THE BEHAVIOUR AND SOCIAL LIFE OF HONEYBEES

by C. R. RIBBANDS

*With a foreword by
Prof. Thomas D. Seeley*

The International Bee Research Association & Northern Bee Books

THE BEHAVIOUR AND SOLCIAL LIFE OF HONEYBEES
by C. R. RIBBANDS

All rights reserved. No part of this publication may be reproduced, stored in a retrieval system, transmitted in any form or by any means electronic, mechanical, including photocopying, recording or otherwise without prior consent of the copyright holders.

ISBN: 978-1-908904-87-4

Published by IBRA & Northern Bee Books, 2016

IBRA
The International Bee Research Association
Hendal House,
Hendal Hill,
Groombridge,
East Sussex,
TN3 9NT.

www.ibrabee.org.uk

Northern Bee Books
Scout Bottom Farm,
Mytholmroyd,
Hebden Bridge HX7 5JS.

www.northernbeebooks.co.uk

C. R. RIBBANDS

Foreword

In 1953, Dr Ronald Ribbands, then a 39-year-old researcher in the Bee Department of the Rothamsted Experimental Station in England, made a landmark contribution to bee research and to beekeeping: his book, "The behaviour and social life of honeybees". It was the first fully comprehensive synthesis of what biologists and beekeepers had learned over the previous centuries about the behaviour and sociobiology of the honey bee, *Apis mellifera*. Most of the studies cited by Ribbands are from the period 1900-1950, but he also reviewed the important contributions from the 1700s and 1800s, including those by Huber on comb building and swarming, and by Dzierzon on sex determination. The breadth of Ribbands's review is also impressive for the way it draws on studies published in German, French, Dutch, and Russian, not just those in his native English. It is, therefore, a wonderful time capsule that shows us what humankind had deciphered about the intricacies of honey bee social behaviour as of 1953.

Ribbands's book provides present-day bee biologists and beekeepers with beautifully clear summaries of the older scientific studies that are the foundations to the topics that are now hotbeds of behavioural research on honey bees: sensory biology, learning and memory, orientation, chemical and acoustical communication, nestmate recognition, queen-worker relations, mating behaviour, caste determination, comb building, individual and colonial thermoregulation, and foraging activity.

Those unable to read German, for example, will enjoy Ribbands's detailed and accurate summaries of von Frisch's papers from the late 1940s in which he reported his step by step tests to solve the mystery of how a worker bee informs her nestmates of the location of a profitable food source: by performing a dance that indicates the bonanza's direction and distance from the hive (Yes), or by flagging the treasure trove with abdominal scent gland secretions (No)? The same holds for Lindauer's papers from the early 1950s on how the scouts in a honey bee swarm conduct a plebiscite to choose the new living quarters for their incipient colony. And to cite just one more example, there is Ribbands's lovely account of his own pioneering studies of how the guards at the hive entrance adjust their acceptance thresholds, becoming less permissive during nectar dearths when making acceptance errors (admitting robbers) is costly, sometimes even deadly.

Another source of pleasure in reading this book is discovering what was not known at the time of its writing. The reader will learn that in 1953 nothing was known about the sex attractant pheromone of queens. Also, that there was a controversy over the mating position: do queen and drone clasp face to face, or does drone mount queen on back? And there were only preliminary indications, from queen caging studies by the German researcher Müssbichler, that the workers know that their colony is "queenright" based on a substance derived from the queen and transmitted by food sharing or antennal contact. Interestingly, Ribbands notes in the section on the glands of the honey bee that the mandibular glands are "well developed in workers and very large indeed in queens" (p. 60), which shows how nicely the stage was set for Butler's pioneering discoveries a few years later on the queen substance pheromone produced in the mandibular glands of queen bees. Also fascinating is Ribbands's statement that "bees are cold-blooded" (p. 70) for it reveals how far we have come in understanding that some flying insects are actually hot-blooded and capable of sophisticated control of body temperature.

In researching his book, Ribbands drew mainly on the findings reported in technical scientific papers, but in writing his book he avoided scientific jargon, for he sought to present the knowledge gained from research in ways that would be useful to beekeepers. He discusses how new scientific discoveries can explain things observed by beekeepers, and thus can help them hone their craft. For example, von Frisch's discoveries of how bees communicate the positions of crops explain how it can be that of seven colonies moved to a heather moor within flight range of a valley with clover, two can fill their hives with "practically pure clover honey without a trace of heather" and the other five colonies "had in the supers a pure heather honey" (p. 166).

This reprint of "The behaviour and social life of honeybees" is welcomed, therefore, both as a guidebook for researchers in bee behaviour wishing to explore the roots of their own studies, and as a reference book for beekeepers seeking a deeper understanding of what is happening inside their hives.

Prof. Thomas D. Seeley,
Department of Neurobiology and Behavior,
Cornell University,
Ithica, NY, USA.
December 2015.

Ronald Ribbands and "The behaviour and social life of honeybees".

Dr Charles Ronald Ribbands was born in 1914. After service during the Second World War, when he worked on the control of malarial mosquitoes, he joined the Bee Department at Rothamsted Experimental Station in 1947. He was thus a key part of an incredibly productive period, working alongside Dr Leslie Bailey, Dr Colin Butler, Dr John Free, and Dr James Simpson, when scientists were well funded and able to devote their time to science rather than filling in grant application forms. Ribbands's studies focussed on various aspects of foraging and communal behaviour of honey bees, which led him to conclude that the evolution of their social life "can be understood in terms of adaptive response to food supplies. The source of these adaptations is food sharing".

In 1952, Ribbands approached the then Bee Research Association with a proposal for a book based on his own work and that of others, which was to become "The behaviour and social life of honeybees". In 1953 he joined the Council of the BRA, and served until 1958, making many contributions, such as setting up the BRA Library of Translations. As well as publishing a number of scientific papers in *Bee World*, he also instigated the publication in that journal of a "beekeepers edition" of long scientific papers, cutting out the detail and emphasising the practical aspects.

In 1956, Ribbands left Rothamsted to take up a post at the School of Agriculture at the University of Cambridge. He lectured on entomology, and carried out research on the aphid vectors of viruses of sugar beet. He was also heavily involved in a number of other organisations such as the Royal Entomological Society. Tragically, he was killed on 31st March 1967 in a car accident near Bishop's Stortford, Hertfordshire, in which his wife and younger son were also injured.

His obituary published in *Bee World* noted: "He produced many original ideas, in his pursuit of research, and also with regard to the administration of organisations he became interested in. His forceful personality gave those ideas wide circulation, and made many of them effective. He was only 53 when he died, but he did more in half a normal working life than many do in the whole span".

The "behaviour and social life of honey bees" was published by the BRA in 1953. A US edition was published by Dover Publications in 1964, but curiously the book has never been reprinted. As Prof. Seeley points out in his foreword, despite the passing of more than 60 years, it remains an important book. As well as being an important historical document, showing the state of understanding of the biology of the honey bee in 1953, it also remains an excellent example of clear writing, making science available to the student and beekeeper. It thus gives me great pleasure that the book is now being reprinted by Northern Bee Books in association with the International Bee Research Association, which will bring it to a new generation of readers.

Norman L. Carreck,
Science Director,
International Bee Research Association,
University of Sussex, UK.
December 2015.

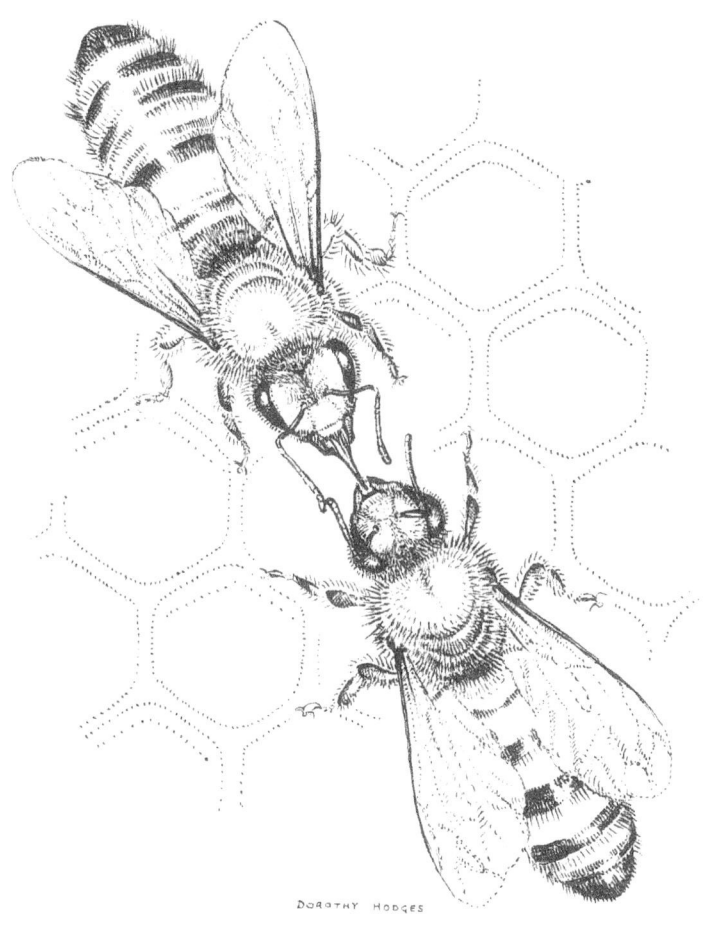

Fig 1 FOOD BEING PASSED FROM ONE BEE TO ANOTHER. FOOD TRANSMISSION WELDS THE COLONY INTO A SOCIAL UNIT (Ch. 38).

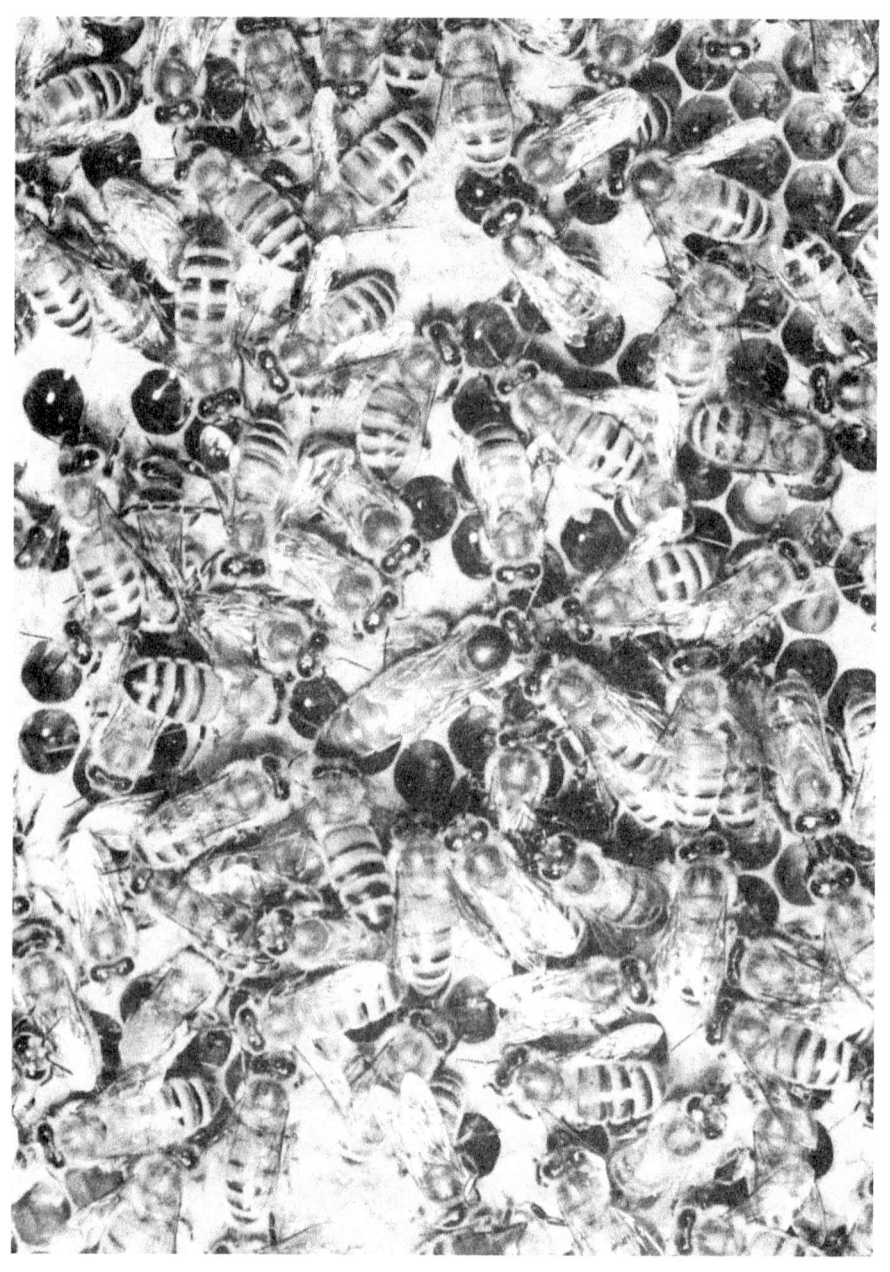

R. V. Roberts

PLATE I *The queen with her attendant workers*

THE BEHAVIOUR AND SOCIAL LIFE OF HONEYBEES

By C. R. RIBBANDS

School of Agriculture, Cambridge
Formerly Principal Scientific Officer
Bee Department
Rothamsted Experimental Station

WITH 9 PLATES
OF PHOTOGRAPHS
AND 66 OTHER
ILLUSTRATIONS

THE COMMON BEE, from a number of peculiarities in its economy, has called forth the attention of the curious; and from the profit arising from its labours it has become the object of the interested; therefore no wonder it has excited universal attention, even from the savage to the most civilized people.

JOHN HUNTER, Esq., F.R.S.
Philosophical Transactions of the Royal Society
LXXXII, p. 128 (1792)

Preface

THIS BOOK IS an attempt to provide a comprehensive review of knowledge of the behaviour and social life of honeybees, to meet the needs of research workers and students as well as practical beekeepers and readers of natural history. It has been written in everyday English, scientific jargon being relegated to footnotes.

The success of *Bee World* and of the Bee Research Association indicates that there is need for books of this kind. British beekeepers have a wholesome interest in relevant scientific research, but new results can seldom be properly valued until they are related to existing knowledge; in this book my aim is to provide that background of information. Most of the book is a review of published research, and facts have been cited wherever possible, little space then being devoted to mere opinion. However, in the less explored portions of the subject few or no facts are available; I have tried to span these gaps with inference and speculation (a necessary prelude to future research), and I hope that the book is thus less unbalanced and more interesting than it would be if it were a mere catalogue of knowledge.

It is a pleasure to record my indebtedness to those who have helped me in this work. Miss Nancy Speirs and Miss Joan Saunders have borne the brunt of the dreary tasks of searching through journals, translating, card indexing and checking references; these things have provided the basis for the book. Miss Saunders has also drawn all the illustrations—except Figs. 1 and 30, for which I am indebted to Mrs. Dorothy Hodges. Dr. H. Kalmus has helped by translating some difficult German papers which would otherwise have been incomprehensible, and I have drawn freely upon the valuable library of translations which is available to all members of the Bee Research Association. On special points the informed criticisms of my colleagues, Mr. L. Bailey, Miss Margaret Ryle and Mr. J. Simpson, have been of great value. Dr. Eva Crane has not only criticized and improved the whole of the manuscript, but she has helped in many other less tangible ways. My wife has shared with me the correction of the proofs.

Permission to copy illustrations has been received from the publishers of the *American Bee Journal, Bee World, Biologisches Zentralblatt, British Journal of Animal Behaviour, Journal of Economic Entomology, Journal of Experimental Biology, Memoirs of Cornell University Agricultural Experiment Station, Nachrichten der Gesellschaft der Wissenschaften zu Göttingen, Naturwissenschaften, Proceedings of the Royal Society, Research Bulletin of Iowa Agricultural Experiment Station, Zeitschrift für vergleichende Physiologie,* and *Zoologische Fahrbücher*; wherever they could be traced, authors have given their permission.

This book is a consequence of the work of Professor Karl v. Frisch, whose discoveries aroused my interest in honeybee behaviour.

RONALD RIBBANDS

Rothamsted Experimental Station
Harpenden, Herts
April 19th, 1953

Contents

Chapter		Page
1	Introduction	11

PART I
THE ROOTS OF BEHAVIOUR

2	*Vision*	13
3	*Taste and Smell*	32
4	*Water and Temperature Perception*	44
5	*Co-ordination; Touch; Perception of Gravity and Distance*	47
6	*Hearing; Vibration Perception; Sound Production*	51
7	*The Glands of the Honeybee*	55
8	*Structural Differences Between Workers, Queens and Drones*	64

PART II
INDIVIDUAL BEHAVIOUR IN THE FIELD

9	*Factors which Influence Flying Activity and Foraging Range*	69
10	*How Honeybees Find Their Way Home*	76
11	*Drifting*	89
12	*How Honeybees Discover Crops, and Return to Them*	93
13	*Crop Constancy; Foraging Areas*	107
14	*The Foraging Method*	115
15	*Pollen Gathering; Division of Labour among Foragers*	119
16	*Foraging Statistics*	127
17	*Time Perception*	133
18	*Mating Behaviour*	142

PART III
COMMUNICATION BETWEEN HONEYBEES

19	*Recruitment to Crops*	147
20	*Selection of a Home*	168

Chapter		Page
21	Recognition of Companions	172
22	The Defence of the Community	179
23	How Honeybees can be Directed to Particular Crops	184

PART IV
LIFE WITHIN THE COMMUNITY

24	Food Sharing	191
25	Clustering	195
26	Wax Production and Manipulation	197
27	Comb Reinforcement; Comb Colour; Propolis	205
28	Nectar Ripening; Hive Humidity	210
29	Temperature Regulation; Wintering	216
30	Brood Rearing	231
31	Queen and Worker Differentiation	245
32	Drone Production	255
33	Swarming and Supersedure	260
34	Seasonal Differences in Physiological Condition and Length of Life	269
35	Laying Workers	277
36	Queen Recognition; Queen Introduction	284
37	Division of Labour	299
38	The Evolution of the Honeybee Community	313
	References	321
	Index of Authors	341
	General Index	346

List of Plates

		Facing page
I	*The queen with her attendant workers*	1
II	*Honeybee gathering pollen from columbine*	128
III	*Honeybee on an apple blossom, using her antennae*	129
IV	*A pollen gatherer approaching a peach blossom*	144
va	*A pollen-covered forager, gathering nectar from a sunflower*	145
vb	*Honeybee gathering nectar from flax*	145
VI	*Returned bees fanning and scenting at their hive entrance*	192
VIIa	*A returned forager at the hive entrance, fanning and scenting*	193
VIIb	*Food transmission*	193
VIII	*A successful forager, in flight*	208
IX	*A forager on sweet clover*	209

Chapter 1
INTRODUCTION

Rather let the naturalist copy the industrious bee, and draw genuine treasures from those flowers of science which have been reared by other hands; and, combining these with his own discoveries, let him endeavour to concentrate all into one harmonious system, with parts curiously formed, arranged and adapted to each other and to the whole, and calculated to preserve the genuine sweets of true wisdom pure and unsophisticated.

W. C. COTTON. Prelude to *My Bee Book* (1842)

THE BEE IS an insect; it lives in the same world with us, subject to the same physical and chemical laws, but insects and vertebrates are built on entirely different plans, and they often respond to those laws in different ways.

In studying insects we can often only appreciate their abilities in terms of our own; thus it is easy to measure their limitations, but very difficult to discover, or to assess the importance of, abilities which we do not possess. Insects are a very successful group of animals, and their success suggests that they have achievements as well as limitations; our study and assessment of honeybee behaviour is handicapped by the fact that we are not insects—aspects of their life we can only guess, and there may be other aspects which we have not yet imagined.

Although the word *anthropomorphic* is used among biologists as a term of derision, scientists are usually (perhaps always and inevitably) man-biassed in their approach to problems of senses and behaviour. An example of this trait is the amount of published research on the senses of honeybees: there are 54 references to original work in the chapter on vision, 14 on taste, 1 each on temperature and water perception, 4 on smell, 2 on hearing. Moreover, despite the many studies of honeybee vision, their perception of polarized light (a very useful ability which man does *not* possess) was discovered only very recently, long after the limitations of their eyes in relation to ours had been carefully assessed.

The plan of this book is derived from the belief that the social life of the community must be restricted by the potentialities of its members, and the behaviour of individuals by their structural limitations. The first section deals with the roots of behaviour (the senses and glands of the individual), the second with individual behaviour in the field, the third with communication between individuals in respect of field and hive-entrance activities, and the fourth with life within the community. In the last chapter evidence from earlier chapters is fitted together in a way which suggests that social life has evolved in a series of adaptations to a restricted food supply. Each chapter is complete in itself and cross-referenced to relevant information given elsewhere, so that the reader need not confine himself to the order given in the book. The first section is intended for reference; the book proper, which is more readable, commences on p. 69.

The honeybee is too often considered in isolation, as if it were an exceptional insect. It is a typical member of the Hymenoptera, that group of higher insects which includes more than 60,000 species of ants, bees, wasps, gall-wasps, ichneumons, chalcids and saw-flies. Most of these species—and most kinds of bees—are solitary insects, but Hymenopteran attributes have on various occasions provided the basis for social organization (found elsewhere among insects only in the termites).

Honeybees are not peculiar in their senses or their individual behaviour, and some of their social life enriches our understanding of these things. We know more about the behaviour of honeybees than of any other insects, and because they are easily kept, marked and observed they will be suitable subjects for further experiments; thus the honeybee can serve as a type from which students of the behaviour of other insects may draw useful analogies.

Studies of honeybee behaviour differ from many other studies of animal behaviour in one important respect—they are much less concerned with sex and mating activities, and more concerned with the everyday life of their subject. Perhaps this is why the student of honeybees obtains an impression of variety and adaptability, whereas even some behaviour studies among higher animals convey an impression of inflexibility. Convention in sexual relations may be advantageous to the species, but rigidity in other activities is a disadvantage.

PART I THE ROOTS OF BEHAVIOUR

Chapter 2

VISION

How large unto the tiny fly
Must little things appear!
A rosebud like a feather-bed,
Its prickle like a spear.

WALTER DE LA MARE, *The Fly*

Visual mechanism

THE VISION OF honeybees is limited by the structure of their compound eyes, but this need not concern us in detail. Very crudely, the eyes can be compared to a bundle of narrowing tubes,[1] each tube having opaque walls and a sensory area at its base (Altenberg, 1926). Each sensory area is stimulated by the light rays which pass down the tube, and these rays include only light from that portion of the visual field which would be enclosed if the tube were projected to the limit of vision. Thus each sensory area receives a stimulus from a different part of the field of vision. The insect perceives the stimuli from all the sensory areas as a mosaic (this was first suggested by Müller, 1826).

Acuteness of vision

The amount of detail in the picture perceived through the eye depends in the first place upon the number of tubes. Each eye of the worker, queen and drone honeybee contains about 6300, 3900 and 13,000 facets respectively (Cheshire, 1886), so the picture which can be seen through each eye is a mosaic formed from not more than that number of differently coloured and shaded spots, each spot being homogeneous.[2]

Still more important, the acuteness of vision varies inversely with the angle of the visual field which is covered by each tube. This angle varies in different sections of the eye, being smallest near the middle in the lower half of the eye. In a vertical section of the eye the angle increases fourfold between the centre and the periphery (Baumgartner, 1928).

[1] Otherwise called 'ommatidia'.
[2] There is no pattern in each such as that depicted by Herrod-Hempsall (1930).

Precise determinations of acuteness of vision in honeybees have been made by measuring the reflex responses of caged bees to a movement in their visual field. Parallel stripes of dark and light stripes were used, and the movement of these stripes to the left or the right caused the bees to be deflected from their course. Hecht & Wolf (1929), who designed this technique, used as an index of acuteness of vision the minimum width of stripe to which the bees responded. They found that it varied with the logarithm of the intensity, giving a sigmoid curve as for the human eye; it was low at low illuminations, and increased rapidly to a constant; its maximum corresponded to a visual angle of 0·98 degrees, which was identical with the measured value for the angle covered by each facet in the region of their maximum density.

Less precise determinations have been made by conditioning bees to fly to syrup which was placed in that one of three small boxes which was distinguished by a mark. The syrup was removed and the boxes rearranged. Repeated tests showed that the bees could just recognize a 0·8-inch square from a distance of 16 inches, which corresponds to a visual angle of a little less than 3 degrees (Baumgartner, 1928). Wolf (1931) demonstrated that bees in the field could be trained to follow a path made of coloured cards each of which made an angle of 1·79 degrees at their eyes.

The higher values for acuteness of vision which were determined by the laboratory method can be appreciated by relating them to our own attainments. The maximum acuity of the bee is lower than the lowest human one; Hecht & Wolf considered that under similar conditions it was about $\frac{1}{100}$ of that of man. Graham & Hunter (1931) criticized Hecht & Wolf's technique on the ground that the bees were responding to a moving pattern whereas acuteness of vision in humans is typically determined on the basis of recognition of a stationary pattern. They therefore tested human subjects with an apparatus similar to that which Hecht & Wolf had used for bees; they concluded that the maximum acuteness of human vision of a moving pattern was about 60 times that of the bee (but they had doubts about the validity of such a comparison).

Despite this unfavourable comparison, Barlow (1952) estimated that facet size is optimal in the centre portion of the bee's eye; if they were smaller and more numerous diffraction of light would prevent any improvement in acuity in that region of the eye.

In the above experiments acuteness of vision was determined in the vertical plane. The angle covered by each facet is more than three times as great in a horizontal plane as in a vertical one, so bees in the field are very astigmatic, and resolve their environment vertically with more accuracy than horizontally (Baumgartner, 1928; Wolf, 1931). The eye of the bee is about four times as long as it is wide, so that despite the increased vertical resolution the visual field is greater in length than in width. Hecht & Wolf (1929) emphasized that the bee's eye is an instrument which functions efficiently only along a narrow vertical belt.

The honeybee is usually considered to be short-sighted, but this is a misinterpretation. The insect eye possesses no focusing mechanism; short-sightedness is a derangement of focusing ability, which can only occur in eyes with focusing arrangements, so the honeybee is not short-sighted in the sense in which we use the term among ourselves. Its indistinct vision, due to lack of acuteness, applies at all ranges. The definition of an object falls off as the distance between object and eye increases, just as it does in the human eye.

Discrimination between degrees of brightness

V. Hess (1916) investigated the discrimination between degrees of brightness by putting honeybees in a dark box which could be lit by two lights, from opposite sides. When the lights were unequal the bees went towards the brighter one. V. Hess concluded that the bees could distinguish differences nearly as small as those perceived by the human eye. Bertholf (1931a) found that honeybees began to distinguish between two squares of light when the brightness of one was reduced to 70% of that of the other.

Wolf (1933a) used reflex responses to vertical stripes in order to study this problem. For each brightness of one of the two sets of stripes he determined the brightness of the second set which first caused the bee to respond to a movement of the striped field. Discrimination was poor at low illuminations, but increased with increasing illumination; the minimum difference detected by the bees was 23%. The discrimination of both the human eye and the eye of the bee reached its maximum at an illumination of 100 millilamberts, when the discrimination of the bee was $\frac{1}{20}$ that of the human eye. The human eye could perceive different brightnesses over a much greater range than the bee's eye.

However, the experiments of Müller (1931), which are reviewed on p. 30, indicate that honeybees were able to distinguish a difference of 14% between two light sources.

Adaptation to darkness

V. Hess (1918) found that the sensitivity to light after varying periods in darkness increased very quickly at first, and then more slowly. After 15–20 minutes in the dark the sensitivity had increased 500- to 1000-fold. Wolf & Zerrahn-Wolf (1935b), again using the reflex response to vertical stripes, obtained a similar result. They pointed out that in similar circumstances the adaptive range of the human eye is about ten times as great as that of the bee.

Perception of movement

Wolf (1933b) studied the response of bees to movement in their visual field by moving a striped pattern underneath them, at different speeds. As the frequency of the flicker increased the intensity of illumination also had to be increased in order to produce a response. In optimum lighting the bees reacted to flicker frequencies up to 54 per second. The maximum flicker frequency to which the human eye responds is between 45 and 53 per second, so the recognition of flicker by the bee was at least as good as that by man.

Autrum (1949) tackled this problem by more refined electrophysiological methods and found that the fusion frequency of the eye of a fly is much higher than that of a mammal; when only 1–4 ommatidia were stimulated the fusion frequency was 60–165 per second, but for larger areas it rose to 265 per second. The bee's eye may be similarly sensitive.

The perception of movement is of great importance in relation to the perception of form and pattern.

Perception of form and pattern

Early experiments which demonstrated the importance of vision in orientation (p. 76) implied some perception of pattern, and Turner (1911) provided evidence of the perception of small patterns. He made small boxes, $2\frac{1}{4} \times 1\frac{1}{2} \times \frac{3}{4}$ inch, and painted them red or green, or with either vertical or horizontal stripes (0·2 inch wide) of red and green. These boxes were all set out, with honey supplied only in those patterned with vertical stripes, and bees learned to enter only boxes with this pattern. Then all the boxes were removed and replaced by a new set, differently arranged, which did not contain and had never contained any honey. In fifteen minutes bees paid thirty visits to the vertically striped box, but none to others. When examining the boxes, the bees hovered within about 0·4 inch of them.

Experiments with blue patterns on a white ground were carried out by v. Frisch (1914); he trained honeybees to select

Fig 2 PATTERNS USED FOR TRAINING PURPOSES BY V. FRISCH (1914).

a gentian-shaped pattern (Fig. 2a) from a sunflower-shaped one (Fig. 2b) but could not train them to distinguish between triangles and squares. Circles were divided into blue and yellow halves and honeybees were trained to distinguish those with the yellow half on the left from those in which it was on the right.

Mathilde Hertz (1929a) showed that honeybees could be trained to associate certain black figures, on a white background, with the presence of food.[1] She was not able to train them to distinguish between such figures as triangles, squares, and circles (Fig. 3, upper row). However, they could distinguish between patterns which were either more or less subdivided (Fig. 3, between upper and lower rows), but they could only be trained to choose the more subdivided pattern, which they always preferred. Hertz had expected that the training would be absolute, but she found that when the bees were trained to choose a more subdivided from a less subdivided pattern, both

[1] Hertz (1937) found that compact white patterns, on a black ground, were not attractive, but that bees could be trained to elaborate patterns of white on black. Deep black areas were as attractive in themselves as solid coloured areas.

patterns being surrounded by a complex environment, the bees continued to go to the training position if the patterns were exchanged—they were also making use of more distant visual marks.

This partially successful attempt to train bees to small patterns was followed by an investigation of the choice of patterns, in which bees were trained to a feeding-place without a pattern, and were then offered alternative models and patterns without feeding. Hertz (1929*b*) gave them the choice

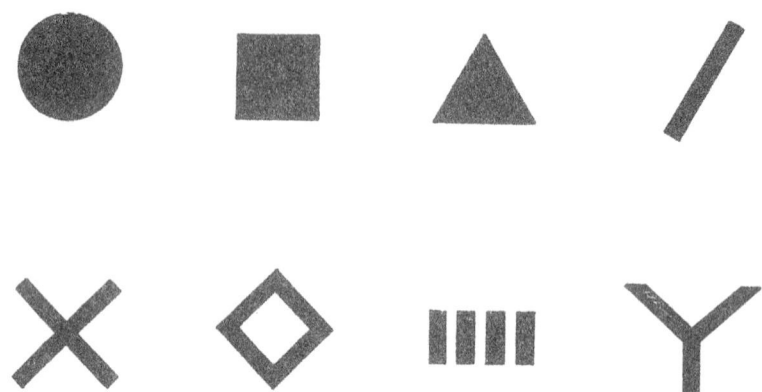

Fig 3 PATTERNS USED BY HERTZ (1929*a*). Those in the upper row were not distinguished from one another. Nor were those in the lower row, but any of the patterns in the lower row were preferred to those in the upper one.

of various black and white patterns, laid horizontally; she concluded that their spontaneous preferences were related to the quantity of outline, irregularity and size of the patterns, and the independence of their components. There was no absolute criterion of attractiveness, but the attractiveness of any pattern depended upon its value relative to that of other objects which the bee could see at that time. The bees always 'decided according to the figural disequilibrium', i.e. they preferred the most contrasting patterns. When there were considerable contrasts hungry bees were attracted without previous training. Bees were not spontaneously attracted to patterns with little contrast, but they could be trained to distinguish these smaller differences and to prefer a pattern with little contrast to one with even less. In consequence of the spontaneous attraction, complicated flower shapes would be more attractive to untrained bees than big simple flowers.

This investigation was continued, using various three-dimensional models of white paper, placed on a white background (Hertz, 1931). The models included rings, cones and imitations of double flowers. The behaviour of the bees towards these faint differences between the illumination of similar models was observed; the model which produced most visible shadow was the most attractive one, the bees being deflected towards

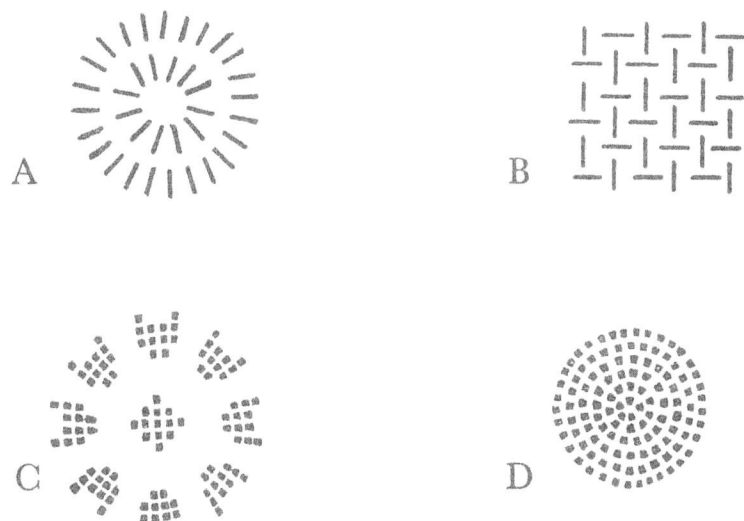

Fig 4 PATTERNS USED FOR TRAINING PURPOSES BY HERTZ (1931).

the region of deepest shadow. Hertz related attractiveness of the deep shadow of her colourless models to Kühn's (1927) observation that untrained bees which were confined in a box always went to its most contrasting portion, whether darkest or lightest.

Wondering whether these results with models could all be reduced to a matter of differences between shades of grey, Hertz compared the attractiveness of chessboard patterns with that of similarly arranged marbles on a white ground. She concluded that the bees strongly preferred three-dimensional white objects to flat grey figures which appeared equally deeply shaded to humans.

Hertz (1931) also compared the reaction of honeybees to several patterns in which identical components were differently arranged. Fig. 4A was preferred to Fig. 4B, for instance, although both contained 36 identical lines, and Fig. 4C was preferred to Fig. 4D. Hertz concluded that the arrangement of the parts of a pattern was important, and that the bees were not merely discriminating between patterns by adding together the simple impulses from their components.

Gertrud Zerrahn (1933), who trained honeybees to one pattern and then presented them with a choice of patterns, concluded that the bees had a spontaneous tendency to alight upon whatever pattern possessed the greatest length of contour or the most variety. Emphasizing that this spontaneous tendency could not be overcome by training, she considered that the response was 'forced' by whatever pattern afforded the greatest visual change, and hence the greatest rate of change of stimulation of the eyes. Wolf (1933b, c) demonstrated the importance of flicker in bee vision, and (1933c) showed that when bees in a dark chamber were allowed to walk towards flickering fields of equal intensity but varying frequency, the numbers which approached the different fields were directly proportional to their flicker frequency. He held that the bees chose spontaneously between different patterns on exactly this basis; because the vision of the bee was not acute the individuality of a flower was not perceived.

Wolf & Zerrahn-Wolf (1935a) gave honeybees a free choice of pairs of chessboard patterns. Chessboards of different size but equal length of outline were equal in attractiveness, so they again concluded that the reaction of bees to patterns depended only on the rate of change in the stimulation of their eyes.

This simple hypothesis does not account for all the observed facts. New evidence concerning pattern perception had been provided by Mathilde Hertz (1933). Having trained honeybees to a pattern, she offered an alternative pattern and noted whether the bee remained attached to the training pattern. Some of the patterns are shown in Fig. 5. In some instances bees forsook the training pattern when some different pattern was offered as alternative. Thus B1 was more attractive than C3, after an hour of training to the latter; bees trained to A1 went equally to C1, those trained at A2 also went to B1 and C2. These results confirmed the previous ones (1929a), which had indicated that pattern perception was not acute.

However, bees trained to C1 or C2 would not visit A1 when it was offered as alternative, and those trained to D2 would not go to A2. In these instances the bees remained attached to patterns with *less* length of outline than the alternatives, so they could not be responding to the one which produced the greatest rate of change of stimulation. In addition, bees trained to A1 would not visit C2 when that was offered as an alternative; thus bees could be trained to either member of this pair. They could also be trained to either member of the pairs B1 and C2,

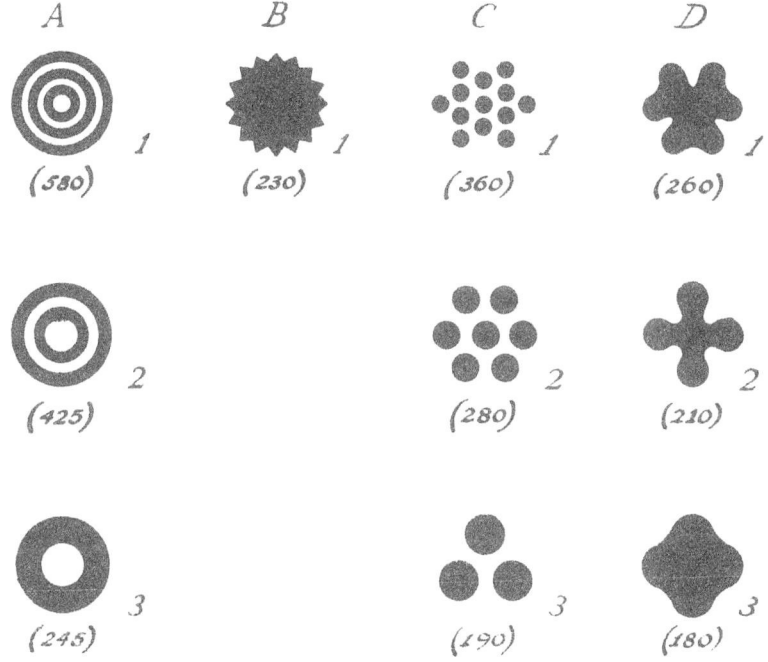

Fig 5 PATTERNS USED BY HERTZ (1933). The numbers in brackets are an index of the length of outline of each pattern.

A1 and C3. Hertz suggested that reciprocal training was possible because different properties of the patterns were attractive. Bees trained to a particular quality of pattern always deserted that pattern for any similar but more complex one, e.g. C3 was deserted for C2, C2 for C1.

Reciprocal training was also possible between alternatives which were of the same shape but which had other different characteristics (Hertz, 1934c). For instance, bees could be trained to choose either one black disc $1\frac{1}{2}$ inches in diameter or a set of seven small grey discs each 0·22 inch in diameter and 0·22 inch apart. They would also choose either of a set of three black discs or a set of seven similar smaller ones (Fig. 6); the area covered by the pattern of the smaller discs was two-thirds of the area covered by the others; the length of outline of the smaller discs was 4½ inches and that of the larger one was 6¾ inches.

Summarizing all these results, Hertz (1935) concluded that the amount of outline possessed by different black-on-white patterns was certainly not the only quality by which they were distinguished; a figure was perceived as a complex thing, which might have several different properties which could be used for alternative training.

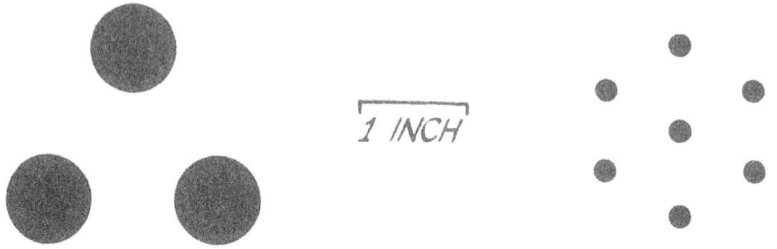

Fig 6 PATTERNS USED BY HERTZ (1934c).

The great limitations of the insect eye in respect of acuteness of vision and extreme astigmatism explain how pattern perception by the honeybee is so very inferior to that of man. Bees possess some ability to perceive small patterns, but there is some doubt concerning the uses of this ability (there can be no doubt of the use of the larger patterns of their environment in orientation).

Flowers in nature usually do not stand singly, but in groups, so there is substance in Wolf's (1935) contention that the quick variation in the stimulation of the facets could cause the bee to alight and discover groups of flowers even when the detail of the flower parts was below the level of visibility. He observed that bees settled more readily on flowers when there was a slight wind than on windless days. Subsequently (1937) he made an 'artificial flower-bed', composed of 56 white cardboard discs, each supplied with syrup. One half of this bed was gently moved backwards and forwards, and Wolf found that about twice as many bees settled on the moving portion as on the still half. He concluded that in nature the visual reaction of honeybees was largely due to the flicker effect produced from the motion of the flying bees relative to the flowers.

Colour vision

Lubbock (1875-76) showed that honeybees could be trained to seek for a paper of one particular colour which they had learned to associate with the presence of food. They would select papers of that colour from papers of other colours. Forel (1910) obtained similar results. There was still a possibility that the bees were differentiating between different brightnesses, as v. Hess (1910, 1916) believed, and not true colours, although Turner (1910) pointed out that bees which responded to differently coloured boxes in shadow would respond to them without hesitation in bright sunlight, and the brightness was very dissimilar in these different conditions.

This doubt was completely removed by v. Frisch (1914), in the first of his classic experimental studies of honeybee behaviour. He selected a series of 15 shades of grey cards, ranging from black to white, and arranged them on a table along with a single blue or yellow card. All the cards were covered with a large sheet of plate glass, and a small watch glass was placed over the centre of each card. The watch glass on the coloured card was supplied with sugar syrup, and bees accustomed themselves to feed there. At short intervals the cards were rearranged, but food was only supplied on the coloured card. Then clean empty watch glasses were supplied at all the cards, so that no food at all was available, but the bees still went unerringly to the coloured card, which they distinguished from all the shades of grey. V. Frisch was unable to train them to select one shade of grey from the others.

Knuth (1891) noticed that the inconspicuous greenish-white flowers of *Sicyos angulata* were visited by many bees and flies, and by comparing the action of floral tints on a photographic plate he showed that these flowers were probably of an ultraviolet colour. V. Hess (1920) demonstrated that bees are very sensitive to ultraviolet, but he believed that this was due to a fluorescence of the outer parts of the eyes, and that they did not have true ultraviolet vision. A technical advance was made by Kühn & Pohl (1921), who trained bees to different wavelengths of the light from a mercury lamp, which was shone as a spectrum on to a table-top and shifted about at will. After an hour of training the bees would go correctly to either ultraviolet or another band of colour.

Richtmyer (1923) and Lutz (1924) analysed the ultraviolet content of many flowers, and Lutz concluded that about 30%

of conspicuous flowers are distinctly ultraviolet. Sensitivity to ultraviolet counterbalances inability to perceive the red end of the spectrum. Ruth Lotmar (1933) also studied flower colours, and she found that many white flowers, such as *Convolvulus*, *Phlox* and roses, reflect little ultraviolet. These flowers should therefore appear blue-green to the honeybee, since Hertz (1939) showed that this is the complementary colour to ultraviolet. Bees probably cannot readily distinguish between some

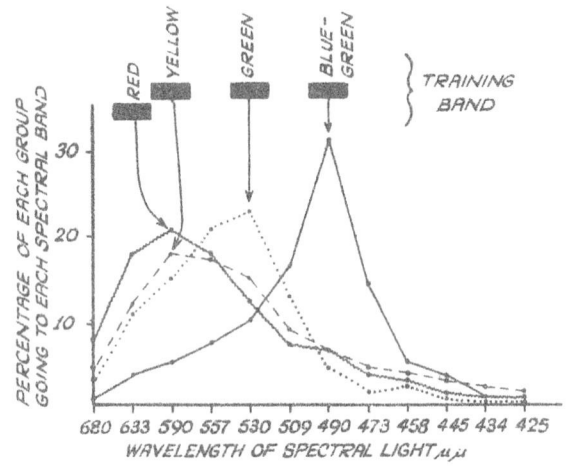

Fig 7 CHOICE OF SPECTRAL COLOURS BY BEES TRAINED TO FOUR DIFFERENT BANDS OF THE SPECTRUM (Kühn & Fraenkel, 1927).

varieties of flowers which appear to us white, pink or other pale colours because the blue-green appearance predominates over such slight differences. On the other hand, Lotmar found that evening primroses and red poppies reflected much ultraviolet, and they will appear to be this colour to the honeybee.

Kühn (1924) and v. Frisch (1924) agreed that the honeybee could be trained to any of the four wave-bands 6500–5300 Å. (orange-yellow), 5100–4800 Å. (blue-green), 4700–4000 Å. (blue-violet), and 4000–3100 Å. (ultraviolet). Both experimenters considered that there was no discrimination within these wavebands—bees trained to yellow went indiscriminately to orange, yellow and grass-green; those accustomed to blue went to purplish-violet or blue.

Bees were therefore believed to lack the capacity to distinguish shades of colour, until the experiments of Kühn & Fraenkel (1927), who divided the light in a spectrum into a series of sections which they shone upon a table. The table was moved about so that the bees did not become trained to any position, and bees were fed upon one section of the spectrum. After training, the sections of spectrum were shone upon a paper bridge which was above the table and at right angles to the direction from which the bees flew. An observer watched the bridge, brushed off with a feather each bee which landed on any portion of the spectrum, and announced the results to a colleague. In the 6500–5000 Å. portion the bees seemed to

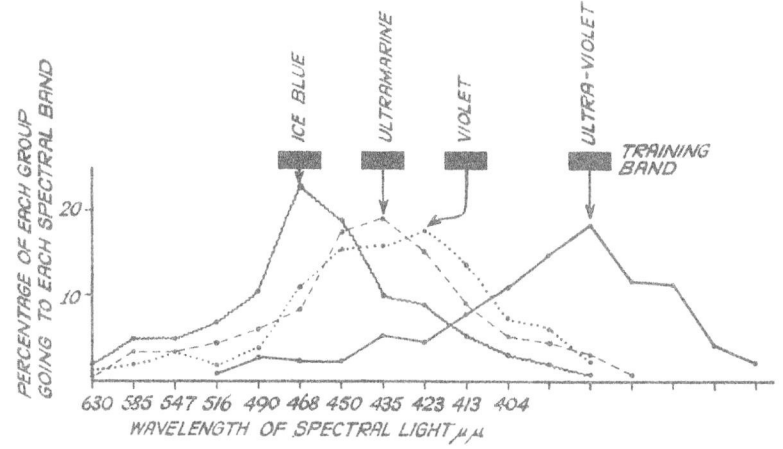

Fig 8 CHOICE OF SPECTRAL COLOURS BY BEES TRAINED TO FOUR DIFFERENT BANDS OF THE SPECTRUM (Kühn & Fraenkel, 1927).

distribute themselves differently when trained to 5400–5200 Å. (green) and 6100–5700 Å. (yellow), but there was no clear difference between the latter bees and those trained to 6500–6000 Å. (red) (Fig. 7). In the 4800–4000 Å. portion the distributions were rather different after training to 4800–4600 Å. (ice-blue), 4420–4280 Å. (ultramarine), and 4150–4050 Å. (violet) (Fig. 8), although the curves were not as sharp as those which resulted from training to 5000–4800 Å. (blue-green). Kühn & Fraenkel concluded that the bees could not only distinguish between orange-yellow, blue-violet, blue-green and ultraviolet, but also between different portions of the first two bands, i.e. different shades of these colours. Different shades of ultraviolet or blue-green might also be distinguished, but this matter was not investigated. Bertholf (1931*a*) found that the longest wavelength to which honeybees would respond was 6770 Å., and (1931*b*) the shortest was 2970 Å.

Kühn (1927), again using spectral colours, was unable to train honeybees to different brightnesses of one colour, because when given a choice they always preferred the brightest. This conclusion is consistent with the inability of v. Frisch (1914) and Hertz (1931) to train bees to a shade of grey.[1]

Kühn (1927) was able to show that blue-violet and orange-yellow were complementary colours to honeybees, and that their eyes showed 'simultaneous contrast phenomena'—a grey

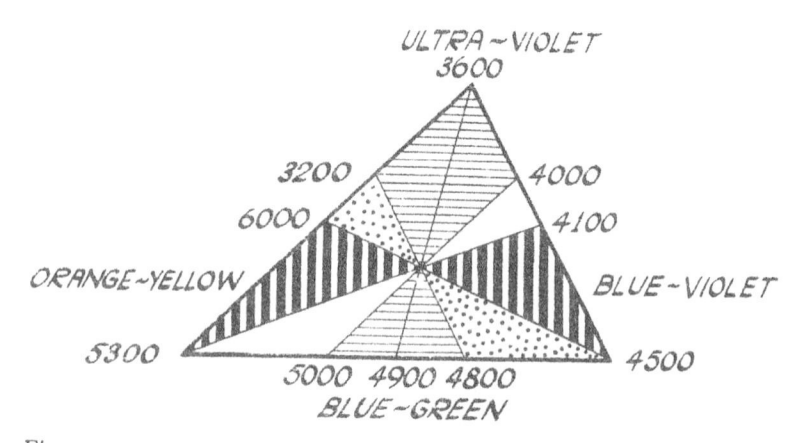

Fig 9 COLOUR TRIANGLE DESIGNED TO REPRESENT COLOUR VISION (Hertz, 1939). Wave-length in μ.

field surrounded by yellow was mistaken for blue by blue-trained bees, and a grey field in a blue-violet background was examined by yellow-trained bees. Thus we may suppose that their eyes would also exhibit 'after-image phenomena', and that the basis of their colour vision is similar to our own in these important respects.

Kühn (1927) suggested that ultraviolet and blue-green were also complementary colours for the bee, and this was established by Mathilde Hertz (1939). Hertz tried to train bees to seven different blue-green papers, having covered the papers with a glass plate which transmitted the whole of the solar spectrum. She found that they could be easily trained to two of the papers, and with greater difficulty to two more, but that prolonged attempts to condition them to either of the other three were unsuccessful. Yet when these latter papers were

---

[1] Bierans de Haan (1928) stated that he had trained bees to choose a dark shade of grey in preference to a lighter one; the papers which he used for his experiments probably differed in their ultraviolet content (Hertz, 1937).

covered with a filter which absorbed ultraviolet the bees could be trained to them. These papers had previously been reflecting both ultraviolet and blue-green, and they had appeared whitish or greyish to the bee. Hertz then showed that bees trained to a blue-green paper would visit with equal eagerness a white paper which did not reflect ultraviolet.

A 'colour triangle' (with each colour represented by a point on one of its sides), can be constructed for the human eye, so that any straight lines which pass through the centre of the triangle (which represents white) will cut the sides at the positions of a pair of complementary colours. Hertz constructed a similar colour triangle (Fig. 9) which might represent colour vision in the honeybee. If this is so, the basis of human and honeybee colour vision might be similar, but with the red receptor of humans replaced by an ultraviolet receptor in honeybees.

Colour preferences

Müller (1882) and Lubbock (1882) both conducted many experiments to determine the colour preferences of honeybees. Müller worked with natural flower colours placed under sheets of glass, Lubbock with coloured papers; they took no account of ultraviolet. Both investigators found a distinct preference for blue, but Therese Oettingen-Spielberg (1949) found that searching bees in a room went to yellow more frequently than to blue. Since bees can be trained to various colours, experiments of this type are not easy to interpret (Lovell, 1910). Contrast with the background might also affect the preference; blue provides the maximum contrast to grass green.

Koch, P. (1934) approached colour preferences from another viewpoint. He reported that he had kept 28 colonies in one apiary for 14 years, and that during this time the hives had always been painted six different colours. The average honey yields from the differently coloured hives had shown consistent differences, thus: dark blue 48½ lb., black 42 lb., brown 40 lb., white 26½ lb., light green 22 lb., pink 21 lb. This result indicates that bees had shown a preference for darker-coloured hives and had drifted to these from the others. In Europe, where colonies are usually kept close together in beehouses, the painting of hives is frequently advocated in order to help the bees to identify their own hive; Koch's results demonstrate that this system can have disadvantages.

Stimulation by colour

Other investigators have sought to compare the power of stimulation of various wavelengths of coloured light by placing bees in dark cages and comparing numbers attempting to escape towards two different sources of light. This technique was used by v. Hess (1920) when he demonstrated that honey-bees were sensitive to ultraviolet light; he found that unfiltered sunlight was eight or nine times as attractive as sunlight filtered through a glass which eliminated ultraviolet rays, but transmitted all the others. Lutz (1924) obtained a similar result, in that the caged bees went to a filter which transmitted only ultraviolet in preference to a filter which transmitted the whole of the spectrum except ultraviolet.

Bertholf (1927) elaborated this technique, using monochromatic light beams of known intensity; he found that the relative attractiveness of the various colours was yellow-green 100, blue 54, yellow 29, violet 23, red 2. Then he investigated the ultraviolet portions of the spectrum in a similar manner (1931*b*), and concluded that there was a peak of stimulation in the ultraviolet (3650 Å.) which was about $4\frac{1}{2}$ times as great as the peak in the yellow-green region (5530 Å.). Bertholf used light of different energies at different parts of the spectrum, and obtained his results by calculating the relative attractiveness of colours of equal energy; he said that the energy content of the ultraviolet portion of sunlight is about one-fourth of the energy content of the whole of the humanly visible portion and that the stimulation given to a bee's eye by the ultraviolet portion of sunlight is therefore nearly as great as that given by the whole of the humanly visible portion.

The results obtained by Sander (1933), who compared the relative efficiency of wavelengths of equal energy, were different. His bees responded most to two different wavelengths, yellow (5700 Å.) and blue (4600 Å.), with a steady decline into the ultraviolet. Sander found that as the intensity increased the efficiency of stimulation increased more rapidly at either end of the spectrum than in the middle. This invalidates some of Bertholf's calculations; it may also explain why Sander, using low intensities of artificial light, obtained no peak in the ultraviolet—contrary to the results of v. Hess, Lutz and Bertholf.

Perception of the plane of polarized light

V. Frisch's work on communication of the direction of crops (p. 157) revealed that honeybees can appreciate the plane of polarization of light.[1] The human eye cannot do this, and our understanding of the problems involved is facilitated by the model (Fig. 10) of the arrangement of polarized light in the sky, designed by Bailey (1953). This model might be constructed from a transparent globe marked with north and south poles and lines of latitude; inside the globe, pivoted at the centre, would be a flat disc of the same diameter. The sun would be at the north pole, and the observer (supposedly capable of detecting polarized

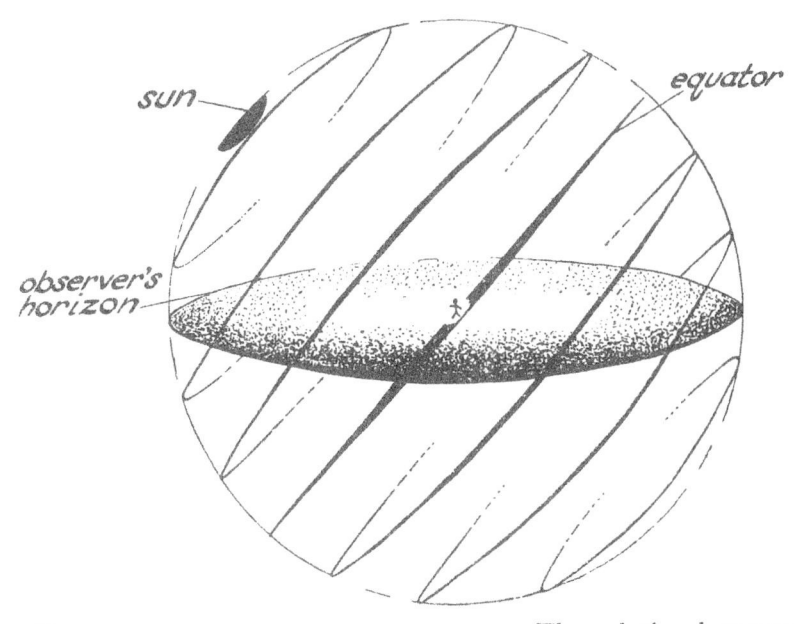

Fig 10 PERCEPTION OF POLARIZED LIGHT. The relation between the sun and the plane of polarization, as it would appear if this property could be perceived. The earth is represented by a flat disc (shaded) which is pivoted at the centre of a transparent globe, representing the sky. See text.

light) would stand at the centre of both disc and globe. The edges of the disc would represent his horizon, and the hemisphere of the globe above the disc would represent the sky. The plane of polarization of sunlight scattered from the atmosphere would then appear to the observer to lie along the lines of latitude of the globe, and the percentage of polarized light in

[1] Light travels in waves, which vibrate transversely (i.e. at right angles to the direction of travel). In ordinary light these vibrations are in an infinite number of transverse planes; in polarized light they are in only one transverse plane. The light reflected from any part of the blue sky is partially polarized, the plane and the degree of polarization depending on its direction in relation to that of the sun.

the light observed would fall away from a maximum (50–70%) along the equator of the globe, and would become zero in the direction of the sun.

Bailey emphasizes that any part of the equator would appear to the observer as a straight line in the sky, and when looking at this line he could immediately indicate the bearing of the sun, which would be at right angles to this line and to his line of sight. Examining the results from which v. Frisch had shown that bees could orient accurately when presented with a small portion of blue sky (p. 157), Bailey found that they had always been presented with a portion on or very near to this equatorial region, or with some artificial equivalent. It would be more difficult to orient accurately in relation to a portion of sky from any other region, and Bailey considers that there is as yet no evidence to determine whether bees possess this ability.

Autrum & Hildegard Stumpf (1950) suggested that analysis of polarized light by the bee's eye is achieved through the eight radially arranged sensory cells of each ommatidium. Menzer & Stockhammer (1951) also suggested a mechanism of analysis by the ommatidium.

The function of ocelli

Like most adult insects, honeybees possess three simple eyes, or 'ocelli', which are on the top of the head (Fig. 16) and which are sensitive to light. Each ocellus is in some respects similar to, but much larger than, one facet of a compound eye; they contain a lens but they could not produce an image, and insects with their compound eyes painted over behave as though blind whether they possess ocelli or not (Homann, 1924). An intriguing problem is presented by the persistence of these simple structures alongside the compound eyes, which seem to be fitted to fulfil all their functions much more efficiently.

Bozler (1926) showed that *Drosophila* reacted less quickly to a sudden exposure to light if their ocelli were painted over. Müller (1931) studied the reactions of honeybees to light before and after he had blacked out their ocelli with a mixture of soot and wax. His apparatus was a rectangular board which could be illuminated by either a 50-watt lamp (e) at one end or similar lamps (r and l) at the two sides; a line was marked down the middle of the board from e to the opposite end (s). The bees were kept in a cold dark box and when they ventured out they were placed at s; e was switched on and the bee moved along the line towards it; e was then replaced by r and l and the reaction of the bee was recorded.

In one experiment r and l were 15·8 and 14·6 inches from the centre of the board, so that their relative brightness at the centre was 1·16 : 1. When the lights changed, untreated bees went 3 times straight on, 84 times towards the nearer source and 13 times towards the other; the results from bees with all ocelli blackened were 1, 74 and 25 respectively; therefore even at this threshold level for brightness discrimination the ocelli did not significantly affect the proportion of responses.

In further experiments the middle ocellus and one or other of the side ocelli were blackened out; with r and l equidistant these bees always went towards the side with the intact ocellus; for equal numbers of these bees to go to either side, the lamp on the side with the covered ocellus had to be at less than half the distance of the other one (i.e. the illumination had to be about 5 times as strong). From this result Müller concluded that the ocelli act as stimulatory organs, their excitation being added to that received through the compound eyes; however, further experiments did not show that the ocelli were of special value in weak light, as one might then expect.

Wolsky (1933) agreed that ocelli are stimulatory organs, and Parry (1947), working with locust heads, provided physiological evidence in support of this view.

Meanwhile v. Buddenbrock (1937) stated that ocelli were usually present in winged insects, absent in wingless ones. Kalmus (1945), who demonstrated a high correlation in this respect, mentioned mutations in *Drosophila* which reduced either wings or ocelli without effect upon the other character; this suggests that ocelli are not vestiges fortuitously tied to wing development, but are usefully concerned with flight. Their use in flight has not to my knowledge been studied, except that Müller (1931) noted that some of his treated bees were still alive after several weeks, with their ocelli covering intact, and that in free flight they behaved exactly as other bees did.

Chapter 3
TASTE AND SMELL

> *They haven't got no noses,*
> *The fallen sons of Eve;*
> *Even the smell of roses*
> *Is not what they supposes;*
> *But more than mind discloses,*
> *And more than men believe.*
>
> G. K. CHESTERTON, *The Song of Quoodle*

NERVE ENDINGS OF many types are sensitive to irritant chemicals, and this common chemical sense has become differentiated into the two senses of taste and smell. There is no very sharp distinction between these two senses, but it is convenient to consider as taste the perception of sweet, sour, salt and bitter qualities,[1] and to include other chemical perceptions in the sense of smell.

Odour and taste are not usually both well-marked properties of the same chemical compound, probably because odours have lipoid-soluble and tastes have water-soluble properties (Moncrieff, 1944). In man taste is a matter of contact with some quantity of a solid or a liquid, while odorous particles are minute, suspended in air, and are present and can be perceived at a distance from their source. Nevertheless, many of the finer distinctions which we call 'flavours' are a consequence of the perception of odorous particles which have passed from the mouth into the nose.

Insects also perceive odours from distant sources, but they may perceive odours through direct contact between their antennae and the source, and smells may then have a recognized distribution in space (Forel, 1908). Insect antennae function primarily as odour receptors, but secondarily also as taste receptors; it has even been suggested that the antennae can perceive sweetness without direct contact with the source (Minnich, 1932).

---

[1] Frings (1951) has recently suggested that these four qualities are not distinct but are due to populations of sense organs susceptible to different quantities of one kind of stimulus.

TASTE

Taste in the mouth

Will (1885) observed the behaviour of honeybees which visited flowers of aubrietia in which he had placed honey, or honey mixed with common salt, bicarbonate of soda or quinine. He noted that the bees collected the pure honey but rejected the contaminated supplies, so he concluded that they possessed a sense of taste in their mouthparts.

A detailed study of the taste perception of the mouthparts of the honeybee was carried out by v. Frisch (1927–34), who used the following as a standard method. Having trained about 12 individually marked bees to drink sugar syrup from a shaded dish, he substituted other solutions for five-minute periods and noted whether the bees refused the solution or drank it either with or without hesitation. This standard method was not very sensitive, because the bees were not at any time presented with a choice of solutions.

Experiments with sucrose demonstrated variation in the concentration of the solution which the trained bees would collect. They almost always rejected $\frac{1}{4}$ molar ($\frac{1}{4}$ M) solution but occasionally they even rejected 1 M solution (30·5 g. in 100 g. water); the threshold depended upon the quality of the natural crops which were available, and it was raised when these were good. The viscosity, osmotic pressure and temperature of the solution, and the presence of flower scents, had little or no effect upon the threshold, but there was variation between individual bees. V. Frisch stated that the threshold varied with the age of the bee (the sense of taste being dulled with age), but the evidence in support of this statement was very scanty (1934, Table 10).

The collecting threshold was raised if the bees had imbibed a very sweet solution during the previous five-minute period, but it was lowered to $\frac{1}{8}$–$\frac{1}{16}$ M (4–2%) in bees starved for the previous hour. V. Frisch equated the collecting threshold to the threshold of taste perception, but probably this assumption was only justified when starved bees were used.

Perception of sweetness

Skramlik (1926) concluded that only one quality of sweet taste occurs in the animal kingdom; sugars may, however, also contain bitter, sour or salt components in their taste (Cameron, 1947).

V. Frisch (1934) tested 19 different substances which are

sweet to man but which are not sugars or sugar-like; none of these was sweet to bees. He also tested 34 sugars and related substances, and he compared his results with estimates of their food value which had been made by Berta Vogel (1931) and which were based on the length of life of groups of differently fed bees. Seven of these substances (sucrose, glucose, fructose, maltose, trehalose, melezitose, α-methylglucoside) were very attractive and of high food value to bees, and were similarly attractive to man. Fucose and inositol were without food value but were rather attractive to bees; xylose and arabinose were of high food value but were not attractive to them; sorbose was unattractive and valueless; lactose, melibiose and mannose were slightly attractive but actually detrimental to bees; all these sugars are attractive to man. Sorbitol was not attractive to bees; surprisingly, Vogel had found that this sugar-alcohol extended their expectation of life more than an equivalent amount of sucrose.

One can see from these results that taste is not a measure of food value. The seven sugars most attractive to bees were also of high food value, but the non-attractive sugars and related substances were not necessarily without food value.

Sucrose, glucose and fructose, the three sugars which occur in quantity in nectars (Ruth Beutler, 1930), are both attractive and valuable. Small quantities of maltose, and probably of melibiose and raffinose, also occur in some nectars (Gwenyth Wykes, 1952a); the first is attractive and highly nutritive, the other two are not. Sucrose, glucose and fructose are the chief sugar constituents of honeydews, in which two other attractive and nutritive sugars—trehalose and melezitose—are usually also found (Vogel).

V. Frisch (1934) found that sucrose was more attractive than glucose or fructose, and that both of these were preferred to maltose. This was not in accord with the earlier results of Kunze (1927), which were based on less adequate data. Gwenyth Wykes (1952b) reinvestigated this problem, using another technique. She placed solutions of the sugars, of equal concentration, in Petri dishes or in the cells of a honeycomb. The solutions were arranged in a Latin square and many bees were allowed to visit them; the quantities imbibed provided an index of relative preference, and the results were analysed statistically. The consistent descending order of preference was sucrose, glucose, maltose, fructose. The relative sweetness probably varied with concentration. Her experiments showed that the observed preferences for solutions of mixed sugars sometimes differed from those predicted on the basis of an additive effect of the constituent sugars. The acceptance of a mixture of equal parts of sucrose, glucose and fructose was significantly higher than that calculated by addition, while the inclusion of maltose appeared to lead to a lower acceptance than the calculated mean. These results were compared with similar discrepancies which have been observed in man (Cameron, 1947; Dahlberg & Penczek, 1941).

V. Frisch (1934) discussed the molecular structure of the sugars which tasted sweet to bees, but he could not determine the relation between taste and chemical constitution; for instance, glucose and mannose differ in structure only in the inversion of an OH group and one carbon atom, but the former is very attractive to bees and the latter is not; they taste equally sweet to man. The sugars which are highly attractive to bees are also attractive to other insects, but the converse is not true; raffinose, unattractive to bees, is as attractive as sucrose to some ants (Annaliese Schmidt, 1938). Comparison between different species of insects shows that marked irregularity in their responses to different sugars is characteristic (Skramlik, 1926); in man similar variations, but of lesser degree, occur between individuals (Cameron, 1947).

Although sugar is a principal item in the diet of honeybees, their mouthparts seem to be less sensitive to small concentrations of sugars than are those of some other insects; perception of $\frac{1}{16}$ M sucrose by starved bees can be compared with $\frac{1}{64}$ M perception by the cabbage-white butterfly *Pieris rapae* (Verlaine, 1927), $\frac{1}{100}$ M by the blow-fly *Calliphora vomitoria* (Minnich, 1931) and $\frac{1}{100}$ to $\frac{1}{200}$ M by ants (Schmidt, 1938). Records of sugar concentration in nectar range from 3% in pear (Vansell, 1942) to 87% in apple (in the dry air of Palestine; Fahn, 1949), but the usual mean in species visited by honeybees is 30% (Ruth Beutler, 1930) to 40% (Park, 1949).

No adequate technique has as yet been developed for determining the minimum difference in sugar concentration which is detectable by bees. V. Frisch (1934) found that they could distinguish between $\frac{1}{8}$ M and $\frac{5}{32}$ M sucrose, but finer discrimination might be of advantage to them. Experienced human tasters can distinguish between 10 and 10·25% sucrose (Dahlberg & Penczek, 1941).

Distasteful substances

The survey of v. Frisch (1934) embraced substances which are acid, sour or bitter to man. These tastes can be grouped together as distasteful to bees (which can probably recognize them as separate qualities). V. Frisch showed that the effects of various acids were additive, and there was also summation of the effects of salts (e.g. NaCl+LiBr) and of bitter substances. NaCl+HCl and NaCl+quinine were also additive, but quinine+HCl produced compensation instead of addition. Octoacetyl sucrose is intensely bitter to man but is apparently tasteless to bees, and bees appeared to be rather less sensitive than man to quinine, but much more sensitive to cocaine. Bees reacted similarly to equimolar solutions of quinine and cocaine, but to man cocaine is only $\frac{1}{40}$ as bitter as quinine. V. Frisch's attempts to measure the threshold of perception of these substances were vitiated by the insensitivity of his technique; the bees were not at any time presented with a choice of solutions, but they were trained to sucrose solution which was then replaced by a sucrose solution to which another substance had been added; the training solution was well above the threshold of acceptance.

V. Frisch's experiments with common salt demonstrate the experimental difficulties. All the bees trained to 1 M sucrose imbibed 1 M sucrose + $\frac{1}{8}$ M NaCl, but one quarter of those trained to $\frac{1}{2}$ M sucrose and one half of those trained to $\frac{1}{4}$ M sucrose were upset by this addition. When 11 bees trained to $\frac{1}{4}$ M sucrose were offered $\frac{1}{4}$ M sucrose + $\frac{1}{16}$ M NaCl, all of them drank it, and only one hesitated. Yet measurements of honeysacs showed that the mean volume imbibed by bees taking 1 M sucrose + $\frac{1}{16}$ NaCl was less than that of bees taking 1 M sucrose only, so $\frac{1}{16}$ M NaCl must have been perceived. In later work Butler (1940b) demonstrated that bees which were collecting water perceived and preferred $\frac{1}{80}$ M NaCl.

V. Frisch found that 31% of bees were upset by the addition to 1 M sucrose of $\frac{1}{10}$ M acetic acid, but none were affected by the addition of $\frac{1}{20}$ M acetic acid (0·3%). Glynne Jones (1952) obtained a similar result when the offered solutions were alternated in time, but he found that 0·01% acetic acid repelled 46% of honeybee visitors to 2 M sucrose when unadulterated 2 M sucrose was offered as an alternative (the roles of taste and odour in repellency were not separated).

Taste perception with the antennae and legs

Dönhoff (1855a) observed that honeybees stretched out their tongues when their antennae were touched with a glass rod dipped in honey or sugar solution, but that they did not do so if the rod had been dipped in water; Briant (1884) and Will (1885) also noted that the antennae of bees could be stimulated

Fig 11 An arrangement of slotted cards round a bee's head to enable the taste thresholds of her legs and antennae to be determined (Marshall, 1935a).

by contact with honey. Minnich (1932) located taste receptors on the antennae and the legs of honeybees. He anaesthetized bees, cut off their wings, and mounted them on small wax blocks. Each bee was allowed to recover and to drink as much distilled water as it would take. Distilled water was then applied to the antenna or leg with a small brush, and if there was no response (extension of the proboscis within 10 seconds) the experimental solution was similarly tested. Minnich compared the responses to equimolar solutions of glucose and lactose, which are alike in viscosity and osmotic pressure although the latter is almost tasteless. Marshall (1935a) adopted a similar technique, but he placed slotted cards round the neck of the bees (Fig. 11), instead of anaesthetizing them and mounting them in wax; he used sucrose and did not compare it with lactose.

Minnich found that $\frac{64}{100}$ M sucrose was clearly distinguished from either lactose or distilled water by either the antenna or the fore leg, but similar stimulation of the hind leg caused a response in only 9% of the tests. The antennal response occurred when only the extreme tip of the antenna touched the solution. Kunze (1933) found that the antennae were sensitive to 2·5–3% (c. $\frac{1}{12}$ M) sucrose or glucose, and to higher concentrations of some other sugars. Addition of weak acids, alkalis or bitter substances inhibited the response. Marshall (1935a) found that $\frac{1}{12}$ M sucrose was the average threshold value to which the antennae were sensitive, whereas the threshold required for stimulation of the foreleg was 1 M (21 bees) or $\frac{1}{2}$ M (6 bees).

SMELL

Site of the receptors

A successful attempt to locate the receptors of smell was made by Lefebvre (1838), who wetted a needle with ether and brought it, from behind, close to a bee which was sucking sugar solution. There was no reaction when he moved it close to the abdomen, stigmata or wings, but when it was placed near to the antennae these were moved towards it.

Subsequently observers of other insects gathered more evidence which indicated that scents are perceived by the antennae, but McIndoo (1914a) claimed that the olfactory sense of honeybees was not located there. However, McIndoo's techniques have been strongly criticized, and they are considered invalid (e.g. Marshall, 1935b).

Precise experiments were carried out by v. Frisch (1921), who trained bees to a scent, amputated their antennae, and found that they would not then respond to the scent; in control experiments he showed that the responses of honeybees trained to colours were not impaired when their antennae were amputated. Then he demonstrated that odour perception was not interrupted by amputation of one antenna only, but was eliminated when the eight terminal segments of both antennae were cut off; when eight segments were removed from one antennae and seven from the other the bees still responded to a scent to which they had been trained. Thus the olfactory receptors were only located on the eight terminal segments of the antennae; this result was confirmed by Frings (1944).

Odour perception. Experimental technique

An experimental technique for studying odour perception was developed by v. Frisch (1919*b*), who measured the ability of honeybees to perceive and distinguish between various scents. Bees were conditioned to collect sugar syrup from a small cardboard or porcelain box supplied with a few drops of a solution of the training scent in liquid paraffin. This box was then replaced by several similar boxes, without syrup, one containing the training scent, others another scent or no scent at all. V. Frisch observed whether the trained bees distinguished between these boxes.

These experiments were carried out with quite large groups of bees, coming during five-minute periods. V. Frisch obtained very valuable results in this manner, but a disadvantage of his method was that the bees tended to be attracted towards any bee(s) which happened to land in the boxes (p. 196). For instance, the bees were trained to a scentless box and given the choice of two similar boxes, and in five successive minutes one box attracted 0, 2, 18, 10 and 11 bees (total 41) while the other attracted 0, 4, 1, 1 and 2 bees (total 8). For this reason such results cannot be usefully analysed statistically.

Ribbands (1953*c*) developed the following technique. Pairs of $2 \times \frac{3}{4}$-inch glass tubes were fastened together with elastic bands, and each pair was placed on a glass Petri dish. Three such pairs were used each time, and were spaced out on a cover placed on a bicycle wheel (Fig. 12), which was slowly rotated.

The training scent was dissolved in ethylene glycol and placed in one tube, the top of which was covered with cotton mosquito netting so that the solution was beyond the reach of bees. The other tube of this pair was filled with sugar solution, which 6 to 10 individually marked bees were trained to collect. One of each of the other pairs of tubes contained ethylene glycol without scent (for threshold experiments) or with another scent (for mixture experiments); the other member of each pair was filled with water. After training, new sets of tubes were put out, with the same scents in one tube of each pair but with nothing in its partner; bees which landed on the top of the pair with the training scent were counted as positive, those which landed on either of the other pairs were negative. There were so few bees that individuals landed one at a time, and were not attracted to one another. Observations were continued until the number

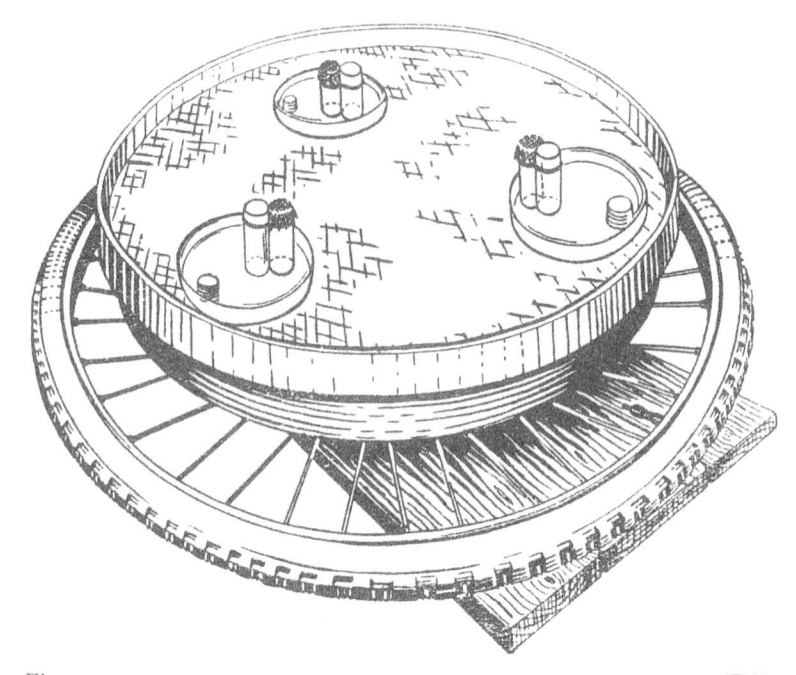

Fig 12 APPARATUS FOR INVESTIGATING SCENT PERCEPTION (Ribbands, 1953*c*). Description in text.

of positives exceeded the number of negatives by five (a significant positive result) or until the number of negatives exceeded the number of positives by six (which implied absence of discrimination).

Recognition of scents

V. Frisch (1919*b*) showed that the bees could readily distinguish between scented and scentless boxes, and that they could be trained to either scented boxes or unscented ones (e.g. bees trained to unscented boxes paid 49 visits to two unscented boxes but only 2 visits to a rose-scented box and no visits to a lavender-scented one).

Bees, conditioned to visit a box supplied with an essential

oil made from the skins of sweet Italian oranges (*Citrus aurantium* ssp. *sinensis*), were then allowed to choose between a box containing this oil and 23 similar boxes containing a variety of other essential oils. In 5 minutes they paid 205 visits to the training scent and 60 visits to an extract of the skins of sweet Spanish oranges, but only 37 visits to all the other 22 boxes together. The training scent was then compared with a different set of 23 scents; this time there were 120 visits to the training scent, 148 visits to an essence prepared from *Citrus medica* and 93 visits to an essence prepared from *Citrus aurantium* ssp. *Bergamia*, but only 29 visits to all the rest of the boxes. Thus the bees distinguished the training scent from 43 other scents, and they confused it only with three scents of very similar origin (which contained the same constituents in different proportions). Subsequent experiments indicated that the training scent was preferred to any of these three scents.

Para- and meta-cresol methyl ether differ only in their molecular arrangement, yet they smell different to both bees and humans. V. Frisch also found that pairs of chemicals (e.g. mirabane and bitter almond), which were of different chemical composition but whose odour was very similar to man, were also similar in odour to bees, which sometimes confused them. Isobutyl benzoate and amyl salicylate could be distinguished by bees but not by man. V. Frisch concluded that, despite the great structural difference between the nose of man and the antenna of the bee, their odour perception had a similar physiological basis.

Bees were trained by v. Frisch to lysol and skatol, and even to carbon bisulphide which was poisonous to them. Yet they could be only imperfectly trained to the camphor-like Patchouli oil, which was readily perceived and definitely repellent.

Bees readily visit certain flowers which are both inconspicuous and unscented to man, but v. Frisch's attempts to train bees to scents from three of these flowers—*Ampelopsis quinquefolia*, bilberry and red currant—were unsuccessful.

Discrimination between mixtures of scents

V. Frisch (1919b) found that bees trained to tuberose scent readily distinguished that scent from a mixture of 10 parts tuberose and 1 part jasmine, but they scarcely distinguished it from a 24:1 tuberose-jasmine mixture. Bees trained to either bromstyrol or methyl heptenone distinguished these scents from mixtures containing 24 parts of the training scent and 1 part of the other scent.

Ribbands (1953c) found that bees trained to 1% phenyl ethyl alcohol distinguished this scent from a 1% mixture containing 119 parts phenyl ethyl alcohol and 1 part geraniol; bees trained to 1% benzyl acetate selected it from 119:1 benzyl acetate-linalol mixture of the same concentration. Bees trained to a mixture of 5 scents, containing 5% scent, were able to distinguish this mixture from a mixture similar but containing also 0·2% geraniol, while those trained to the 5 scents + 0·5% geraniol selected the training mixture from a mixture without geraniol.

Moreover, bees distinguished between mixtures which contained only the same two scents, but in different proportions: bees were trained to select a mixture of equal parts of 1% linalol and benzyl acetate from a mixture which contained $\frac{9}{20}$ 1% linalol and $\frac{11}{20}$ 1% benzyl acetate.

This ability to discriminate between mixtures of scents helps honeybees to find the crops to which their companions have directed them (Ch. 19), and enables them to recognize the individual odour of their comrades and distinguish it from the odours of strangers (Ch. 21).

Threshold of scent perception

The threshold of scent perception should ideally be expressed in terms of the quantity of scent contained in some standard volume of air, but there has been no successful attempt to measure the antennal acuity of the honeybee in these terms. Instead various scents have been mixed with a solvent, and the threshold has been expressed in terms of the minimum dilution in the solvent to which the bees will respond. Human scent perception has been used as a yardstick.

V. Frisch (1919b) measured the threshold of perception for two pure scents which were dissolved in liquid paraffin. Bees conditioned to bromstyrol readily recognized it in 1:2,000 dilution; in 1:20,000 dilution v. Frisch concluded that it was no longer perceptible to most of the bees, but some seemed to recognize it. Methyl heptenone was recognized in 1:2,000 dilution, but not in 1:20,000. Comparing these results with the perception thresholds for himself and his wife, he concluded that the thresholds of perception for bees and humans were similar.

Using a different technique, with scents dissolved in ethylene glycol, Ribbands (1953c) found that bees perceived methyl heptenone in 1:40,000,000 dilution; the threshold for perception by the most sensitive 25% of humans who were tested was 1:1,000,000. Honeybee thresholds for perception of various scented chemicals varied from 1:100,000,000 (phenyl ethyl alcohol, benzyl acetate, methyl benzoate) to 1:5,000,000 (linalol). Honeybee thresholds varied from one-tenth to one-hundredth of the thresholds for humans; honeybees were attracted by odours in concentrations which were imperceptible to man (cf., p. 176) and Ribbands concluded that although comparison between humans and honeybees was of doubtful value, yet it was certain that honeybee threshold values were substantially lower than those of man.

Chapter 4
WATER AND TEMPERATURE PERCEPTION

Water perception

EXPERIMENTS WHICH WERE reported by Krijgsman (1930) and Krijgsman & Windred (1930) showed that the blood-sucking flies *Stomoxys* and *Lyperosia* would react to water at a distance, the response being most marked in flies which had been deprived of drinking water for a day. Buxton (1932) pointed out that there was little other precise evidence of a tropism towards water, but he supposed that many insects possessed organs which enabled them to find water from a distance and that these might be presumed to exist in mosquitoes, for instance, which lay their eggs in water.

Soon afterwards the water perception of honeybees was studied by Mathilde Hertz (1934*a*). Having trained a group of bees to collect from a gauze-covered dish of sugar syrup, she took away this dish and offered a series of alternatives. The bees did not gather over gauze-covered empty dishes, but they congregated over gauze-covered dishes which contained distilled water beyond their reach. They stretched their tongues toward water or moist earth which was 1½–2 inches away, but not toward water which was 3–4 inches away.

Hertz partly filled dishes with several different liquids, each to the same level below the wire gauze; the bees gathered over distilled water, concentrated salt solution or tap water with one drop of 70% alcohol, but they did not gather over 70% alcohol or glycerine. Ribbands (1953*c*) used a scent-perception technique (p. 39) to show that honeybees could be trained to seek sugar syrup in the vicinity of water, but his prolonged attempts to train bees to associate syrup with ethylene glycol were ineffective.

Hertz concluded that there was no doubt that the honeybees' reaction was caused by the molecules of water itself, and that they could distinguish water-vapour concentrations in the air, but she pointed out that we could not deduce from this that they could smell it. She did not know whether the effect on the receptors was chemical or physical, but she was confident that there was true perception by sense organs, which were presumably in the antennae and which were directly dependent on the water content of the air.

Temperature perception

A study of temperature perception was carried out by Heran (1952), who devised an apparatus which enabled him to offer

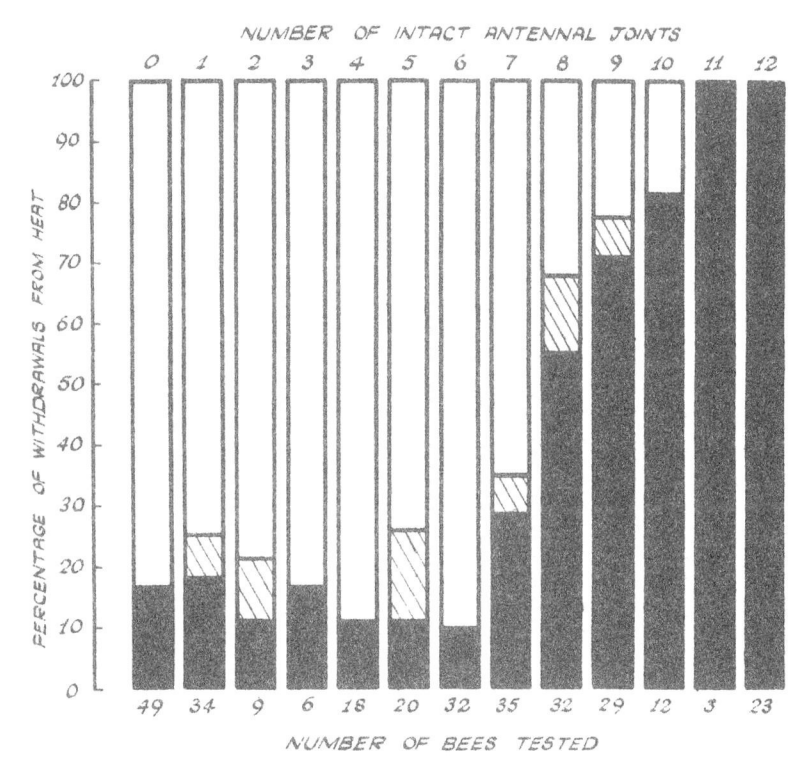

Fig 13 REACTIONS OF BEES TO HEAT (40° C.) AFTER VARIOUS ANTENNAL AMPUTATIONS. Withdrawals in black, doubtful reactions shaded (Heran, 1952).

a choice of three different temperatures at which trained bees could take sugar solution from a dish. They were trained to collect at various temperatures (20, 25, 32, 36° C.) and they remembered the training temperature and distinguished it from temperatures 2° C. above or below it. The bees were thus able to adapt their temperature-perception ability to a quite unnatural purpose.

Heran then studied the resting temperature which they preferred. A long glass-covered box was heated at one end, and the temperature gradient along the box was measured from a row of thermometers, placed horizontally and close to the floor. Five to seven bees were placed in the apparatus at a time, and the temperatures of their resting places were noted. Young bees, up to 7 days old, rested at broodnest temperature (35–37·5° C.); older bees showed more latitude (31·5–36·5° C.), and the thermal preference of winter bees was lower after confinement at a low temperature (32·8° C. after 2 hours at 30° C., 27·8° C. after 5 days at 13·7° C.).

By cooling the apparatus slowly Heran showed that the bees responded to temperature changes of 0·25° C. They were less sensitive to increased temperature and receded from it only when it exceeded 40° C. These experiments were repeated with bees which had suffered various degrees of antennal amputation. All the bees with complete antennae or with only one joint cut off responded to temperature increase, but sensitivities decreased with increasing amputation and were uniformly low when more than five segments had been cut off (Fig. 13). In other insects the most sensitive heat receptors are also found on the terminal segments of the antennae (Wigglesworth, 1950).

Chapter 5
CO-ORDINATION; TOUCH; PERCEPTION OF GRAVITY AND DISTANCE

An important but inadequately studied group of senses responds to displacements, stresses and pressure changes. These changes may be within the body of the insect, when the sensory cells (called proprioceptors) register the relations between one part of the body and another. Alternatively, the sensory cells (then called exteroceptors) may register such changes outside the animal. The former senses help to co-ordinate the parts of the body, and enable the insect to appreciate the pull of gravity; the latter ones include touch, the perception of air movements and perhaps of distance travelled, and hearing (Ch. 6, p. 51).

These senses have been studied mostly by physiological techniques, for which larger insects, such as locusts and cockroaches, are more convenient. For this reason we have to infer their qualities in the honeybee. These techniques involve electrical detection of impulses in the nerve from an isolated sense organ exposed to the stimulus which is being investigated.

The receptors vary according to the type of stimulus. The simplest receptors of external stimuli are the hairs of the body surface[1] (Wigglesworth, 1950); in modifications the hair may be very short or it may become a minute canal surrounded by a small oval dome of thin cuticle sensitive to pressure changes; these domes[2] may be protected by being sunk in small pits overhung by the more rigid surrounding cuticle. In another modification[3] the receptor is wholly embedded within the body and has become a sensory rod lying along the axis of an elastic strand inserted into some pliable region of the cuticle. In more complex modifications receptors of this last kind are arranged in groups to form sense organs, which are sunk into the surface of the body ('Johnston's organ' at the elbow of the antenna, 'subgenual organs' in the tibia of each leg).

---

[1] Hair sensillae.   [2] Campaniform sensillae.   [3] Chordotonal sensillae.

Co-ordination

Pringle (1938) found that domed sense cells were frequent in those parts of the cuticle of cockroaches which are subject to stresses (e.g. the mouthparts and legs), but that they did not occur in any part of the cuticle which was not subject to them; then he showed that these cells responded to stresses. Pringle concluded that, in insects, the force exerted on the limbs by the weight of the body and the contraction of muscles is measured as compression in the skeleton; this is in direct contrast to the vertebrate plan, where it is measured as a tension in tendons. The insect sense cells served only to record the position of a *whole* limb or palp, not that of any one muscle; Pringle considered that this was an important limitation on the variety of insect movement and behaviour. However, he also demonstrated the function of groups of small hairs at the joints of the limbs, which registered the bending of the joints.

Perception of gravity

Orientation in relation to gravity is important for an insect which spends most of its life on a vertical surface. Moreover, this orientation has an important role in communication between foragers (p. 158).

The mechanism of orientation to gravity in ants was discovered by Vowles (1953*b*). A small particle of soft iron was attached with shellac cement to various parts of the bodies of ants, which were compelled to walk in a dark room on a vertical board which could be subjected to a magnetic field. In the presence or absence of the magnetic field these ants did not behave differently from normal ants, which tend to run in a straight line when on a vertical surface. When the magnetic field was switched on or off the ant usually hesitated, but if iron had been glued to the head, thorax or abdomen the ant subsequently continued normally; however, if iron had been glued to the second joint of both antennae the ant either stopped and cleaned her antennae or smoothly changed direction and then continued in a straight path. Results of gluing iron to only one antenna suggested that the gravity receptors function independently in the two antennae and are used singly, the receptor on one side being dominant at any one time and change from one antenna to the other often occurring during a single straight run.

Johnston's organ, situated within the second joint of the

antenna, is the only organ present in that region which could act as a gravity receptor; Vowles sectioned the antennae of ants and found that Johnston's organ had a similar structure in ants to that which had been described in the honeybee.

Vowles also tested ants on a vertical turntable, and concluded that they orientated to gravity with an error of $\pm 33°$. They were confused between pairs of orientations symmetrically placed on either side of the vertical.

Touch

Sensitive hairs which act as touch receptors are probably scattered over the whole surface of the body, but there is a concentration of them on the antennae (studied by Vogel, 1923) and on or near the mouth (McIndoo, 1916).

Perception of distance

Perception of distance travelled is important for orientation (p. 80) and for communication of crop sources (p. 150). The experiments described on p. 160 led v. Frisch (1948) to suppose that the energy expended on a journey was used as a measure of the distance travelled. However, it seems more likely that some modification of a proprioceptor would serve this purpose. Very slow or incomplete adaptation to a continued stimulus is a distinguishing characteristic of proprioceptors (Lissmann, 1950); this property could enable the insect to record the duration of displacement of a structure which was bent during flight, and this duration would be a measure of the distance travelled.

Eggers (1923) suggested that Johnston's organ might register changes in tension caused when the antennae were either actively bent back or passively bent by air currents. Hollick (1940) mounted a fly in a wind tunnel and showed that whenever the insect was exposed to an air stream it pulled up its legs in a characteristic attitude, as in flight; this response ceased when a wax coating was applied to the junction of the second and third antennal joints, where Johnston's organ is situated. Hollick noted that in an air stream the antennae of untreated insects were held a little more erect, and the third joint was rotated relative to the second one; he decided that stimulation of Johnston's organ clearly influenced the tone of the leg muscles.

The experiments of Bethe and Wolf (pp. 80–4) have demonstrated the importance of antennae in honeybee orientation, and the disturbance which is caused by rotation. Direction is not perceived by the antennae (p. 29), so these results may well be due to interference in the perception of distance.

For these reasons I should expect to find the site of distance perception in hairs on the antennae, or in Johnston's organ. Moreover, some of these sense organs might be stimulated only in certain conditions, e.g. if the hairs or the antennae were erected when the insect became excited. This postulate might explain why foragers only dance on their return from good crops, and the dance itself might be in some respects comparable to the discharging of a condenser which had received impulses during the foraging flight.

Chapter 6

HEARING, VIBRATION PERCEPTION, SOUND PRODUCTION

*As secret as the wild bees song
She lay there all the summer long.*
JOHN CLARE, *Sacred Love*

SOUND WAVES ARE conveyed by very small vibrations of solids, liquids or gases. Organs which are sensitive to such vibrations will serve as organs of hearing, and any distinction between sound perception and vibration perception would be arbitrary.

There is little direct evidence of sound perception by honeybees, but analogy with other insects suggests that they are not deaf. As with the senses mentioned in the previous chapter, the most fruitful approach has been physiological.

Hearing: the kind of sounds which might be perceived

Sound waves consist of very small air movements, and sense organs which are so refined that they respond to such movements will serve as organs of hearing.

Kröning (1925) was among those who have tried, unsuccessfully, to train bees to associate sounds with the presence of food; it has been supposed that bees are deaf, and Snodgrass (1925) lent weight to this view by stating that 'from an anatomical standpoint it would seem that bees must be deaf'. More recent opinion has veered in the opposite direction; Pumphrey (1950) concluded that hearing is a well-nigh universal attribute of animals.

This changed attitude has followed from a realization that insects and humans appreciate quite different aspects of sound. Mayer (1874) showed that the long hairs on the antennae of male gnats resonated to a tone of the same pitch as the hum of the female in flight. Having demonstrated that impulses occurred in the nerve from the anal cerci of crickets in response to sound waves, Pumphrey & Rawdon-Smith (1936) concluded that the significance of a sound to these insects could not be judged by its apparent loudness to the human ear; a response from a cercus was readily elicited by a low-frequency stimulus which was quite inaudible to man.

Pumphrey (1940) pointed out that sounds produced both changes of pressure in, and small displacements of, the conducting medium. From anatomical evidence he considered that, whereas the human ear was essentially a pressure receiver, it was probable that all insect auditory organs recorded displacements rather than pressures. Moreover, whereas the human ear discriminated between different frequencies by the method of harmonic analysis it was highly improbable that any insects could do this, or that they could discriminate between pure tones to which they were sensitive except on the basis of differences in intensity. However, Pumphrey & Rawdon-Smith (1939) experimented with apparatus which provided pure tones of any desired frequency, modulated in amplitude to a second chosen frequency; nerve impulses from the tympanum of a locust indicated that this structure responded to the modulation frequency. This was in sharp contrast with human hearing; humans are quite insensitive to large variations in the modulation frequency providing that the frequency is not so low that discrete beats are perceived. Pumphrey (1940) concluded, 'It is therefore easy to understand how insects may recognize as similar sounds which are quite different to human ears, and conversely, may distinguish between sounds which to human ears are identical.'

There have been no comparable investigations in respect of the honeybee, which possesses neither anal cerci nor tympanal organs; the above experiments are relevant because they illustrate the kind of properties which any hearing organ in the honeybee might possess. It is important to remember that sounds used for communication are less likely to be discovered if they are inaudible to the human ear.

Hearing: perception through solids

Sound waves may be transmitted through solids, liquids or gases, and there is evidence that vibrations transmitted through solids can be perceived by honeybees.

In each leg of the honeybee Schön (1911) described groups of innervated elastic strands, stretched between the walls of the tibia to form a 'subgenual organ'. These structures occur in the legs of ants, bees and wasps, butterflies and moths; they are especially developed in grasshoppers and locusts, where they are provided with a tympanum and are known to perceive airborne sounds.

Autrum & Schneider (1948) have shown that these organs are very sensitive to mechanical vibration; in Fig. 14 the threshold for response of the honeybee leg is compared with that of the locust leg and the human ear drum. The honeybee leg only responds to sounds of a higher order of amplitude; against this

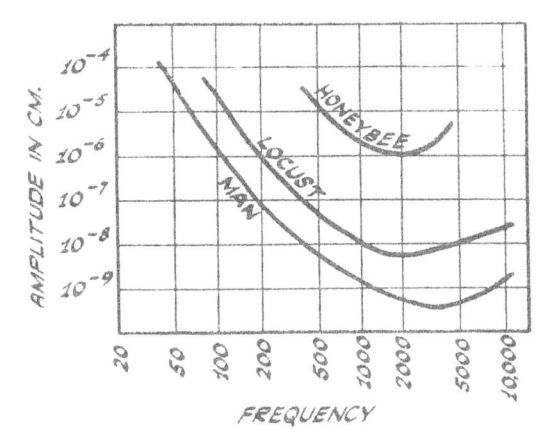

Fig 14 THRESHOLDS FOR RESPONSES TO VIBRATION. Data for human ear drum from Wilska (1935). Data for middle leg of locust and honeybee from Autrum & Schneider (1948).

disadvantage, sound vibrations can travel greater distances through solids with less energy loss than through air.

Airborne sounds are produced by the stridulating organs of grasshoppers and detected by the subgenual organs; the work of Regen and of Faber (cited Pumphrey, 1950) has proved that these sounds provide a very important method of communication, especially in relation to mating—Pumphrey concluded that these insects possess a symbolic emotional language. I think it possible that the subgenual organs of honeybees also assist in communication; the dancing forager expresses distance travelled in terms of the waggle of her abdomen (p. 154)—these vibrations might be transmitted through antennal contact and perceived in Johnston's organ—situated at the angle of the antenna—which is somewhat similar to the subgenual organ in structure, and they might also be transmitted through the comb and perceived in the subgenual organs of potential recruits.

We have seen that insect auditory organs have a different structural basis from those of vertebrates, and that insects distinguish different characteristics of sound waves; 'auditory organs' developed to perceive sounds transmitted through solids may provide another contrast with the human airborne sound receptor (itself a refinement of the lateral line system of fishes, which perceives water-borne sounds).

Hansson (1951) failed to train bees to fly to any sounds, but he was able to train them to associate food with tones (400–1200 c.p.s.) if they were compelled to *walk* into sounding boxes fitted with a loud-speaker; he could not train them to distinguish between different tones.

Sound production

Queens can make a piping noise which is audible to the human ear, and to which other bees are said to respond (p. 287). The pipe of a young virgin is 'a strikingly pure tone, with only about 10% of total harmonic, of 320–340 cycles per second' (Woods, 1950). Workers can produce a more feeble note.

Landois (1867) said that the sound is produced by the spiracles, and is not affected by cutting off the wings. Woods (1950) constructed a model whistle from a metal tube of the size of the queen's abdomen, with 10 holes, 0·007 inch diameter, to represent spiracles; a good imitation of a queen's pipe was obtained by blowing a very small quantity of air through this whistle. Woods suggested that the pipe of mated queens was less effective because their air sacs were restricted by their ovaries.

The tones produced by wing beats were investigated by Hansson (1945), using a cathode ray oscillograph. Stinging bees produced a tone of 300 c.p.s., and lower tones were produced, in order, by virgin queens, by workers flying to good crops, to gather water, to sparse crops, or those flying home. The lowest flight tone (180 c.p.s.) was that of the drone. However, there was much overlapping because individual variations were considerable, and Hansson considered that this was evidence against their use as signals. Lowering of the tone was, in general, a consequence of heavier loading.

Chapter 7

THE GLANDS OF THE HONEYBEE

> *Now what delight can greater be*
> *Than secrets for to knowe,*
> *Of sacred Bees, the Muse's Birds,*
> *All which this booke doth show.*
>
> CHARLES BUTLER, *The Feminine Monarchie*

THE SALIVARY GLAND SYSTEM

A COMPLEX SALIVARY GLAND system is present in all Hymenoptera (Bordas, 1895*a*), and some of the components of this system have been modified for social purposes. A thorough physiological and biochemical study of these glands might yield valuable results, but information at present available is often scanty and sometimes contradictory.

The glands were discovered at various times during the nineteenth century; the functions of some of them were investigated subsequently, but those of others are still unknown. In the head and thorax there are six sets of glands.

The pharyngeal glands

In 1811 Ramdohr announced the discovery of a pair of salivary glands in the thorax of honeybees, and two other pairs of glands were found in the head by Meckel (1846). Fischer (1871) made the first contribution to our knowledge of their function.

The pharyngeal glands are the largest glands in the worker honeybee. They are situated in the forepart of the head (Fig. 15) and when unravelled they are two long ducts, about $1\frac{1}{2}$ times the length of the bee's body, to which 1,000 or more berry-shaped bodies are attached (Cheshire, 1886). The ducts join and open at the back of a narrow trough or feeding groove on the tongue, so that the secretion of the glands can be passed on to either a larva or another bee. There is no closing mechanism to the duct, so the secretion must be sucked out of the gland by the action of the pharynx (Kratky, 1931). The pharyngeal glands are absent in drones and vestigial in queens. Fischer

(1871) noted that the pharyngeal glands were only turgid and well developed in nurse bees. Leuckart supposed that bee milk[1] might therefore be produced in these glands, so he set his pupil Schiemenz (1883) to study this question. Schiemenz confirmed Fischer's observation, and noted that royal jelly[2] and the contents of the pharyngeal gland were both acid (pH about 4·8). Langer (1912) found that the proteins of royal jelly and of the glands were identical. These results, taken together, leave no doubt that the bee milk is produced in these glands.

The waxing and waning of the glands in worker bees has received further attention. Soudek (1927) carried out a detailed study and arbitrarily defined stages in the development of the glands, thus I (full), II (diminished), III (empty), IV (atrophied). This classification has been adopted by others. Soudek found that the glands were small and empty in newly emerged bees, but feeding with pollen (although not with some pollen substitutes) caused rapid development to full condition. At a later age the glands regressed through stages II to IV.

Kratky (1931) confirmed that the pharyngeal glands did not develop in newly emerged bees which were fed on sugar solution only, but they did develop when pollen was added to the diet, if the bees were then less than 10 days old. Bees fed on syrup for 11–20 days, and then on syrup +pollen, did not develop their glands. Young bees with full glands were fed on syrup until the glands retrogressed, and then they were fed with syrup +pollen; no redevelopment occurred.

Kratky noted that there was often considerable variation in the gland development of individuals which were of the same age, and a strong nectar flow tended to speed up retrogression. Gerstung's brood food theory (p. 264) postulated that the duties of the individual are determined by the state of her glands, but both Rösch (1930) and Kratky (1931) observed some young foraging bees with full glands. Kratky found 15–20% of foragers with such glands. It is obvious that nursing duties can only be carried out when the glands are in a suitable state, but it seems that cessation of those duties causes gland retrogression, rather than vice versa. There may be a mutual relationship.

Most of the bee milk produced by the pharyngeal glands is fed to the larvae, but bee milk also provides all or most of the

---

[1] 'Bee milk' is more exact than the more usual term 'brood food'.
[2] The term 'royal jelly' is reserved for bee milk in or from queen cells.

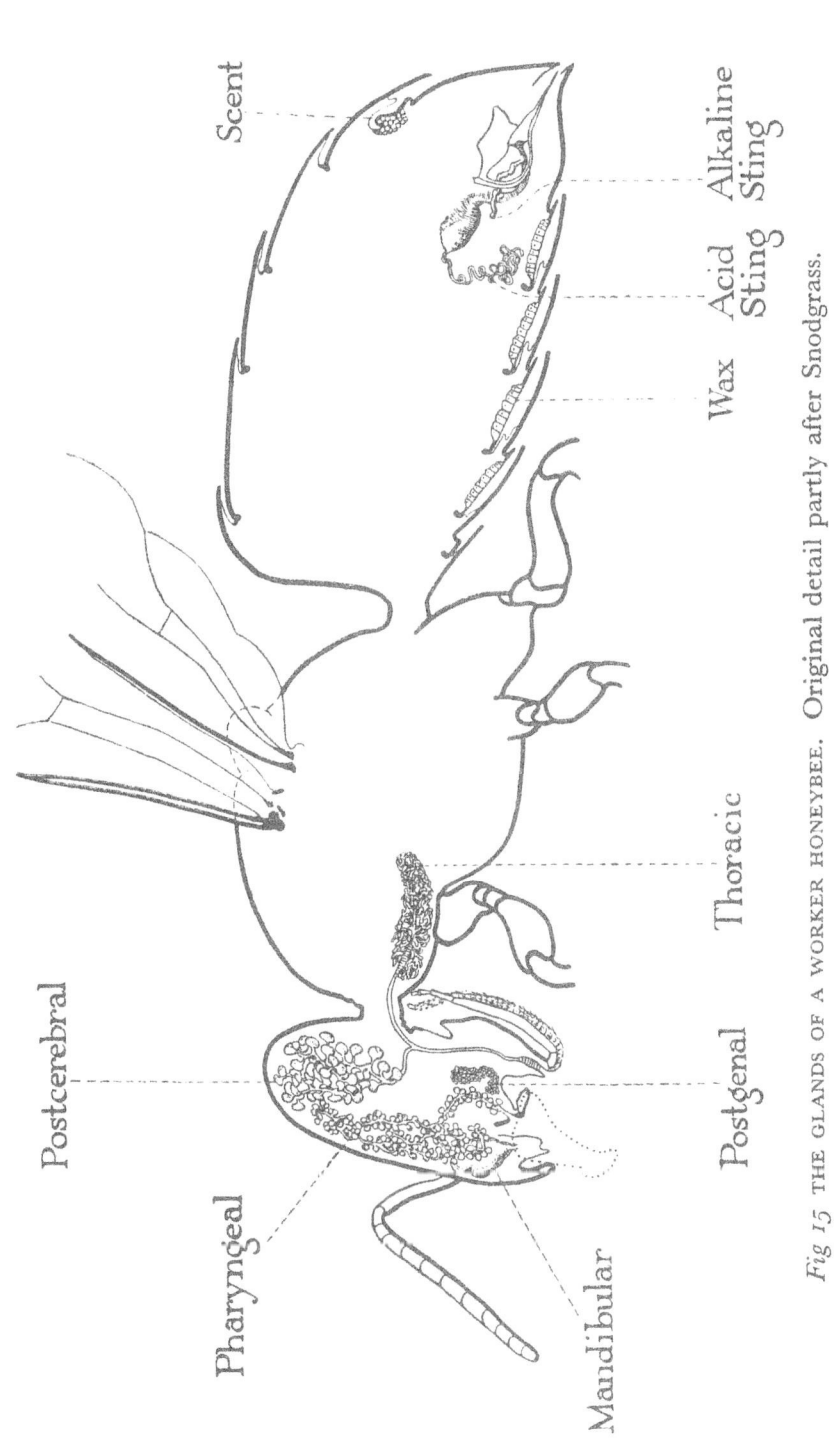

Fig 15 THE GLANDS OF A WORKER HONEYBEE. Original detail partly after Snodgrass.

diet of the laying queen. Feeding of the queen is usually carried out by nurse bees (p. 284), which are most numerous in the broodnest. The weight of a queen, without eggs, is about 0·1 gram, and Cheshire (1886) calculated that 2,000 eggs, which a good queen might lay in one day, would weigh four times as much. Yet the queen's stomach, which is smaller than that of a worker, contains no pollen grains but is filled with a substance microscopically indistinguishable from royal jelly. Her egg-laying capacity, like the very rapid growth rate of the larvae, is dependent upon the receipt of the rich and immediately available protein supply which bee milk provides. There is also evidence (p. 259) which suggests that bee milk is the diet of the drones.

As there is such clear evidence that the pharyngeal glands are full in nurse bees and retrogress in foragers one might suppose that the production of bee milk would be their only function, but biochemical analysis has shown that they are also an important source of digestive enzymes. Kratky (1931) detected no enzymes in the pharyngeal glands of newly emerged bees, or those fed from hatching on sugar syrup only, but he found a strong production of invertase and diastase by the full glands of nurse bees. They contained no protease. Kosmin & Komarov (1932) pointed out that invertase plays an important role in the metabolism of the bee because sucrose cannot be absorbed through the gut wall until it has been broken down or inverted into glucose and fructose. They agreed that there was no invertase in the pharyngeal glands of newly emerged bees, but they observed a slight inversion capacity in bees 24 hours old. The capacity gradually increased, but in 18-day-old bees it was only about one-eighth of the inversion capacity of their midgut (whole midgut, including contents). The pharyngeal glands of many 18-day-old bees had commenced to retrogress, but their inversion capacity rose very sharply. In 20- to 40-day-old bees the inversion capacity of the pharyngeal gland was about one-half of the capacity of the midgut at that time, but it exceeded the inversion capacity of the midgut of nurse bees. Thus it appeared likely that the atrophied pharyngeal glands were secreting a salivary juice which had an important digestive role.[1]

Homologues of the pharyngeal glands are well developed in most bumble bees and in *Andrena*; they are usually present in

---

[1] Inglesent (1940) could find no invertase in this gland.

other Hymenoptera but they are often small, and in true wasps they are rudimentary (Bordas 1895a).

Postcerebral glands

The post-cerebral glands are a pair of much-branched glands which occupy much of the back of the head (Fig. 15). Their ducts join with the ducts of the thoracic glands and form a tube which terminates on the tongue in a salivary valve. Cheshire (1886) considered that the tube was placed so that the secretion was pumped out only when the tongue was extended as for sucking.

These glands are well developed in workers and in the queen, but they are atrophied in drones.

They do not vary with the age or occupation of the workers, and they developed satisfactorily in newly emerged bees which were only fed on sugar suspension (Kratky, 1931). A suspension of them is alkaline (pH 5–8·6) and Kratky could find no invertase, diastase or protease in it. Inglesent (1940) provided results which suggested that it might contain a lipase, but as his enzyme determinations of both the pharyngeal and the thoracic glands are inconsistent with the more detailed analyses of both Kratky and Kosmin & Komarov this work needs confirmation.

The postcerebral glands are said to produce a fatty secretion, which might be concerned with the handling of wax (Heselhaus, 1922), but this is speculation based on the development of the glands in the wax-producing species.

The development of these glands in bumble bees is very variable, e.g. in *B. campestris* they are rudimentary in males but large in queens and workers, but in *B. pomorum* they are much bigger in males than the other sex; they are absent or rudimentary in most other Hymenoptera (Bordas, 1895a).

Thoracic glands

These much branched glands occupy part of the front of the thorax (Fig. 15). They unite into two ducts which swell into two large salivary sacs and then join to form a common duct which unites with the two ducts of the postcerebral glands. These glands are derived from the cocoon-spinning gland of the larva (Schiemenz, 1883) and they are well developed in workers, queens and drones, but least so in the latter (Bugnion, 1928).

These glands, like the postcerebrals, do not vary with age or occupation; they develop satisfactorily in newly emerged bees fed only on sugar solution (Kratky, 1931). They also are said to produce an alkaline fatty secretion (Heselhaus, 1922). Kratky (1931) found no enzymes in them; Kosmin & Komarov (1932) found a trace of invertase in bees less than one week old, but none at all in older bees; Inglesent (1940), ignorant of their results, reported the presence of much invertase in the thoracic glands of bees of all ages.

Thoracic glands are well developed in bumble bees and wasps, and they are present in some solitary Hymenoptera, but their salivary sacs are only found in the honeybee (Bordas, 1895a, b).

Mandibular glands

The mandibular glands are a pair of oval sacs, described by Meinert (1860), which open at the base of the mandible (Fig. 15). They are very small in drones, well developed in workers (Bordas, 1895b) and very large indeed in queens (Snodgrass, 1925). They do not vary in size with age or occupation but they become reduced when the newly emerged bees are fed on a protein-free diet, although there is apparently no protein in their acid secretion (pH 4·1–5·1), which contains a finely divided whitish substance (Kratky, 1931). Kratky found no enzymes in it.

Wolff (1875) thought that because their secretion had a strong smell the mandibular glands were olfactory in nature, but Schiemenz and Heselhaus (while agreeing about the strong smell) thought that the glands were concerned with the intake of food. Forel (1908) said that the mandibular glands had a stinking secretion and were analogous to the anal gland of certain ants, the secretion of which had the same odour for humans.

Heselhaus's (1922) opinion of its function was based on the observation that honeybees have an acid saliva (blue litmus paper with a drop of syrup was turned red when worker bees sucked it clean); he said that this also happened with a queen, which has no pharyngeal glands (the only other glands with an acid secretion).

Katharina Pasedach-Poeverlein carried out an interesting and relevant experiment (1940). She divided newly emerged bees into two equal lots, one to be fed for 20 days with honey only, and the other to be fed with honey +pollen. Preparations of the glands of some of the bees confirmed Kratky's conclusion that the pharyngeal and mandibular glands were reduced by the former treatment but the postcerebral and thoracic glands were unaffected. Both groups of bees were then starved for a short time and afterwards fed with a small quantity of 60% sugar solution. Thirty minutes later their honeysacs were dissected out and examined; the concentration of the sugar solution had declined 15·3% in the bees which had lived on honey and pollen, but it had declined only 1·6% in those which lived on honey only. Thus the dilution was caused by a secretion from the pharyngeal and/or mandibular glands. Drones, in which these glands are rudimentary, were starved and fed on 60% sugar solution, the concentration of which was only reduced 0·6% during 30 minutes in their stomach.

She dissected out the honeysacs of worker bees fed on solutions of known sugar concentration; ligatured at both ends and placed in Ringer solutions of various concentrations, they did not change in volume. Thus the honeysac wall was impermeable. In addition, normal bees were fed on sugar solutions of various concentrations, and their honeysacs were examined at intervals. In those fed on 60% sugar syrup the concentration[1] had declined 0·5% by the time their feeding was completed and 2·5% after they had flown 570 yards to their home. In those fed on 15% sugar syrup there was no change, but in those fed on 10% syrup the concentration was increased about 0·3% after they had flown home. These results indicated that salivary juices had a refractive index approximately equivalent to 15% sugar syrup, that they came from the mandibular and/or pharyngeal glands and that they were added to the food during and after its intake.

These results of Heselhaus and Pasedach-Poeverlein, considered with those of Kosmin & Komarov, suggest that the mandibular and pharyngeal glands together produce the salivary secretion; they may be responsible for different ingredients of the saliva.

Mandibular glands are well developed in nearly all Hymenoptera (Bordas, 1895a).

Postgenal glands

Bordas (1895a, b) described a small pair of sac-shaped glands, homologous with similarly sited glands in some wasps, bumble

[1] Measured by refractive index.

bees and other Hymenoptera, which were against the postgenal plates. A canal from them opens at the base of the mentum. They are rudimentary in drones. Bordas called them internal mandibular glands, but Heselhaus, who confirmed their presence, called them postgenal glands. Their function is unknown.

Sublingual glands

These structures were studied in detail by Bordas (1895*b*). They are situated at the base of the tongue between the two paraglossae and said to open on each side of the mouth.

They are well developed in drones, but they are rudimentary in worker bees. Their function is also unknown. Snodgrass (1925, p. 155) states that they are not glands, but groups of fat cells. Bordas (1895*a*) recorded their presence in wasps, and in some solitary Hymenoptera.

GLANDS IN THE ABDOMEN

Four kinds of glands in the abdomen are concerned with processing by-products of the workers' diet. The alkaline and acid sting glands are present in many other Hymenoptera (Bordas, 1895*a*). The other two kinds, the wax and the scent glands, are characteristic of honeybees and play an important role in their community life.

Wax glands

Butler (1609) stated that wax was carried by bees in the form of little scales, which they masticated and moulded into comb. Wax scales on the underside of the abdomen were afterwards rediscovered by others, including Hunter (1792), who thought they had been secreted. A review and a description of the glands was given by Dreyling (1903–5). The sternal plates cover most of the underside of the abdomen, and they overlap each other. On the anterior overlapped portion of each of the last four of these plates (of segments IV to VII) there are a pair of oval polished surfaces, the wax plates; the glands are thickened portions of the hypodermis of these plates. The wax is secreted in liquid form through the plates and it hardens into wax scales in the pockets formed by the overlapping sterna; the wax scales therefore have the same outline as the wax plates from which they were produced.

The glands develop, and shrink again, according to the age and the duty of the worker bee. Dreyling noted this, and the process was subsequently studied in greater detail by Rösch (p. 303). Wax production is dealt with later (p. 197).

Scent glands

Worker bees sometimes raise their abdomen, turning the tip downward so that they expose a whitish membrane which connects the last two abdominal terga (Pl. VIIa). According to Zoubarev (1883), dissection by Nasanov revealed glands which supply this area.

Sladen (1901–02) discovered that the whitish membrane was a scent-producing organ. He recognized a distinct odour coming from a group of bees which were exposing this membrane, and then found that the same odour was sometimes given out from exposed membranes dissected off from newly killed bees.

The structure of the organ was described in detail by McIndoo (1914b) and Jacobs (1924). These investigators agreed that it is not present in drones. McIndoo said that the queen possesses gland cells in the same position and arrangement as in workers, although the articular membrane is only visible externally at the instant when the queen bends her abdomen to sting, but Jacobs insisted that the queen possesses nothing suggestive of a scent organ and that this area of the integument is almost entirely lacking in gland mouths.

Jacobs reported that single gland cells, of the same structure as the 5–600 which are concentrated together in the scent gland, are scattered all over the bee. This observation would fit with experimental results (pp. 176, 177) which indicate that the colony odour always comes from workers in some degree and that the odour is derived from scented waste products of metabolism. Concentration of many of the cells in one site enables the giving out of odour to be regulated.

Jacobs found that most solitary bees possess integumentary glands, which are sometimes concentrated in groups (e.g. in *Andrena albicans*) and which seem to occur most often on the ventral surface of the abdomen. He suggested that the scent organs originally evolved in female solitary bees as a means of attracting males—and that the development of social life changed their function (see pp. 172–83).

Chapter 8

STRUCTURAL DIFFERENCES BETWEEN WORKERS, QUEENS AND DRONES

Strange all this difference should be
'Twixt Tweedledum and Tweedledee.

BYRON, *Feuds between Handel and Bononcini*

IMPORTANT STRUCTURAL DIFFERENCES are associated with the highly organized social life of the honeybee. Differentiation of the sexes is more extreme than would be possible in a non-social insect; the queens and drones never have to fend for themselves, and they have thus become modified into complementary kinds of reproductive machinery, with reduced non-reproductive functions. The workers, however, carry out all the activities of the colony except mating and egg laying, and their adaptation to the varied tasks is aided by the suppression of reproduction.

To permit this primary division of labour the workers feed the reproductive castes with prepared food, and these reproductives have smaller digestive systems, some of their food-processing glands being reduced or absent (pp. 55–63); their probosces, jaws and legs and the hairs on their bodies are not specialized for collecting and manipulating food, and they have smaller brains than the workers. Their reproductive organs occupy a very large portion of their bodies.

The sexual organs of the two reproductive castes are so arranged that the drone can pass his whole complement of sperm into the queen in one instant and the queen can store the sperm throughout her lifetime. The very long life-span of the queen is one important aid to community existence, but the drone lives for only a short time. Single matings which suffice for life are an attribute of many other insects, but the amplified egg production and longevity of the queen honeybee have increased her attainment.

The sense organs of the drone are highly developed to enable him to locate the flying queen, and his stronger wings enable him to catch her, but the sense organs of the queen are less efficient than those of the worker.

Differences in the brain and sense organs affect behaviour, but studies of sense physiology have almost all been carried out with worker honeybees, so the sensory capacity of queens and drones has to be inferred from comparative anatomy.

Sense organs

Cheshire (1886) found that the antennae of workers were on average $\frac{1}{70}$ inch diameter and $\frac{1}{8}$ inch long, while those of queens were only $\frac{1}{125}$ inch diameter and $\frac{1}{11}$ inch long. The antennae of drones were $\frac{1}{52}$ inch diameter and $\frac{1}{8}$ inch long. Thus the sensory surfaces of the antennae were approximately in the ratio—queen 1, worker 2, drone 3. The minute structure of the antennae was also examined. On the antennae there are small hairs which are usually regarded as organs of touch, and Cheshire estimated that there were 6–8 times as many of these hairs on worker as on drone antennae. The probable smell receptors are small elliptical plates on the antennae; Cheshire counted about 1600 on each queen antenna, 2400 on each worker antenna and 37,800 on each drone antenna; Schenk (1902) gave a somewhat similar estimate. Thus one may infer that the drone's sense of smell is more delicate than that of the worker, and that the sense is used by him in his search for a mate.

The eyes of drones occupy more of the head than those of queens or workers, and they meet in the middle of the head (Fig. 16). Cheshire estimated that the eye of a queen contained 4920 facets, while that of one of her sons contained 13,090, and that of a worker daughter contained 6300 facets. The superior eyesight of the drone also serves to equip him for mating.

Brain

Differences in the brains of workers, queens and drones were described by Jonescu (1909). The relative sizes of the three brains are shown in Fig. 17. The optic lobes of the drone are much larger than those of the worker, which in turn are rather bigger than those of the queen; this corresponds to the structure of the eyes themselves. The antennal lobes are of similar size in workers and drones, but considerably smaller in queens. Jonescu observed that the internal structure of the antennal lobes was much less complicated in the drone than in the worker;

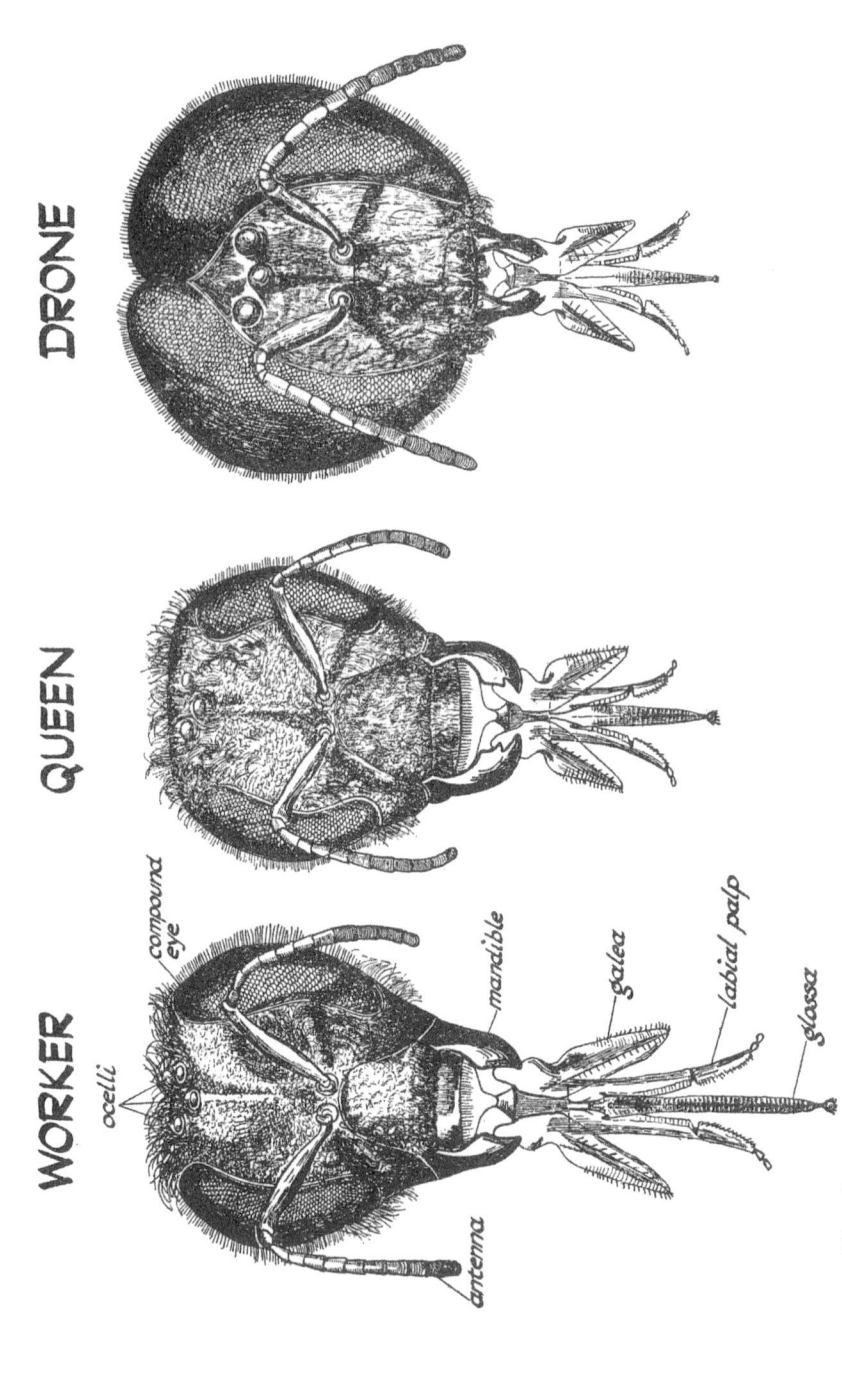

Fig 16 HEADS OF A WORKER, A QUEEN, AND A DRONE. These drawings show differences in eyes, antennae, ocelli, and mouthparts.

this difference could be associated with a difference of function, for the functions of the worker antennae are more diverse.

The third prominent swelling of the brain, the mushroom bodies, are not associated with a particular sense organ but are

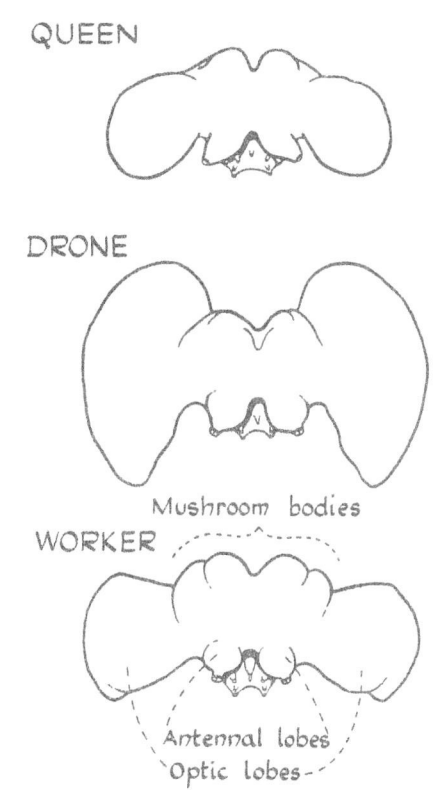

Fig 17 BRAINS OF A QUEEN, A DRONE, AND A WORKER (after Jonescu, 1909). The large optic lobes of the drone correspond with his large eyes. The mushroom bodies, principal centres of co-ordination, are largest in the worker.

the principal centres of co-ordination, to which come a large number of sensory and motor nerves from all parts. These are believed to be the centres of instinct and memory and of associated actions. The mushroom bodies are much smaller in the queen than in the worker, and they are still smaller in the drone.

One may note, in passing, that although the brain development of the social Hymenoptera seems to be superior to that of the solitary species, yet the perfected social life of the honeybee has not been accompanied by a greater brain development than that of other social species. Armbruster (1919) devised an arbitrary 'brain index' based on the dimensions of different parts of the brain, and he calculated that the brain index of both bumble bees and wasps was superior to that of honeybees.

Worker adaptations

Differences between the glands of the workers and of the reproductive castes are discussed elsewhere (pp. 55–63). The workers also possess other distinctive features which are relevant to a study of their behaviour.

Their most important external modifications will be discussed in relation to their task of pollen gathering (p. 119). Their stings have now lost all functions connected with their original role in egg laying, and have become adapted solely for defensive purposes.

PART II INDIVIDUAL BEHAVIOUR IN THE FIELD

Chapter 9

FACTORS WHICH INFLUENCE FLYING ACTIVITY AND FORAGING RANGE

Blaw, blaw ye wastin' winds, blaw soft
Amang the leafy trees,
With gentle gale from hill and dale
Bring hame the laden bees.

ROBERT BURNS, *O' a' the airts*

THE EFFECTS OF weather on flying activity are not readily analysed; weather is itself a complex—temperature, light intensity, wind, rain—and flying activity can be directly affected by any of these, or their interaction, and also indirectly through their influence on the crop.

Nectar concentration and abundance are largely determined by weather factors, but that problem is beyond the scope of this book. Nevertheless, the different ways in which various plants react to the weather introduce a further complication to any analysis of flying activity.

Comparisons between weather statistics and records from weighed hives, the interpretation of which suffers from these and other difficulties, cannot contribute very greatly to an analysis of the effects of the weather components on bee activity.

The effect of nectar supplies

Lundie (1925) made an electrical device which recorded the exits and entrances of bees from their hive; in the wide range of conditions in which weather was not a direct limiting factor, flying activity at the entrance was greatly influenced by nectar supply. This happened mainly because loads are gathered more quickly during nectar flows (p. 130); we do not know how many more bees participate in foraging in such conditions, so the extent to which activities upon the crops are augmented cannot be decided.

The effect of temperature

In winter, spring and early summer, temperature is the important limiting factor for flying activity. Park (1923b),

Corkins (1930) and Walker (1945) have recorded winter flights at an air temperature of 0·5–1·5°C.; more usual temperatures are 8–10°C., and on cloudy or windy days a much higher temperature is required (Wilson, 1922). Higher temperatures are usually necessary for foraging (except for water, which is quickly gathered), although Parks (1925) reported pollen and nectar gathering at 5°C.; Lundie (1925) found that flight commenced at 12–14°C. on clear days in April and at 16–18°C. in May, but that on dull days the temperature had to be 2°C. higher. Lundie found no evidence that a heavy nectar flow induced bees to fly in numbers at a lower temperature than they would have done in dearth conditions at the same time of year.

As bees are cold-blooded, their activity increases with temperature; this is not necessarily reflected in increased flying activity at their hive entrances, since at lower temperatures they gather smaller loads (p. 129). The usual positive relation between temperature and foraging activity may be reversed in tropical conditions, probably due to indirect effects of temperature on crops (Ratnam, 1938; Cherian et al., 1947).

The effect of light intensity

The effect of light intensity on exits from the hive in summer was measured by Butler & Finney (1942), who found that within any one day there was an average increase of 10–20% in the number of bees leaving the hive when the radiation rate increased by 0·1 calorie/sq. cm./min. (the mean radiation rate during the experiment varied from 0·38 to 0·74 calorie/sq. cm./min.). These workers disagreed with the conclusion of Cameron (Brittain, 1933) that the intensity of ultraviolet light was more important than the intensity of the rest of the spectrum.

Changes in light intensity probably enable honeybees to return home quickly in advance of approaching storms. Lundie (1925) gave several examples of this happening, and pointed out that flight activity before and after storms was more closely correlated with changes in light intensity than with any other factor.

The effect of wind

Their own flight speed of about 14 m.p.h. (p. 127) suggests that flying activity is likely to be greatly inconvenienced by winds of similar magnitude. Surprisingly, Lundie concluded that they are not necessarily seriously affected; in one of his records the wind speed was 9–10 m.p.h.[1] on 19 May and 1–4 m.p.h. on the following day, but there were 54 more flights on 19 May. However, a study of his graphs shows that the experimental hive lost 200 grams on 19 May and gained 1200 grams on 20 May. On windy days I have seen marked bees repeatedly make short trial flights and return without foraging—Lundie's method would record such trials equally with foraging trips, and this could account for his result.

Counts of honeybees on apple blossom, made by Brittain (1933), showed a maximum number at a wind speed of 1 m.p.h. and a steady fall to one-seventh of this number at wind speeds in excess of 7 m.p.h. Brittain concluded that even very low speeds had a decided influence on foraging activity.

The effect of rain

Heavy rain inhibits foraging, and wet blossoms after rain may be a serious inconvenience; bees will fly short distances in light rain to some crops (e.g. rhododendron, Brittain, 1933; lime, Ribbands, *unpub.*).

Intervals for lunch?

Hambleton (1925) took hourly records of changes in hive weight through two seasons, and noted that there was usually a temporary decline in the rate of gain of the colony at midday. He pointed out that this decline was not closely correlated with changes in physical factors, and recalled that Bonnier (1879b) had found that plants produced less nectar at this period and fewer bees left the hive. Hambleton felt, however, that this explanation was not entirely satisfactory.

Lundie (1925) did not notice any midday drop in flying activity, but Lindauer (1949) observed that honeybees danced less readily in the early afternoon (p. 162). The experiments of Schuà (1952a) involved training individually marked bees to collect syrup from capillary tubing in artificial flowers; he observed that between 12 a.m. and 1 p.m. their rate of visiting temporarily declined (to about one-half), and he also could not determine its cause (which was not light intensity or temperature); the quantities of syrup absorbed by the bees were not altered during this period.

[1] Taken from an anemometer 5 feet above ground level.

The effect of colony size on flying activity

Lundie (1925) noted that strong colonies in an apiary commenced to fly at somewhat lower temperatures than weak ones. Woodrow (1932) worked in an apple orchard and compared the flighting of ten colonies at temperatures above and below 18° C.; he reached the same conclusion, but this was mainly buttressed by comparison between the largest (8·2 lb.) and smallest (1·6 lb.) colony—eight other colonies (3–6½ lb.) were not obviously affected. Data from seven more colonies (Woodrow, 1934) lent no support to the original conclusion.

The theme of Woodrow's papers was that strong colonies are consistently superior to weak ones for fruit pollination, but the strong colonies were not better *per bee* (my calculations from data in Table I, Woodrow, 1934). A different result might be obtained later in the season, when the larger colonies have reached a peak of brood rearing (p. 235).

The effect of foraging range on flying activity and honey production

The effect of foraging range on flying activity can be conveniently measured in terms of honey production.

Eckert (1933) recorded colony gains from crops growing 7 miles away and reported that during three seasons he had placed groups of colonies in an arid region of Wyoming, devoid of crops, at various distances from an irrigated area; colonies within 2 miles of the irrigated area gained as much as those upon it. Sturtevant & Farrar (1935), in the same locality, compared a group of 30 colonies sited 1 mile within an irrigated area with a similar group sited 1½ miles beyond the edge of the area (2½ miles from the first group). In 1931 the average gain of the distant colonies was 59% of the average gain (73 lb.) of those among the crops; 1932 was a better year, and between 22 June and 16 August (after making allowances in respect of stolen honey) the distant colonies gained 91% of the average gain (218 lb.) of the others. Sturtevant & Farrar concluded that 'apiaries located out some distance from the nectar source may not always do as well as those surrounded by a considerable area of nectar-secreting plants', but 'if necessity requires the placing of apiaries appreciable distances away from the nectar source . . ., other things being equal, the beekeeper can secure a quite satisfactory honey crop'.

These experiments were conducted in a dry district where it is likely that flying conditions were uniformly favourable. Moreover, the colonies were sited in an area and not upon a crop, so the most valuable crops may have been at some distance from the on-sited colonies, thereby reducing the differences between the distances flown by the various groups.

TABLE 1. The effect of foraging distance upon colony net gains

Crop and date		At crop	$\frac{3}{8}$ mile away	$\frac{3}{4}$ mile away
Apple	1949 (2–12v)	13·3±2·7	10·5±2·6	2·0±1·4
	1950 (28iv–18v)	3·0±2·2	−4·4±1·6	−12·2±1·1
Lime	1949 (28vi–15vii)	55·9±6·0	54·0±9·5	32·5±4·2
	1950 (1–15vii)	35·3±5·9	24·3±3·4	20·1±4·8
Heather	1949 (4viii–20ix)	59·6±5·8		51·2±2·1
	1950 (2viii–5x)	25·6±1·2		2·4±0·7
Cabbage	1949 (28iv–2v)	−0·4±0·6	−4·8±0·1	−2·9±0·3
Onion	1949 (7–30vii)	38·0±5·7		30·4±7·5
Total gain per colony, all experiments together—1949		166·4		113·2
1950		63·9		10·3

A series of experiments carried out by Ribbands (1951) compared the gains in weight of a group of colonies sited on the edge of crops with those of groups sited $\frac{3}{8}$ and $\frac{3}{4}$ mile away from the same crops. Three of the experiments were repeated on the same sites in two successive years, under different weather conditions. The results, which are summarized in Table 1, showed that an increase in foraging distance was consistently associated with a decrease in colony gain.

The food consumed by bees in flying such short distances (p. 216) is unimportant in relation to their average nectar load of about 40 mg. (p. 128). Yet the time so spent is important. Bees flying at 14 m.p.h. (p. 127) can make a return trip of $\frac{3}{4}$ mile in 6·4 minutes, so the effect of flying time upon the weight of forage brought in, which varies inversely with the time required to harvest and unload the crop, can be calculated (Table 2). At the height of the lime flow in 1950 Ribbands found that individually marked bees were taking 40·3 ± 5·9 minutes for a complete trip, so a reduction of 14% in their nectar intake at ¾ mile could be attributed solely to extra flying time; any greater reduction was due to reduced flying activity. Pollen loads are gathered much more speedily than nectar loads (p. 131), so the economic flight range for pollen gathering must be less than for nectar collecting.

TABLE 2. Calculated effect of foraging distance on the weight of nectar brought into the hive (Ribbands, 1951)

Time for complete trip, including unloading (min.)	5	15	30	60	120	240
Reduction of nectar intake at ⅜ mile (%)	39	18	10	5	3	1
Reduction of nectar intake at ¾ mile (%)	56	30	18	10	5	3

The effect of foraging distance was very variable; on one occasion (apple, in 1950) the effect even at ⅜ mile was considerable, although on another (heather, in 1949) the gain at ¾ mile from the crop was 85% of the gain at the crop. A foraging distance of ¾ mile reduced the mean total gain by only 32% in the good honey season of 1949, but by 83% in the poor season of 1950. Weather was the most important factor in the production of the distance effect, which was greatest on those days when low temperatures, little sunshine, or high winds were recorded. The differences in weather accounted for the big differences in the magnitude of the effect upon the same crop in the two years.

These results illustrate another complexity in the effect of weather on flying activity; any detrimental effect increases with extension of the foraging range. Ribbands (1952a) pointed out that each colony required a considerable quantity of honey and pollen for maintenance during these experiments; the amount, independent of the distance from the crop, probably ranged between ½ lb. and 1½ lb. per colony per day. Table 3 suggests the relation between incoming forage, the recorded colony gains, and the honey surplus ultimately available for the beekeeper. It shows that the effect of distance upon incoming forage produces a magnified effect upon colony gains, and a still greater effect upon honey surplus.[1] Any increase in the rate of incoming forage will diminish the

[1] Other effects, e.g. those produced by drifting or by disease, are similarly magnified in terms of honey surplus.

proportion required for maintenance, so reducing the effect of foraging distance upon colony gain.

TABLE 3. Suggested relation between incoming forage, colony gain, and honey surplus (Ribbands, 1952a)

	Estimated quantity of incoming forage (daily maintenance allowance, May–July 1 lb. Aug.–Sept. ½ lb.)	Actual colony gains (Table 1)	Estimated honey surplus (allowing 40 lb. honey for winter)
1949			
At crop	235 lb.	166 lb.	126 lb.
¾ mile	182 lb.	113 lb.	73 lb.
Reduction with distance	23%	32%	42%
1950			
At crop	130 lb.	64 lb.	24 lb.
¾ mile	76 lb.	10 lb.	—30 lb.
Reduction with distance	42%	84%	100% (heavy additional feeding required)

These results illustrate one disadvantage of the practice of concentrating large numbers of colonies into a few favourable apiary sites, often several miles apart. This disadvantage is maximal in unfavourable flying weather (including early spring) when a large proportion of the flowers in the area may be left unworked, while the concentration of bees in the immediate vicinity of the apiary exceeds the capacity of the forage there.

Walstrom *et al.* (1951) studied the relation between foraging range and production of red clover seed. Colonies having been placed along one edge of a belt of red clover, ten sample areas of one square yard were harvested at various distances from the bees; the seed yields fell off with increasing distance, but it was realized that more detailed and controlled investigations would be necessary before the extent of the effect could be measured.

Chapter 10

HOW HONEYBEES FIND THEIR WAY HOME

*Oft have I wondered at the faultless skill
With which thou trackest out thy dwelling cave,
Winging thy way with seeming careless will
From mount to plain, o'er lake and winding wave.*

THOMAS SMIBERT, *To the Wild Bee*

THE FORAGING HONEYBEE has to find her bearings in relation to her colony and her crop. Accuracy, the essential feature of orientation to the hive, is ensured by constant repetition during early exercise flights. Colour, which is often a prominent feature of a natural crop, is less likely to be usable as a recognition mark at the hive, but in homecoming the honeybee may make extensive use of landmarks at some distance from the hive entrance, and these may be coloured. Orientation involves the integration of various perceptions, some of which—colour, form, odour—have already been studied separately (Ch. 2, 3).

Landmarks

Fabre (1879) described experiments with marked mason bees (*Chalicodoma*), which often returned to their home after he had taken them up to 3 miles away from it. He held that they possessed a sense of direction which enabled them to do this, but critics pointed out that his results could also be explained in terms of orientation to landmarks. At the suggestion of Darwin he tried to disorientate the bees by rotating them during their journey or by affixing small magnets to them (Fabre, 1882), but these techniques did not affect their homing abilities.

Romanes (1885) took a colony of honeybees to a house which was several hundred yards away from the seashore and separated from it by a wide flowerless lawn. Having allowed the bees to fly for a fortnight, he closed their hive entrance. At intervals it was then partly opened, and groups of 20 bees were collected as they tried to get out. The front of the hive was covered with a sticky substance in order to catch all returning bees. Groups were liberated out at sea, on the shore, and on the lawn to seaward 200 yards from the hive; none of these bees returned. Groups were then liberated at intervals at various points in the garden. These bees returned to their hive, although many were liberated at distances greater than 200 yards, so Romanes considered that the bees were orientating in relation to recognized landmarks and were not making use of a special sense of direction.

Wolf (1926) carried out a more detailed version of Romanes' experiment. A colony was moved to a new and distant site, and groups of bees were caught, marked and liberated out of sight of their hive at various distances and times. Three groups of six were caught before any bees had flown from the new site; they were liberated 165, 440 and 770 yards away from it, but none returned. After 1 hour of flying, 6 bees were liberated 60 yards away; only 3 returned. After $3\frac{3}{4}$ hours, 4 bees released 60 yards away all returned; 5 were released 115 yards away but none came back. After 5 flying hours, 6 bees released at 115 yards returned; 6 more were released at each of two points 165 yards away in different directions; 3 of one group returned, but none of the other. Other groups were then released, in various directions and at distances up to 1030 yards; the results showed that both distance and place of release were important. The distance from which the bees returned increased slowly with the number of flying hours; they returned much more readily from cleared areas and did not return at all from dense wood (where they orientated back to the site of release). Wolf therefore considered that the bees gradually obtained an accurate knowledge of their environment by orientation flights, that they flew along paths which were firmly marked by optically fixed points, and that they could register and remember the angles which they described on these paths. He said that after an extra broodchamber is added to a hive, if the bees were accustomed to fly over its roof, they will drop through the usual distance and attempt to enter the hive too high up.

Bethe (1898) experimented with displaced hives. He moved a hive backwards for 20 inches; some of the bees flew round in circles and then entered, but others congregated at the site of the old entrance. Still more of them congregated at the old site when the hive was moved $6\frac{1}{2}$ feet backwards, but when it was returned to its original position again they entered unhesitatingly. Another hive was turned through 90 degrees; the bees again clustered at the old site, but after a few days they were returning normally to the new position. Bethe also liberated small groups of bees from boxes at a distance from their hive; they flew away but they eventually returned to the position from which they had been liberated. This happened either in the presence or the absence of the box. Bethe thought that if vision had been the determining factor in orientation the bees would have found the displaced hive or box, so he concluded that the bees were obeying some unknown force which led them to return to the position in space from which they had come. Because the bees did not return to their hive from distances greater than a mile or two, he held that this force did not act at long distances.

His conclusions were vigorously criticized by both v. Buttel-Reepen (1900) and Forel (1908). V. Buttel-Reepen repeated Bethe's box experiments, and attributed the results to memory for locality. Forel also considered that the results demonstrated orientation by visual landmarks; he said that honeybees with varnished eyes did not return, although they would have done so if the unknown force had been effective.

Kathariner (1903) repeated Bethe's experiments with slightly displaced and with rotated hives. When their hive was rotated through 45 degrees the bees found the entrance correctly; when it was turned through a further 45 degrees so that the side wall faced the front, the bees landed at the old position, were confused, and eventually ran round the corner to the old entrance. From the first result Kathariner concluded that when the visual stimuli were altered the bees modified their return flight so that they landed correctly; when these stimuli were unaltered the bees landed, and Kathariner believed their subsequent running round to the old entrance to be a response to the smell of the nest.

Colour and Odour

In another experiment Kathariner sited two hives on a terrace, with their entrances 51 inches apart. One hive, containing a colony, was painted yellow, and the other, empty, was painted green. After two months the colony was divided between the two hives, the queen remaining in the yellow one which contained fewer bees; the green hive was placed on the site of the yellow one and the latter was moved farther along the terrace, so that the two were in the same relative position. Soon the yellow hive contained about three-quarters of all the flying bees, a result not in agreement with Bethe's hypothesis; Kathariner considered that this was due to vision rather than to the attractiveness of the odour from the queen.

Two days later 15 foragers from each hive were captured, marked, and liberated 110 yards away; 19 were seen to return, 16 of which entered the hive from which they had been taken. Another pair of samples was taken, and these bees were not liberated until the fronts of both hives had been covered with cardboard of the opposite colour. Kathariner noted that the bees which then returned only went into the hives 'after much hesitation', which 'was not understandable unless the eyes played a prominent part in the choice of hive'. Nevertheless, 13 out of 14 marked bees eventually entered the hive from which they had been taken, despite the misleading colour of its front. This result might now be interpreted in terms of distinguishable odours, becoming effective when the visual stimuli were confusing.

A rather similar technique was used by v. Frisch (1914), with a row of hives in a beehouse. The fronts of three of them were covered with zinc sheets, one painted white and opposite sides of the other two painted yellow and blue. The sheets were arranged as in Fig. 18(i), and bees were placed in D; the other hives remained empty. Some days later sheet D was reversed, sheet E was reversed and transferred to C, and sheet C was transferred to B (Fig. 18(ii)). The bees left D as usual, but nearly all of them returned to the empty hive C, thus demonstrating that the colour of the hive front was playing the predominant part in their orientation.

Wolf (1926) carried out a series of experiments on some aspects of orientation. He demonstrated the importance of colour by standing a populated hive and an empty one side by side, the former on a yellow and the latter on a blue sheet; when the sheets were exchanged the rate of homecoming was temporarily reduced. In addition, bees were trained to colour at their hive, for three days, and hive and coloured sheet were then displaced and a sheet of another colour placed on the original site; the disturbance was much less than that which occurred after the same displacement without coloured sheets. Nevertheless, trained bees sometimes flew into hives of complementary colour, so Wolf held that colour was not a very important factor in home-finding orientation; when yellow and blue sheets under a hive were repeatedly exchanged either colour quickly lost its influence on orientation.

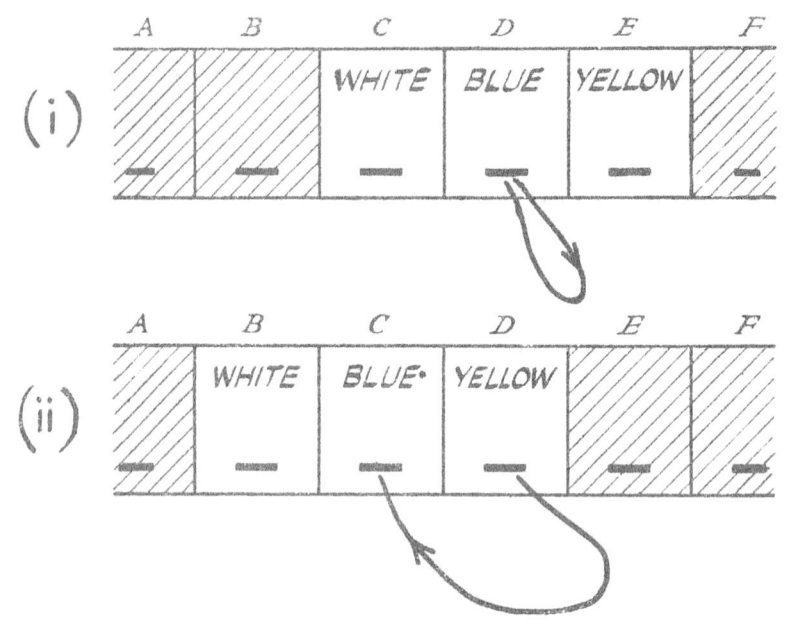

Fig 18 ORIENTATION TO COLOUR (v. Frisch, 1914). (I) First arrangement of coloured hives. (II) The bees remain in the same hives, but the coloured fronts are rearranged; the bees return to an empty hive of the colour to which they had become accustomed.

Wolf then conducted experiments with both scent and colour, and with scent alone (blotting paper was impregnated with plant scent and pinned out of sight inside the hive entrance). The influence of the scents was independent of their quality but dependent upon their intensity; Wolf concluded that scent was more important than colour in orientation to the hive. In all his displacement experiments, bees congregated at the normal position of the hive despite the use of both scent and colour, so he considered that the bees were orientating with the aid of landmarks at a distance from their home.

Perception of distance and direction

Bethe's critics had been primarily concerned to discredit his main thesis (that insects are merely reflex machines) and they paid no attention to one significant experiment. Bethe (1898) had caught a number of returning bees, cut off their antennae, and put them back into their hive, which was moved a short distance backwards or forwards. When the bees flew out the normal ones collected at the old position, but most of those without antennae flew back to the displaced hive. Thus the antennaless bees were orientating only by vision, which brought them to their hive, but in the normal bees some other faculty was dominant.

On an open piece of ground, which was devoid of clear landmarks and at a distance from their hive, Wolf (1926) liberated groups of 20–50 bees from a small box; he found (cf. Bethe) that the bees returned to the position of liberation, whether or not the box was there; there was little or no interest in the box if it was displaced 2–3 yards from this position, but if a bee did find it she exposed her scent gland and attracted others. This experiment was repeated with bees whose antennae had been amputated; when the box was displaced these bees almost never returned to its original position, but after a short time they found the box and clustered round it (cf. Bethe). Hence Wolf concluded that the antennae had a particular function in orientation; he surmised that when a bee flew out she constructed a definite number of orientation curves which were registered by her antennae and that her memory of these enabled her to return to her starting point.

To test this hypothesis Wolf captured returning foragers, marked them, took them to a spot which was known to them and 490 yards from their hive, released them one at a time, and noted the time of their release and of their arrival at the hive. Four groups of 70 bees were used—normal bees (carried carefully from the hive without shaking), rotated bees (carried from the hive in a cage which was continually rotated), bees from which the antennae were completely cut off, and antennaless rotated bees. Their journey times are recorded in Table 4.

There was no significant difference between the time taken by normal bees and antennaless bees in returning, but the rotated bees returned more slowly than either. Rotation had no effect upon antennaless bees, but Wolf found that the time taken in returning was still increased by rotation when only the last seven segments of the antennae were removed. He rested all groups for 10–20 minutes before liberating them, without effect on the result; it made no difference whether the bees were carried in an open or a light-proof receptacle. Similar results were obtained in a repetition of the experiment during the following year.

TABLE 4. Returning times of foragers captured at the hive and released 490 yards away (Wolf, 1926)

Bees	\<1 min.	1–2 min.	2–3 min.	3–4 min.	4–5 min.	5–6 min.	6–7 min.	7–8 min.	\>8 min.
Normal	6	30	18	8	4	2	2	0	0
Rotated	0	11	23	9	9	2	2	6	8
Antennaless	3	24	19	13	9	3	0	0	0
Antennaless rotated	3	23	18	11	8	3	1	0	0

This work was followed up by other experiments (Wolf, 1927), which are widely known. Wolf carried a colony to the middle of an extensive area of open waste land, *'without optical orientation marks'*, and trained the bees to a feeding-place 165 yards away. He repeated earlier experiments on colony displacement for short distances, and found that the bees were still confused at first; forward or backward displacement was much less confusing than a displacement to the left or right and a 90-degree rotation of the colony was more upsetting than a 10-foot sideways displacement, but in all instances normal rate of incoming was quickly resumed.

TABLE 5. Bees trained to a feeding-place 165 yards from their colony, on a site devoid of optical marks. Mean return times (in seconds) of groups of 50 bees captured at the feeding-place, marked and liberated (Wolf, 1927)

Bees	Released at feeding-place, 165 yds. from colony	Released 165 yds. to right of colony	Released 165 yds. to left of colony	Released 165 yds. away on opposite side of colony
Normal	32	88	102	168
Rotated, with antennae	97	184	185	213
Unrotated, without antennae	114	212	196	256
Rotated, without antennae	117	221	214	252
With only 1 antenna	130	216	208	247

Then he captured and marked bees at the feeding-place, confined them in a cage and liberated them one by one, either at the feeding-place or at the same distance from their hive in three other directions from it. The bees were either normal, rotated, antennaless, antennaless and rotated, or deprived of one antenna. Fifty bees of each kind were liberated at each position. There were striking differences between the journey times from the four different points of liberation (Table 5). Among the normal bees those liberated at the feeding-place returned most quickly, those liberated to either side of the colony took three times as long, and those liberated on the

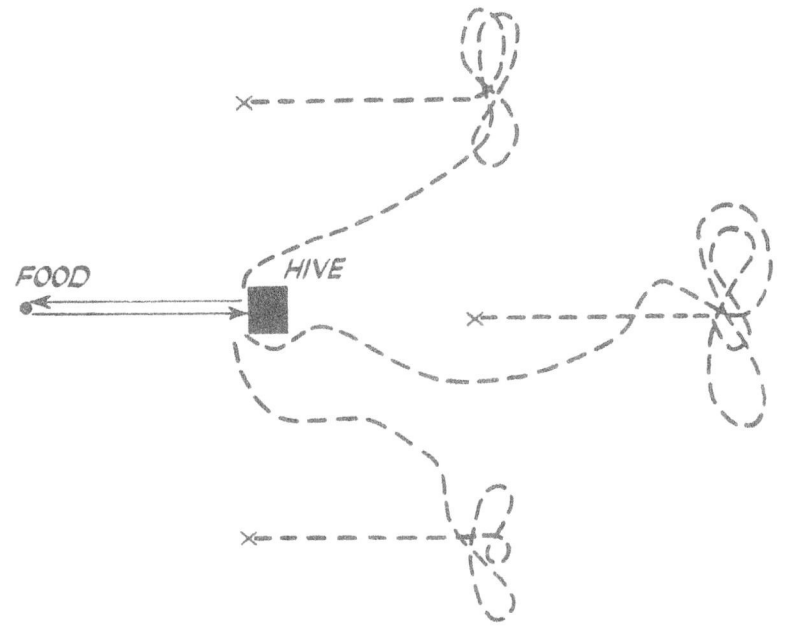

Fig 19 RETURN OF TRAINED BEES CAPTURED AT THEIR FEEDING-PLACE AND LIBERATED IN OTHER DIRECTIONS FROM THE HIVE. X: points of liberation; dotted line, supposed route. The experiment was carried out on a site without landmarks (after Wolf, 1927).

opposite side took five times as long. Wolf accounted for the differences by supposing that when the bees were liberated they flew a definite distance along their original path before they orientated themselves and discovered their whereabouts (Fig. 19). In this experiment the antennaless bees returned less quickly than the rotated ones. This result conflicted with Wolf's earlier experiments (Table 4), so he repeated the new experiment in a situation where optical orientation marks were abundant. In the latter circumstances the antennaless bees returned more quickly than the rotated ones. Thus after rotation the antennae were of limited value in the absence of visual stimuli, but in their presence the abnormal impulses from the antennae were more confusing than no impulses at all.

Wolf said, 'Whereas the normal bees flew off immediately and hardly constructed any orientation curves, always adhering to a straight flight path, the behaviour of the rotated and antenna-less bees was different. The rotated bees with antennae always

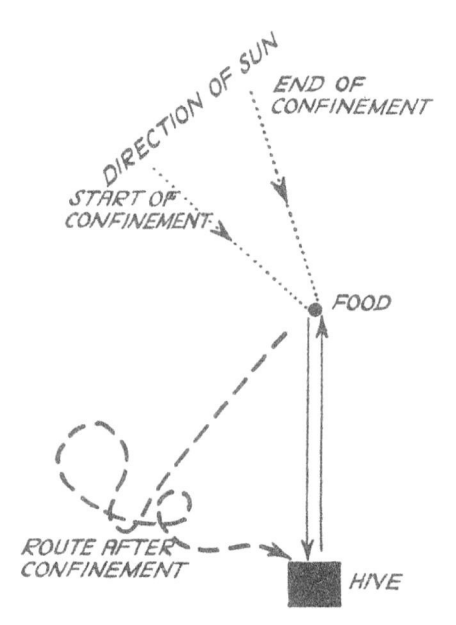

Fig 20 ORIENTATION IN RELATION TO SUNLIGHT, ON A SITE WITHOUT LANDMARKS. Bees confined at the feeding-place for an hour took longer to return to their hive; they returned more quickly if the hive was displaced by an angle equal to that through which the direction of the sun had moved (after Wolf, 1927).

constructed orientation curves for a longer time after their take-off, and then they flew forward in a straight line. The antennae-operated bees flew similarly but behaved unsteadily in their flight, swaying to and fro so that their path was more helical than straight. . . . The antennae apparently regulate the flight and hinder deviation from the prescribed direction.'

Orientation in relation to sunlight

Santschli (1911) and Brun (1914) had shown that ants were able to orientate themselves and travel at an angle to the sun, and Wolf (1927) surmised that his bees were also making use of this ability. To test this hypothesis he captured 50 trained bees at their feeding-place and confined them for 1 hour. When liberated their average time for return to their colony was 107 seconds (range 60–162). A second set was similarly treated, but the colony had meanwhile been displaced by an angle equal to that which the sun had traversed in the hour of confinement; these bees took only 44 seconds (range 10–80) to find their way back. Hence the bees were orientating in relation to sunlight (Fig. 20).

In another experiment Wolf (1927) demonstrated that homecoming bees used optical marks along their track if these were available. Between the hive and the feeding-place he laid down

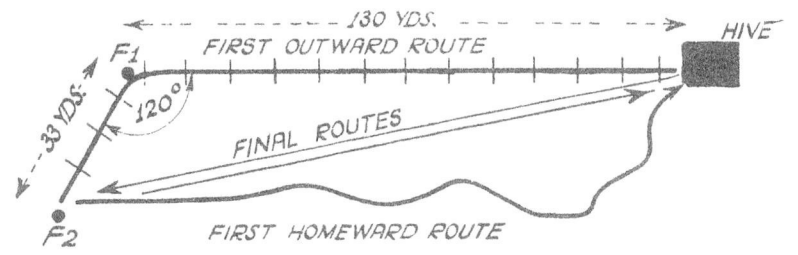

Fig 21 ORIENTATION THROUGH ANGLES (after Wolf, 1931). Trained along a line of coloured cards to F_2, bees flew through the same angle and distance (*not* the same route) on their homeward journey. They quickly learned to disregard the cards and to fly to and fro in a straight line.

parallel tracks of blue and yellow cards, 14 inches square, placed 23 feet apart. When he displaced either the hive or the cards the homecoming bees were confused, but when he displaced both together there was less interruption.

This experiment led to another one (Wolf, 1931) which was carried out on the same waste land, free from optical marks. Bees were trained along a line of blue cards 26 feet apart to a feeding-place F_1, 130 yards away. The food was removed from F_1 and another line of cards was added, at an angle of 120 degrees to the first line and for a distance of 33 yards, from F_1 to a new feeding-place F_2. Within 30 minutes the bees were trained to F_2; they went and returned along the marked track. This continued for about half an hour; then the bees still went along the marks but they returned along path *b* (Fig. 21) and in doing so 'they often curved away to the side, the left side being preferred'. The bees returning along *b* were adjusting their flight, including the angle through which they had to turn, relative to the incident light. However, the corners of the route were soon cut off and within an hour the bees were flying direct to and from the hive along route c. They were orientating entirely by incident light, disregarding the cards which had previously guided them. Variations of this experiment were repeated several times, with the same result.

Speed of adaptation to change of home; the role of distinctive colony odour

The adaptability of homecoming honeybees to changes in the position of their entrance was studied by Ribbands & Nancy Speirs (1953a). They made a tier of three broodchambers, each with a floorboard and entrance, kept a nucleus of bees in the middle broodchamber, and added individually marked bees of known age at intervals. Then the middle and upper broodchambers were exchanged and the entrance turned through 90 degrees. Two days later the upper and lower broodchambers were exchanged and the entrance was turned through another 90 degrees. Thus the bees flew from three different entrances in succession, each at a different height and facing in a different direction (Fig. 22). They could only return by the entrance from which they had flown, and the speed of adaptation of each individual was observed.

The ages of the individuals did not affect their adaptability. The bees re-orientated quickly and completely after the first change, and still more rapidly after the second one. After the first change some of the bees even returned correctly to the new entrance from their first flight, and after the second change 50% of them did so. Orientation circling on exit apparently did not affect the accuracy of the return flight (cf. Opfinger, p. 98), and it appeared that colony odour had played an important part in re-orientation. Fanning[1] at the new entrance, with exposure of the scent glands, facilitated re-orientation, the number of scent-producing bees being proportional to the difficulties of re-orientation; the scent-producing bees, which were of all ages, had not necessarily themselves encountered great difficulty (one-third of them had returned correctly). The speedier adaptation to the second change of entrance probably occurred because the first change had conditioned the foragers to greater reliance upon colony odour as an orientation mark.

Solitary bees also use their own distinctive odour for orientation purposes. Skaife (1952), who kept African carpenter bees (*Mesotrichia caffra* L.) in rows of bamboo tubes, noted that sight

---

[1] Rapid vibration of the wings, producing a directed current of air.

played 'but a small part' in recognition of the home. If a group of tubes were displaced a yard or so away the bees ignored them and returned persistently to their original site. Alterations in differently coloured bands placed at tube entrances did

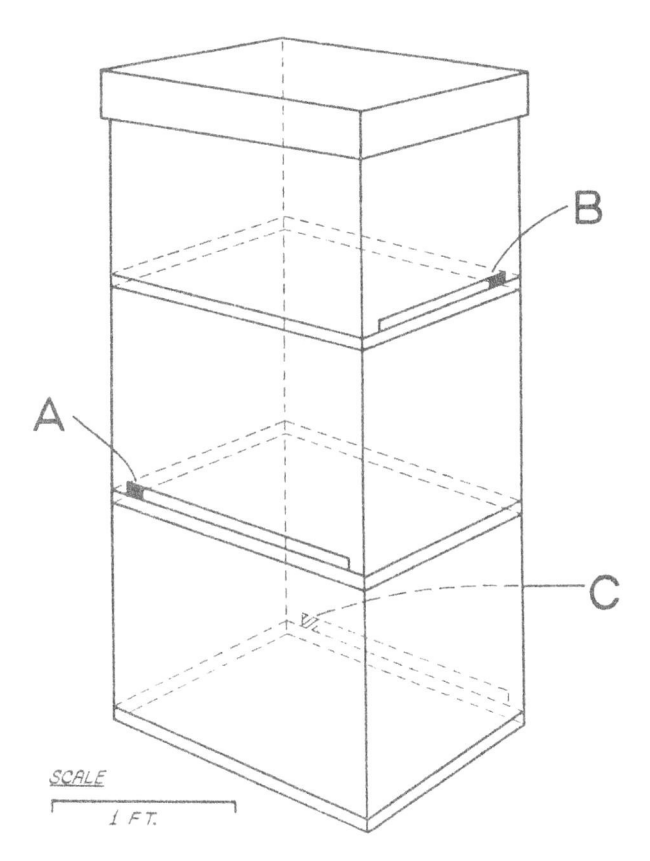

Fig 22 ADAPTABILITY TO CHANGE OF HIVE ENTRANCE (Ribbands & Speirs, 1953a). The bees were made to fly, in succession, from entrances A, B, and C. Results are given on p. 86.

not hinder the bees at all. If the position of the tubes in a group was changed the bees were not greatly troubled; they flew from tube to tube until they found their own, which they entered without hesitation. A female might sometimes be seen turning slowly round in her tube and pressing the tip of her abdomen against its mouth, a performance which presumably served to impress her own scent upon the entrance, and which was kept up for half an hour or more.

Conclusions

The experiments quoted in this chapter have together demonstrated that the eyes and the antennae both play an important role in homecoming orientation.

Colours and shapes provide visual landmarks in the vicinity of the hive and also on the route back to it; we can readily appreciate the importance of these factors, because these perceptions of honeybees are related to, although different from, those which we possess (pp. 16–27). As we cannot perceive the plane of polarization of light, the honeybees' ability to do so is less easily appreciated; Wolf's experiments (p. 84), with the more recent ones of v. Frisch (p. 157), have demonstrated one of its uses, showing that it enables the honeybee to orientate itself in any required direction.

Honeybees probably perceive the distance through which they have travelled with their antennae (p. 49). Wolf's work demonstrated this ability and agreed with that of Bethe (pp. 80–4) in suggesting that this factor sometimes over-rules vision; v. Frisch has demonstrated its value in recruitment to crops (Ch. 19). The ability is used in conjunction with the above-mentioned perception of direction.

The sense of smell is located in the antennae (p. 38). The use of colony odour is complementary to the use of distance and direction; the latter will bring the bee into the vicinity of its goal and the former will then enable it to locate the exact position. Thus these abilities taken together and carried to perfection could provide an alternative to colour and pattern vision, and we can only guess the extent to which they supplement or supplant the capacity for recognition of shapes and colour. One may wonder to what extent the sense of distance and direction is well developed in other animals for which scent plays an important part in orientation.

The relative importance of the different factors used by the honeybee in orientation will obviously depend upon circumstances. The variety of the factors makes an important contribution to adaptability.

Chapter 11

DRIFTING

> *Show me the way to go home,*
> *I'm tired and I want to go to bed,*
> *I had a little drink about an hour ago,*
> *And it's gone right to my head.*

Drifting of workers and virgin queens

BEEKEEPERS HAVE a practical interest in the orientation of the homecoming honeybee, because errors in orientation may lead a bee to enter, or try to enter, some hive other than her own. This process, called 'drifting', is more likely to occur when hives are kept very close together, as in most European beehouses; in such circumstances Rauschmayer (1928) found that nearly half the flying bees in a colony drifted into adjoining colonies when no orientation marks were provided, so he advocated adorning the entrances with marks made up of combinations of various colours; he found that different patterns of the same colour were useless. Also, bees took no notice of orientation marks placed more than 8 inches above their entrance, but they took note of objects below it.

Drifting by virgin queens or by diseased bees has obvious disadvantages, and in the robbing season any drifting bees might be killed. King (1932) marked some hundreds of young bees and reported that in a row of hives four feet apart drifting was so severe that by the time the bees were a fortnight old one in every thirteen had entered another hive. He claimed that in such circumstances production records from individual hives were of little value.

Corkins (1932b), who had advocated such records, arranged an experiment in three apiaries, in which black Caucasian and yellow Italian colonies were equally distributed. Random samples of 1500 bees were shaken from the combs of each colony for ten consecutive weeks, 237,000 bees being counted during the season. The method measured only the drift from colonies of one strain to those of another, but doubling the measured value should somewhat overestimate the proportion of drifting. So doubled, the mean percentage of drifted bees for all colonies for the season was 4·42 ± 0·06. Corkins observed no tendency for bees to drift from strong colonies to weak ones, or vice versa.

In all three apiaries the colonies had been placed 2–3 feet apart, but in one there was a row of only 6 colonies, in another a group of 10, and in the third the 6 colonies which were sampled were in the middle row of a group of 60. In the last, Corkins considered that conditions for maximum drift had

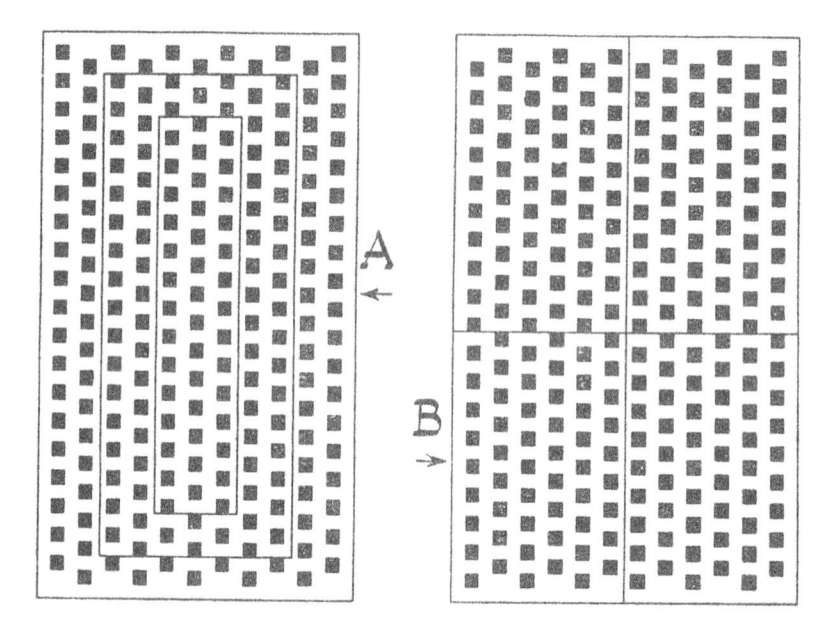

Fig 23 PLAN OF APIARIES FOR EXPERIMENTS ON DRIFTING (carried out by Nekrasov, 1949).

been provided; however, he found no significant differences between drifting in the different apiaries.

Corkins considered that the amount of drifting was not sufficient to impair the value of the records of individual colonies in an apiary. His conclusion should, however, not be accepted unreservedly. His samples were so taken that a high proportion of very young bees, of pre-drifting age, would have been included in them, and the experiences of Koch are relevant (p. 27).

Any effect of drifting upon the foraging population will be magnified in terms of honey yields (p. 74), and the recent work of Nekrasov has shown that drifting can affect these yields.

Nekrasov (1949) studied drifting in seven Russian apiaries, each containing about 250 colonies placed in rows. For analysis the colonies in each apiary were divided into either three rings (Fig. 23a) or four blocks (Fig. 23b). The colonies were removed from their winter quarters and placed on their summer sites, and a few days later (15 April) the weight of the bees in each colony was obtained. The result indicated that during re-orientation there had been substantial drifting towards the centre of the apiaries; in five apiaries (1178 colonies), the weight of bees, expressed as a percentage of the average

TABLE 6. Honey production from colonies in different parts of the apiary (Nekrasov, 1949)

| Location (see Fig. 23b) | Average production per colony in lb. ||||||
	Apiary 1 251 colonies	Apiary 2 203 colonies	Apiary 3 275 colonies	Apiary 4 218 colonies	Apiary 5 256 colonies	Apiary 6 251 colonies
Block A	77·7	49·4	16·3	39·6	49·0	55·8
B	74·8	47·9	15·4	39·8	49·9	63·9
C	59·8	50·8	25·5	50·1	62·2	56·5
D	59·6	49·3	23·9	43·1	52·2	62·7
Difference between the best and worst blocks	27%	6%	51%	24%	24%	13%

weight per colony within each apiary, was 95, 101·2 and 106·4% in the outer, middle and inner rings respectively. Despite this initial drift, weights of bees in the same colonies were not significantly different in the three rings on 20 June (99·4, 100·4 and 99·6% of mean, respectively), and there were no differences between the eventual honey yields from these three groups.

The honey-production records indicated another kind of drift. Analysis of six apiaries, divided into four blocks (Fig. 23b), showed that in some instances there were substantial differences between the yield from the different blocks (Table 6). The difference between the best and worst blocks ranged from 6% to 51%. Nekrasov considered that these differences had arisen because some foragers had returned to colonies nearer to their crop, and he produced evidence that the mean yield in each apiary fell off row by row as the distance from the crop increased.

In four apiaries Nekrasov found that on 1 August the bees in the front row weighed 3·3% more per colony than those in the back row, but the honey yield was 6·8% greater. The colony surplus is only a small proportion of the total amount of food carried into the hive, so a small percentage difference in the foraging force of a colony would exercise a disproportionate effect upon its surplus (cf. Table 3, p. 75), and explain this result.[1]

These results offer no evidence concerning Taranov's (1946) opinion that drifting weakens some colonies and so lowers the productivity of the whole apiary, but they do demonstrate that drifting will tend to mask the performance of individual colonies. To counteract this, many Continental beekeepers advocate the painting of hives with various colours; landmarks and irregular arrangements within the apiary have also been recommended.

Drifting of drones

Livenetz (1951) carried out experiments which suggest that the proportion of drones which drift is not markedly different from the proportion of workers which do so. He also marked groups of 100 drones and placed them in colonies other than their own, in the same apiary. They were introduced into 5 queenless colonies, 4 colonies with unmated queens and 10 with laying queens; in these groups 14, 14 and 33 remained (2·8%, 3·5% and 3·3% of introductions respectively). Thus the drones were not attracted to stay either in queenless colonies or in colonies with virgin queens; this agrees with the observation of Butler (1939), but conflicts with that of Risga (1931).

Anaesthesia and drifting

V. Buttel-Reepen (1900) stated that anaesthesia of honeybees with chloroform or saltpetre fumes destroyed their existing memories, and that bees so treated would make their way into any hive. Evidence to support this statement was not provided; it may have been derived from observations of bees which had not fully recovered from the treatment, since Ribbands (1950b) demonstrated that anaesthesia with chloroform, carbon dioxide or nitrogen had no effect upon the memories of treated bees.

However, Gontarski (1950) stated that nitrous oxide anaesthesia eliminated memories and had been regularly used in his apiaries as a device for preventing the return of bees from newly made nuclei to their original colonies; no evidence was provided. Controlled experiments have now shown that nitrous oxide anaesthesia does not impair the memory of treated bees; a few will orient and return to their new home if they leave it before complete recovery, but this proportion is so small that the treatment is not of practical value (Ribbands, 1953d).

[1] Nekrasov incorrectly deduced from it that 5% of the drifting bees had subsequently returned.

Chapter 12
HOW HONEYBEES DISCOVER CROPS, AND RETURN TO THEM

While Honey is in Every Flower, no doubt
It takes a Bee to get the Honey out.

ARTHUR GUITERMAN, *A Poet's Proverbs*

SOME OF THE components which are involved in orientation to crops have been studied separately in experiments which have involved training honeybees to colour, form and scent (Ch. 2, 3). Their manner of integration has been a matter of some controversy. Investigators did not at first distinguish between the honeybee's return to a known crop and her discovery of a new one. They supposed that each individual found a crop for herself, and they inferred that the properties of the crop which were initially attractive would equally influence orientation and return. The discovery of communication among foragers now suggests that searching bees and collecting bees may be influenced according to a different scale of values.

Scent

A review of early work on crop attractiveness was given by Clements & Long (1923). Plateau (1876) started a long controversy when he observed that honeybees and other insects seldom visited artificial flowers, and that this was evidence that they were not attracted only by sight. He added (1897a) that honeybees were indifferent to the diverse colours found in the same species, since they went indiscriminately to blue, rose, white and purple cornflowers, or to dahlias or scabious of various colours; he listed 27 species of plants with greenish or brownish flowers which are visited by honeybees and pointed out that they visit inconspicuous extra-floral nectaries, e.g. on the stipules of field beans and the petioles of *Prunus* (1897b). Plateau found (1910) that conspicuous but nectarless flowers (e.g., *Lilium candidum*) were quickly visited when scented syrup was added to them, but not when they were supplied with unscented syrup. Therefore he concluded that smell played the predominant role in attraction to flowers and that sight was of little or no importance.

Kerner (1902) stated that the scent of lime trees was readily perceptible to humans at a distance of 30 yards, and that in the Danube valley one could smell vines which were 300 yards away. Knuth (1898) suggested that odour was mainly responsible for attraction from a distance, while the colour of flowers attracted bees when they were 1–2 yards away.

Colour

On the other hand, Andreae (1903) found that bees were attracted towards flowers which had been covered with an inverted bell-jar so that they were visible but scentless, but that they were not attracted to flowers which had been placed in a beaker and covered with dark paper so that they were not visible but their scent could escape. Andreae concluded that colour attracted honeybees from a distance, scent only when they were close to them. Forel (1904) made a series of variously coloured artificial flowers, put a drop of honey on each one, and placed them among dahlias. At first they were neglected, but three honeybees from the dahlias were marked and introduced to them. These bees deserted the dahlias and returned regularly to the artefacts, which were red, white, blue, blue with a yellow centre, rose-coloured, and green. Forel concluded that the olfactory sense of the bees was poor, because they did not find the green artefact but they did find all the others. As each bee had only been introduced to one of the others, and had found the rest for herself, he considered that she appeared 'instinctively to draw inferences from analogy' and that vision had played a predominant role.

In further experiments Giltay (1904), who found that flowers from which the petals had been removed attracted fewer bees than normal flowers, observed that habit was important, because honeybees could learn to visit mutilated flowers freely. Joséphine Wery (1904) found that depetalled flowers attracted only one third as many honeybees as normal flowers, and also that artificial flowers of *Eschscholtzia* and *Dahlia*, placed in natural foliage, attracted almost as many honeybees (19; 22) as the same number of real flowers; the imitations attracted twice as many honeybees as depetalled real flowers (11; 6) and nearly five times as many as real flowers hidden in foliage (19; 4).

Wery prepared two bouquets of equal size, one of the conspicuous flowers of *Helianthus, Dahlia, Aster, Coreopsis* and *Helenium*, the other of strongly scented but inconspicuous mignonette; the former attracted six times as many honeybees (35; 6). When the conspicuous flowers were replaced by imitations they still attracted four times as many bees as the mignonette (25; 6). From these results Wery concluded that colour and form were responsible for about 80% of the attractiveness of flowers to bees, and pollen, nectar and perfume together were responsible for only 20%.

Another type of experiment was carried out by Lovell (1912). Having trained bees to collect honey or syrup from a feeder, he placed a large quantity of honey on an inconspicuous object a short distance from the feeder, and a smaller quantity on a coloured object the same distance away but in another direction. Then he allowed the feeder to be emptied so that the bees searched in its vicinity; in six experiments only the conspicuous object was found, and in others it received two or three times as many visits as the inconspicuous one.

Scent and colour

V. Frisch (1919b) trained bees to visit a small blue jasmine-scented box and then offered them the choice of either a yellow jasmine-scented box or a blue unscented one; in three experiments six times as many bees visited the scented box (76; 12). The bees were re-trained and then given the same choice, but the boxes were placed at the rear of a cage made from partitions 39 inches square (Fig. 24); in ten experiments only 25% of the bees now went to the scented box (10; 29). V. Frisch concluded that the bees could see colour from a greater distance than they could perceive scent—even when the scent was intense.

This conclusion is based upon good evidence, but it is not beyond all criticism: for instance, the quantities of scent and colour used were such as might appertain to a flower or a small bouquet, but flowers often grow in large masses, and their perfume may then be perceptible from some distance, as Kerner (1902) recorded. Moreover, the relative importance of colour and scent varies from species to species. For these reasons one may suppose that there are exceptions to v. Frisch's suggested sequence. Honeybees can be attracted to either inconspicuous scented objects or unscented coloured ones: in some aspects of insect behaviour there is evidence of an inflexible chain of instinctive reactions which can only be elicited in a definite order, but in their crop-finding activities honeybees are certainly more adaptable.

In other experiments v. Frisch found that bees trained to a blue mignonette-scented box also chose the training scent in preference to the training colour. Yet bees trained to a yellow box scented with orange-flower oil subsequently chose the

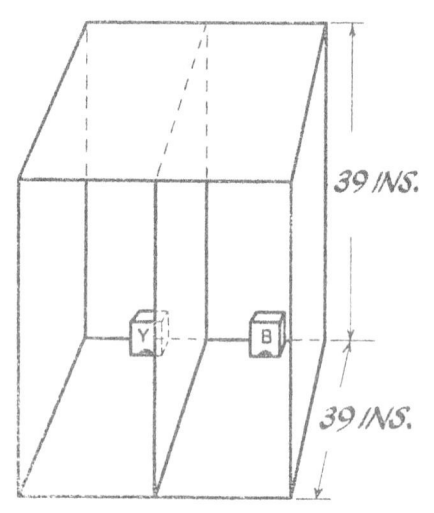

Fig 24. TRAINING TO SCENT AND COLOUR (after v. Frisch, 1919*b*). Bees preferred correct scent with incorrect colour to the opposite combination, but this preference was reversed when the training boxes were caged so that the choice had to be made from a distance of 39 inches.

coloured alternative; to humans orange-flower oil is weakly scented. V. Frisch could not decide whether quality or intensity played the more important part in determining the relative attractiveness of scents. However, he demonstrated the importance of intensity by training bees to blue mignonette-scented boxes and showing that 75% of them chose the scented alternative when the scent was intense but only 40% did so when the scent was sparse. Even when more bees chose the coloured boxes they always hesitated more at the entrance of an unscented box.

Memory of scent and colour

Memory of scent and colour were also tested by v. Frisch (1919*b*). Honeybees were trained for 5 hours to a tuberose-scented blue box, and then immediately presented with alternatives; 60% chose the scented box. Then they were fed for 2 days in an acacia-scented yellow box and subsequently offered the choice of a grey tuberose-scented box or a blue unscented one. They went only to the scented alternative (51; 0), so they had remembered their original training to the scent but had forgotten the training to the colour. Memory of the scent persisted, despite counter-training, for at least 5 days.

The learning of the recognition marks

Elisabeth Opfinger (1931, 1949) studied the phase of the visit to a crop at which learning occurs. In her experiments

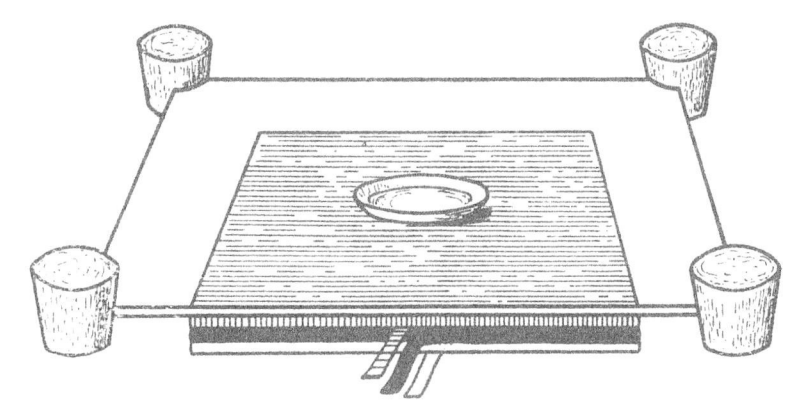

Fig 25 DEVICE FOR STUDYING THE PART OF THEIR VISIT IN WHICH BEES LEARN TO ASSOCIATE COLOUR WITH FOOD (Opfinger, 1931). Cards can be withdrawn in turn.

honeybees which visited a feeding-place were individually marked, and the prominent feature of the feeding-place— either colour, form, or scent—was changed as soon as they arrived and changed again at the moment of departure. Several feeding-places were then arranged, featuring the characteristic which had been visible during either the approach flight, the feeding time, or the departure flight, or some similar neutral characteristic. Opfinger then observed which of these feeding-places the honeybee returned to.

Her first experiments (1931) involved training the bees to a feeding-place on a pile of several coloured cards, with a device for withdrawing the top card and leaving a card of another colour (Fig. 25). The bees made thirty times as many return trips to whatever colour had been under the feeding-place during arrival as to the colour which had been present during feeding or departure; as many bees went to a neutral colour as to either of the other colours. In a refinement, Opfinger divided the arrival period into two parts, by pulling away the card during the descent of the bee; the visits to the first and second colours of the arrival in relation to visits to a neutral colour were then in the ratio 34: 35: 3.

These results showed that the approach flight was *alone* responsible for the learning of the colour of the feeding-place. Other experiments showed that the form of the feeding-place (whether a flat pattern or solid), and optical marks at a distance up to 8–14 inches away from the feeding-place were also learned only during the approach flight. Opfinger observed that this learning occurred although the mean duration of the approach flight was only 3 seconds, while feeding occupied 73 seconds and the departure circling $10\frac{1}{2}$ seconds. She suggested that the explanation might be that during her approach flight the bee devoted her attention solely to the properties of the food source, as during feeding she attended exclusively to that occupation.

The departure flight is usually referred to as the orientation flight, and its unimportance in these experiments was surprising, but Opfinger demonstrated that it provided information about the location of the feeding-place in relation to more distant optical marks.

A 14-inch-square chequer-board of alternating blue and yellow squares was used for one experiment. The bees flew to one corner, the chequer-board was rotated so that they flew off from the opposite corner; the bees returned in about equal proportions to either of these two corners, but much less frequently to either of the other two corners; thus the approach and departure figures were similarly effective.

The effect of more distant landmarks was then investigated. On sites plentifully supplied with potential orientation marks two small white tables (32 × 20 inches) were placed 11 yards or more apart and equidistant from a colony, which was from 12 to 28 yards away in four successive arrangements. Then 123 newcomers to either table (Fig. 26) were individually marked and transferred to the other table; 101 returned to the table they had departed from, but only 22 returned to the table at which they had at first arrived. Hence most of these newcomers had learned the whereabouts of the table during their departure flight.

The two tables were then placed 20 yards apart and equidistant from a colony, while an observer sat at each table and another observer sat on a stool (C) midway between the two tables. Bees going to either table were individually marked,

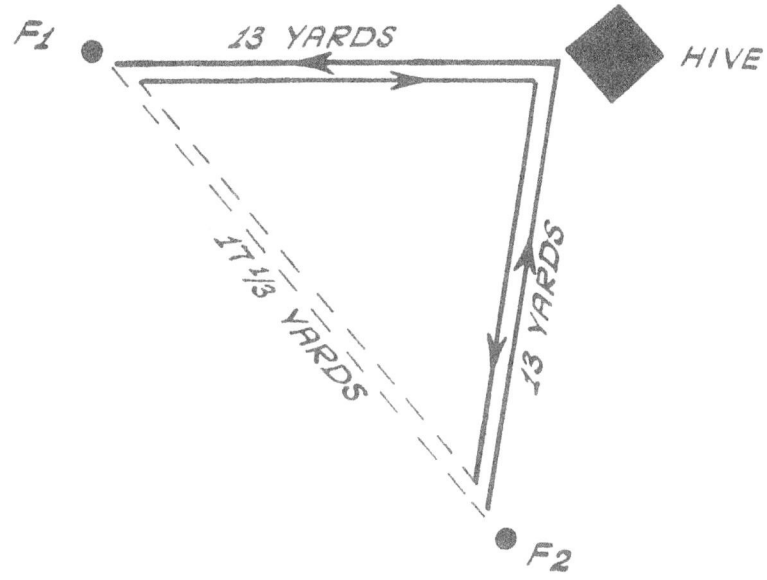

Fig 26 THE LEARNING OF LANDMARKS NEAR TO THE FOOD SUPPLY. Bees were transferred from F_1 to F_2 or vice versa. 80% of them returned to the feeding-place from which they had departed (Opfinger, 1931).

carried to C, and liberated; the observers timed them and noted their behaviour on their return. Ninety-nine bees returned thus:

to the table previously visited	31
to C and then to that table	39
to C only	13
to C and then to the table not previously visited	10
to the table not previously visited	6

Thus about two-thirds of the bees had marked the whereabouts of the feeding-place during their departure flight, but about half of these bees, as well as most of the other ones, had probably acquired some knowledge of the whereabouts of the feeding-place during their first approach flight.

For her final experiment Opfinger (1931), having carried a colony into one of two adjoining quadrangles in the old academy at Munich, established a feeding-place in each quadrangle, in opposite corners. Newcomers were again carried from one feeding-place to the other, but the flying route to one of the two feeding-places was significantly longer because the bees had to fly over the building which separated the quadrangles. In those circumstances about four-fifths of the bees returned to the nearer feeding-place, whether they had previously arrived or departed there; the simpler route had been more readily learned.

Opfinger (1949) carried out a similar investigation to determine when the honeybee learns the scent of the visited crop. Chemical scents and covered natural flowers were used in different experiments. The sugar syrup was placed on a glass slide on top of a dish which contained one scent, and when the bee had settled the slide was raised and the dish was exchanged for one which contained another scent. It appeared that the scent of the feeding-place was also learned only during the approach flight.[1]

Searching and collecting bees

Bonnier (1905, 1906) showed that when a bee became attached to a crop it was not easy to lure her from it even by placing richer sources of nectar in its immediate vicinity. He distinguished between 'collectors', who went straight to a crop to do work which had been determined in advance, and 'searchers' who went towards different plants or any objects whatsoever from which they hoped to obtain food: the latter flew in a different manner, their behaviour was more wasp-like, and the sound of their flight was different. Bonnier noted that when dishes containing honey were put out at sunrise they were found and emptied soon after bees began to fly, but that when they were put in new places towards the end of the morning the bees took longer to find them (between 30 minutes and 3 hours if they were at a distance of 110 yards); if they were put out after midday usually no bees found them. He therefore concluded that 'searchers' were numerous early in the morning and that their numbers dwindled rapidly as the day wore on.

Bonnier showed that when a searcher bee discovered a suitable source of food she became a collector.

---

[1] The scent experiments were carried out after the bees had made *several* journeys, on each of which one scent was present at arrival and the second scent was substituted during feeding. An alternative but less likely explanation of the results would postulate that the bees had learned to disregard the second scent because on arrival they had repeatedly failed to find it.

The proportion of scout bees

Therese Oettingen-Spielberg (1949) has recently shown that very few of the foragers in a colony search for new crops; nearly all wait to respond to the dances of successful colleagues (Ch. 19). Her important work must be reviewed in detail. A queenright nucleus, prepared from honeybees which had emerged in an incubator and had never flown freely, was confined within a large empty room. Blue cards and yellow cards, $5\frac{1}{2}$ inches square, were then placed in the room in the vicinity of the flying bees, and those which visited the cards were observed and killed; in 22 flying hours only one scout came, to a yellow card. Coloured cut-out figures, rich in outline, were substituted for the square cards, but in 10 flying hours only one scout came, again to yellow. Cards with artificial scents were supplied, with no more success, and when a normal nucleus was substituted for the prepared one the result was not materially different.

The cards were replaced with bouquets of lime, cornflower, or delphinium. These attracted more scouts, but in 40 experiments, mostly lasting 1 hour, only 57 scouts from the prepared nucleus were attracted to the bouquets, although the room was filled with buzzing bees. More searchers from a normal nucleus went to the bouquets (94 in 12 experiments), but again only a small proportion of the flying bees acted as scouts.

In further experiments lime, mignonette or phlox flowers were placed in a box in the room, so that they were not visible but their scent could escape from a vent. The natural scents from these flowers, unlike the artificial scents previously used, attracted bees from both the normal and the prepared nucleus, but they arrived at the vent only in small numbers.

These experiments had shown that flowers were much more attractive than coloured and scented cards, and that more scouts came from the normal nucleus than from the prepared nucleus which contained only inexperienced bees.

The next experiment sought to determine the age of the scouts. A nucleus, prepared from eighteen differently aged groups of 200 marked bees, was kept in the experimental room; a bouquet of cornflowers stood on a table in the room, and every bee which searched these flowers was killed and recorded; on another table, scented with thyme oil, concentrated sugar syrup and soya flour were both provided, and all the bees which collected from this feeding-place were distinctively marked. Each day the bouquet was provided 1–2 hours earlier than the food, and both were removed at the same time. In these circumstances the feeding-place (initially much less attractive) attracted many more newcomers than the bouquet; 1009 individuals flew to the feeding-place, while only 53 flew to the bouquet without having flown to the feeding-place. In addition, 207 collectors subsequently went to the bouquet as scouts, but this usually happened during the hours before the feeding-place was provisioned (188 collectors searched the bouquet during 54 hours when food was not provided, but only 19 did so during 27 hours when it was available). Thus 80% of the scout bees had collecting experience. The mean age of the 53 inexperienced scouts was 31·5 days (range 7–47), while the mean age at which the 1009 collectors began their task was 29·0 days (range 9–46). Oettingen-Spielberg concluded that the scout bees were not a special age-group, and that both inexperienced and long-experienced collectors could act as scouts.

The field experiments of Oettingen-Spielberg were still more important. In August, when natural forage was sparse, she trained bees to collect sugar syrup from a coloured, scented card (A) in her garden. A differently coloured and differently scented card (B), not provided with food, was placed 10 inches away from A. Food was provided on four successive days; on each day 70 recruits were marked and allowed to forage, and all additional ones were killed. On the fifth day food was provided at B instead of at A. At first the bees flew around A, but soon one of them found the food at B and within a short time as many bees were there as had previously been at A. The dancing and/or scenting of the successful ones soon attracted the others. This experiment having been frequently repeated, with the same result, it was modified—each newcomer to B was killed. In each of two such experiments only 2 scouts came to B (from among 280 marked collectors).

A refinement of this experiment showed the importance of dancing in facilitating the change from A to B. A and B were arranged as before, 6½ yards away from the colony, and another feeding-place C was arranged 27 yards away from it in another direction. A was a blue card scented with apricot, while B and C were bergamot-scented yellow cards. For 4 days food was supplied at A and C only, and bees foraged at both sites; each day 70 recruits were marked at A, and any surplus unmarked bees were killed. At the same time on the fifth day food was supplied at B and C only, and all bees from A which visited either B or C were killed. During one hour 255 out of 278 of these bees responded to the dancing of the bees from C; 121 were killed at B and 134 at C. Moreover, the tendency to search for the new scent at the old source increased with the duration of the training at the old source; among bees trained to A for 1, 2, 3 and 4 days, 24, 38, 48 and 64%, respectively, went to B.

The commencement of foraging of marked bees was studied by Lindauer (1952). Whereas young bees in the hive fled from dancers the older ones tried to follow them, with increasing excitement, and only 9 out of 159 individually marked bees commenced to forage without being seen to follow a dancer first. Ninety-one of the 150 dance-followers were studied in detail; 79 brought back loads not visibly different from those of the dancers whom they had followed, and 42 of these indicated the same locality by their own dances; of the bees which brought back different loads, 4 did not dance, 4 indicated the same locality as the dancer whom they had followed, and 4 a different one.

Oettingen-Spielberg's conclusion that few bees act as scouts is doubtless valid in normal circumstances, and we may guess that during a good nectar flow all the foragers might become attached to crops, so there may be no scouts at all. At the other extreme, there are times in late autumn and early spring when the weather is suitable for foraging but no crops are available; numbers of searching bees may then be seen in the vicinity of an apiary.

Elisabeth Opfinger (1949) found considerable variation in the speed of learning of different individual bees, and she suggested that foragers might be divided into three categories, based upon this variation. These were (*a*) bees which adhered faithfully even to very sparse crop sources, (*b*) the majority of bees, which remained true to a source so long as the food supply was normal but were re-trainable to another source if the crop failed, (*c*) a very few bees which were especially restless and were always searching here and there and discovering new sources of food. She suggested that these distinctions might be biologically significant and that there might be a division of function between searching bees and collecting bees. However, Bonnier and Oettingen-Spielberg have clearly shown that searchers become collectors and vice versa.

The attractiveness of nectar

Darwin (1876) observed a wall covered with flowering ivy-leaved toadflax (*Linaria cymbalaria*), which no honeybees visited for a fortnight, but which was then abundantly visited on one warm day. Bonnier (1879b) observed a similar sudden spate of visits to lungwort (*Pulmonaria officinalis*) on a hot day after rain, and Pérez (1894) recorded sudden and quick recruitment to *Salvia splendens* warmed by the sun. Bonnier (1879a, b) considered that abundance of nectar was the best guide to the attractiveness of a crop. All three writers postulated that odour could contribute to the attraction; this odour might have come from the secreted nectar, or alternatively the secretion of nectar and the dissemination of perfume might have been correlated processes, dependent upon similar external factors. V. Frisch (1919b) demonstrated that the smell of honey was only slightly attractive to bees, and the sugars in nectar do not attract bees from a distance. When Gubin & Romashov (1933) placed dishes of sugar syrup of different concentrations outside a colony which was being fed they found that the syrup concentration made little difference to the number of bees attracted to each dish; with dishes of honey the differences were somewhat greater, but quite insufficient to account for any direct attractiveness which the small quantity of nectar within a plant might be supposed to possess.

Nevertheless, Oettingen-Spielberg's work has emphasized the indirect importance of nectar abundance in determining crop attractiveness, by proving that nearly all the bees on a crop are recruited to it through the dances of comrades; the vigour of those dances, which controls the extent of the recruitment, is determined by the abundance and concentration of the food supply (p. 161). Thus, indirectly, this factor is much more important than any other.

The discovery of crops by scout bees

Oettingen-Spielberg showed that scout bees could occasionally be attracted to models by their colour or scent, and that they went most readily to natural flowers; this result may imply that optimum attractiveness is attained through an integration of perceptions. Since the scout bees are usually established foragers, their searching may be conditioned by inferences from previous experiences.

Butler (1951) carried out field experiments in early spring, when scout bees are most readily available. Either coloured or scented discs ($\frac{1}{4}$-inch diameter) attracted them in small numbers, and they were more readily attracted to unbaited coloured discs impregnated with the perfumes of flowers (not in bloom at that season) than to similar unscented discs. The number of scouts attracted to the vicinity of coloured scented discs was only just significantly greater than the number attracted towards coloured unscented ones (63; 39) but the number which settled on the scented ones was much greater (42; 1). These results conformed with v. Frisch's conclusion (p. 95).

The searching activities of scout bees may sometimes be a direct consequence of their previous experience. An interesting example was provided by Cooper (1952). Seven stocks of his bees, which had been foraging on mustard, were taken to a heather moor. The hives were well spaced, so that drifting was unlikely, and they found the heather on the day of their release. Three weeks later the colonies were visited: all were also bringing in loads of pollen from white mustard, which was growing one mile away. Twenty yards away there were four other colonies, whose owner knew of no mustard within their previous foraging range; in 30 minutes no mustard loads were taken into these hives, at a time when they were being carried into Cooper's hives at the rate of 20 loads per colony per minute. This result could be attributed either to previous experience of the individual scouts or to their appreciation of a scent which was pervading their stores.

Discussion and conclusions

The problems of crop-finding can be studied in a new perspective now that Oettingen-Spielberg's results have shown that nearly all the newcomers to a crop have left their hive with a knowledge of its whereabouts and of its scent, but with no knowledge of its colour or form (pp. 101-3). This fact, together with v. Frisch's demonstration that scent-training is remembered while colour-training is quickly forgotten (p. 96), with the lack of specificity of the few flower colours, and with the well-developed sense of smell (Ch. 3) and limited vision (Ch. 2) of the honeybee, emphasize the role of scents in attracting bees to crops.

For the foraging bee the properties of scent and colour are different and complementary. The most valuable attribute of colour lies in the fact that the bee can orientate straight towards it, a more direct and therefore a quicker process than proceeding up a wind or diffusion gradient towards a scent. Thus it is possible that three phases might sometimes be distinguishable in crop-finding, (i) searching for a distinctive odour, and then (ii) orientation towards any distinctive colour or pattern, followed by (iii) landing on the selected object if the presence of scent is confirmed. Such a relation between colour and scent, and great variety in the emphasis of the three phases, would help to account for the diverse results and varied opinions of the investigators who have studied the process.

Forel noted (p. 94) that honeybees appeared to draw inferences from analogy when they were foraging upon coloured models; they also exploit differently coloured varieties of the same species of plant. Honeybees can distinguish between the odour of their companions and that of other honeybees, but they are nevertheless attracted also to the latter (p. 178), and it is likely that they will sometimes be attracted to flower odours which differ from, but are similar to, those of flowers which they have hitherto exploited.

Chapter 13

CROP CONSTANCY; FORAGING AREAS

Out upon it! I have lov'd
Three whole days together;
And am like to love three more,
If it proves fair weather.

SIR JOHN SUCKLING, *Constancy*

LONG AGO ARISTOTLE wrote that on each expedition the honeybee confined herself to one kind of flower. Darwin (1876) postulated that the instinct of flower constancy was established by natural selection and supposed that constant bees might work more quickly than inconstant ones, and Loew (1886) pointed out that flower constancy only occurred among those flower-visiting insects which were most completely adapted for pollination.

The constancy of bees is not complete, as Bulman (1892) emphasized when he recorded several instances of repeated changes from crop to crop. At that time he did not differentiate between honeybees and bumble bees, but subsequently (1902) he recorded 48 observed changes of crop by honeybees. Ribbands (1949) observed a marked bee which returned again and again to work two pollen crops at once (p. 116), and some of the bees observed by Bulman may have been so employed.

Observations of mixed visiting by honeybees which were working on a mixed crop containing equal numbers of swede and cabbage plants, evenly distributed, were published by Bateman (1951). He found that the bees possessed 'a high but incomplete constancy to species and a lower but still highly significant constancy to varieties within a species', and suggested that the lack of constancy was due to poorly developed powers of discrimination, especially on the part of younger bees. These results could have been explained just as readily on the supposition that some of the older visitors had at some time become temporarily discontented with their first crop, had inspected the adjoining crop, and had subsequently visited both.

Interpretation of observations of foraging constancy should

allow for the different sense perception of honeybees and humans (Ch. 2, 3). An experiment of Mather (1947) may be relevant. Artificial crosses of *Antirrhinum majus* and *A. glutinosum* are easily made and, when either species was pollinated simultaneously with pollen from both, hybrids were as common as maternal types; yet when plants of these two species were randomly distributed at 1 ft. intervals in a large plot their seed produced only 1·2 and 2·9% of species crosses, respectively. Mather reported that the genetic isolation had been produced by the adherence of the bees to one species or the other, and added that flower colour differences were *not* the cause, since bees freely transferred pollen from a first cross with coloured flowers to *A. glutinosum* with white ones. Mather noted that *A. glutinosum* has a more prostrate habit and slightly smaller flowers than *A. majus*. Is it possible that it also has a different scent, inconspicuous to the human experimenter but detectable by the honeybee?

Mixed pollen loads

There have been attempts to measure the extent of foraging constancy, by determining the proportion of mixed pollen loads brought into the hive (pollen and nectar gatherers are not likely to differ fundamentally in respect of constancy). Annie Betts (1920) captured pollen loads from more than 3,000 honeybees and examined them microscopically. Of these loads 6·8% were mixtures of two or more pollens; in 39% of the mixtures the different pollens were segregated into different parts of the load, but in the others they were intermingled. This work may have exaggerated the proportion of mixed loads (especially segregated ones), since all the mixed loads which were seen were captured. Nearly 15% of all the loads examined contained a few grains of some other species of pollen.

Another analysis was undertaken by Brittain & Newton (1933), who examined microscopically 162 honeybee pollen loads from various crops. They found that only 62% were from single species; one load contained pollen from 5 species. However, during apple bloom the proportion of pure loads rose to 80%, so Brittain & Newton decided that apparent constancy depends greatly upon availability, and that almost any result could be obtained by choosing selected periods for the tests. Later, they pointed out that their method often showed only a single grain of the second pollen (Brittain *et al.*,

1934); it is therefore likely that the proportion of mixed loads was raised by contamination from previous trips.

Other results have tended to the opposite extreme; Mary Percival (1947) recorded 256 mixed loads, which she stated represented only one-thousandth of the total pollen income; this is only the proportion of obviously visible mixtures, since the loads were not examined microscopically.

A more realistic proportion was recorded by Annie Betts (1935), who examined 916 unselected loads microscopically and found among them 18 mingled mixtures (different pollens mingled throughout the load, produced by working two crops at once) and 7 segregated mixtures (produced by collection from first one species and then another). Thus 3% of these loads were mixed. In a more extensive analysis of the same kind, in which 4519 mixed pollen loads were seen, Anna Maurizio (1953) found that the proportion of mixed pollen loads was usually between 1 and 3%. There were only two kinds of pollen in 96% of these mixed loads, but three or four kinds were found occasionally. Maurizio also found a proportion of loads (about one-fifth of the number of mixed pollen loads) in which pollen was mixed with fungus spores or foreign bodies.

Betts and Percival both concluded that mixed pollen loads were obtained from neighbouring plants, which might differ widely in size, colour, scent and structure. Maurizio concluded that they might be obtained from any plants in bloom at the same time.

Foraging areas

Honeybees which were working in a garden upon *Salvia aethiopsis* were marked with coloured paints by Müller (1882), who observed that for ten consecutive days two of these bees returned repeatedly to the same plant. Giltay (1904) also marked honeybees and noted that the same ones returned day after day to the same plot.

Then Bonnier (1906) noticed that many bees were working upon a long belt of buckwheat, and he marked all the bees which were foraging upon a 16-foot strip near to one end of this belt. When he examined the buckwheat on the following day he found that all the honeybees on this strip were marked ones, and that no marked bees were foraging more than 16 feet away from the edge of the strip. These bees were therefore confining their activities not only to one crop but to a particular portion of that crop.

This important conclusion has been confirmed by a number of subsequent investigators, who have studied honeybees working upon many different crops. Estimates of the areas ranged over by bees working in fields of clover, dandelion and other plants were made by Minderhoud (1931), who marked all the bees which were foraging upon areas of 1 square yard and then plotted the subsequent movements of these bees on squared paper. The bees returned to within 10 yards of the square upon which they had been marked. When Buzzard (1936) marked a dozen bees out of about a hundred which were working upon three bushes of *Cotoneaster horizontalis*, each bush covering an area approximately 2 yards square, he found that over a period of 5 days these 12 bees kept to their chosen bush; only three strayings were observed (where branches intertwined) and these were followed by almost immediate return. About 100 honeybees which were working on a 5 × 8 yard patch of rosebay willowherb situated in the middle of a larger area were marked by Butler, Jeffree & Kalmus (1943), who found that during that day these bees seldom moved more than 5 yards beyond this patch.

In all these instances the observed foraging areas were not more than 10 yards in diameter; both Minderhoud (1931) and Butler (1945) related their conclusions to suggestions concerning the cross-pollination of self-incompatible fruits, since it appeared that individual bees might usually confine themselves to single trees. Mommers (1948), however, produced evidence that the foraging areas of honeybees in apple orchards could extend up to 40 yards in diameter. Having marked many bees visiting one tree, he counted bees on neighbouring trees and determined the proportion of marked bees on each tree. His results suggested that when trees were planted in rows, with wider spaces between rows than between trees in a row, the bees tended to fly along the rows more readily than from row to row (Fig. 27). Hence he concluded that pollinators should not be planted in separate rows.

Ribbands (1949) studied individually marked bees which were foraging in a specially planted plot of flowers. Their foraging areas were very variable in size: one bee was temporarily attached to a single flower of Shirley poppy, while another, collecting from *Eschscholtzia*, ranged over the whole length and breadth of the plot (75 × 30 feet). The chief factors affecting the size of the foraging area were the number of

CROP CONSTANCY; FORAGING AREAS

flowers available, the nectar and pollen content of these flowers, and the amount of competition provided by other foragers. This work also showed that a bee might prefer different portions of her foraging area at different periods of the day; moreover, the bee might move or extend or reduce her foraging

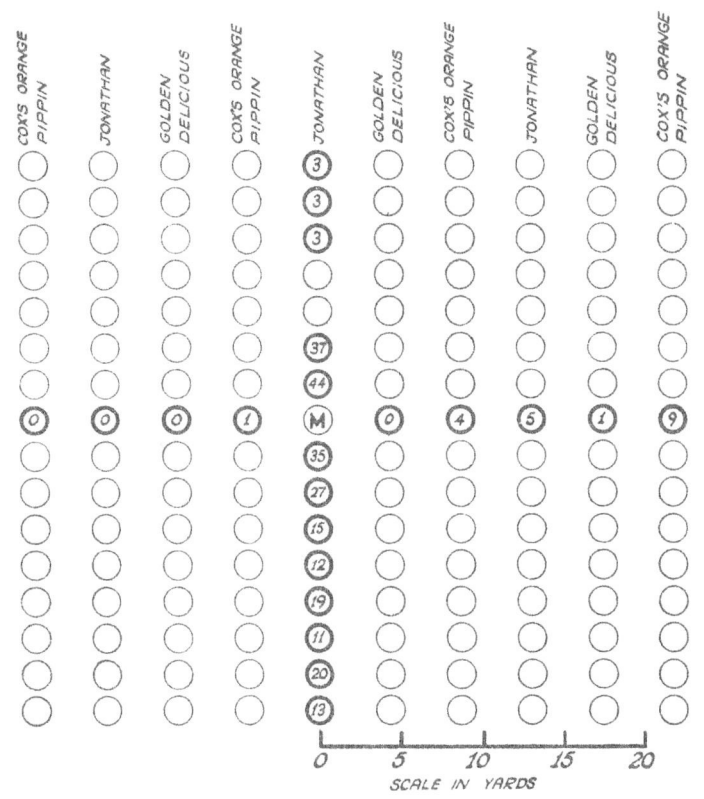

Fig 27 THE FORAGING OF HONEYBEES IN AN APPLE ORCHARD. The bees on tree M were marked; on the same day there were twelve counts, at 30-minute intervals, of the numbers of both marked and unmarked bees on trees in different directions from this tree. The percentage of marked bees seen on each tree is given (data from Mommers, 1948).

area. The attachment to foraging areas was found to be a consequence of the method used when foraging and, like other features of the foraging behaviour, a consequence of the exercise of choice and memory (p. 115).

Sardar Singh (1950) observed individually marked bees which were foraging upon six different field crops. Their foraging areas were 'fairly elastic', and increased as the crops waned. Both size of foraging area and constancy varied in different individual bees upon the same crop (e.g. on buckwheat one individual area was 120 × 110 feet, and another was 30 × 30 feet), and there were differences between the average

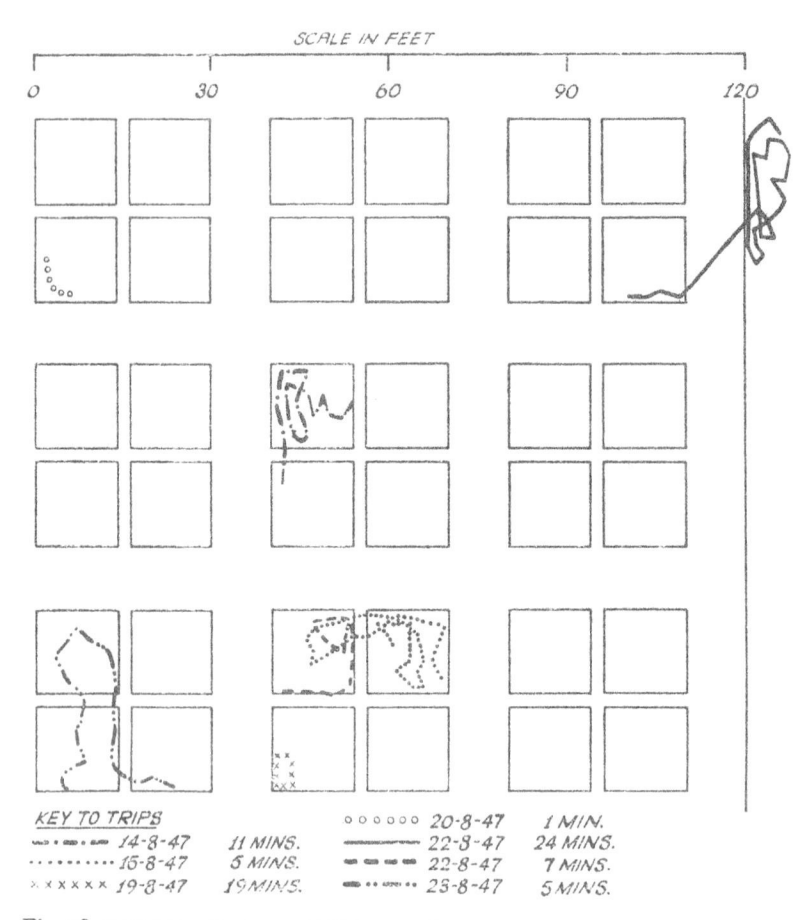

Fig 28 ROUTES OF ONE MARKED BEE WORKING ON BUCKWHEAT PLOTS. The foraging area was either very large, or it varied from day to day (Sardar Singh, 1950).

size of the foraging area upon different crops (e.g. golden rod areas, which were smaller than buckwheat ones, ranged from 40 × 30 feet to 17 × 10 feet). The wide variations in the quality and quantity of different crops would lead one to expect such differences. Single trips were usually restricted to smaller areas (Fig. 28), so the recorded sizes were partly attributable to movement of the foraging area.

One should remember that all observers are likely to underestimate both the average sizes of foraging areas and the variation in size, because a number of bees are marked and only a few of these are followed. The bees which work smaller or more constant areas are easier to watch, and those with very large foraging areas cannot be followed.

Foraging areas and pollination problems

The implications of the work on foraging areas are important in many pollination problems. Attachment to a particular crop is an aid to pollination because it prevents the useless carrying of pollen to other species. Attachment to a particular area of a crop is advantageous to the bee, and it sometimes helps seed growers to keep their strains pure. It has disadvantages for the fruit grower if his orchards are planted in blocks and the distances between self-sterile but compatible varieties are too great, and it vitiates simple calculations of the potential pollinating capacity of honeybees—which might otherwise be based on the number of flowers visited per bee per hour. The number of flowers in the foraging area of a bee is a more correct index of her maximum pollinating capacity, and this index is very variable and less readily determined.

Seed growers' problems have been investigated by experimental plantings of self-incompatible varieties and studies of their progeny. Crane & Mather (1943) observed no intercrossing when red and white radish varieties were grown in large blocks 120 yards apart and without intervening isolated plants. Intercrossing was reduced to about 1% at a distance of 15 feet from the contiguous edge of large plots of the separate varieties.

When isolated plants of one variety were spaced out at intervals from a large block of the other variety, the intercrossing occurred over longer distances. In such conditions intercrossing varied from 8 to 70% among isolated plants 47 feet away from a block of another variety, and one instance of intercrossing was observed in an isolated plant 479 feet away.

These results would fit a supposition that the foraging areas of the bees were small when there was a profusion of flowers but extended over a much greater distance when the blossoms were widely separated. Bateman (1947*a*) experimented with the same two crops and again obtained considerable contamination among single white radish plants grown at intervals of multiples of 20 feet from a block of red ones. The intercrossing steadily decreased with distance from 60% at 20 feet to 13% at 80 feet and about 1% at 160 feet; there was no detectable decrease between 160 and 580 feet, which suggests that this small proportion of intercrossing was caused by unattached foragers. In another experiment Bateman grew red radishes in blocks, each with a small central patch of white ones; he found that the plants could be spaced at densities ranging from 1½ to 11 plants per square yard without effect upon the amount of intercrossing at any distance. This result suggests that within these limits of density the sizes of the foraging areas did not vary with the sparseness of the crop.[1]

The work already accomplished has uncovered new problems which still await solution. For instance, we know that foraging areas vary greatly in size and that their size depends upon many circumstances, but there is no adequate evidence concerning the effect of competition between bees upon the sizes of their foraging areas. The foraging areas of honeybees on waning crops tend to increase; as increased competition reduces the amount available for each individual, one might suppose that it would also tend to increase the sizes of their foraging areas. Yet in the absence of competition any confinement to a small area—leaving the rest of the crop untapped—would usually be disadvantageous, but in the presence of competition a bee may find that the blossoms first visited are better than those which she visits subsequently (because the latter have been robbed by rivals), and so she may be encouraged to return and to create a definite foraging area for herself. In this way increased competition might produce smaller foraging areas; the largest foraging area recorded by Ribbands (1949) occurred in the absence of competition, while the smallest one (on a different crop) was associated with intense competition.

One may suppose that increased competition, by itself, tends to decrease foraging areas, but it also tends to cause crop deterioration, which has the opposite effect. The effect of increased competition upon size of foraging areas under any particular circumstances is therefore not at present predictable. Knowledge of this would help to solve many pollination problems.

[1] Bateman himself (1947b) did not know of the existence of foraging areas and thought that foragers wandered at random.

Chapter 14

THE FORAGING METHOD

And then pell-mell his harvest follows swift,
  Blossom and borage, lime and balm and clover,
On Downs the thyme, on cliffs the scantling thrift,
  Everywhere bees go racing with the hours,
For every bee becomes a drunken lover,
  Standing upon his head to sup the flowers.

V. SACKVILLE-WEST, *The Land*

IN A SUMMARY of the foraging behaviour of honeybees Butler (1945) said that honeybees became 'fixed' to their crops; these 'fixed' bees confined their attention to a small 'fixation area' of a particular crop, some 4–5 yards in diameter, and they foraged there either until they died or until the crop failed. He suggested that a bee would 'fix' in any area from which she could obtain a full load of nectar within 30–60 minutes and that on each trip any 'fixed' bee would almost always alight first on one particular plant and that after moving elsewhere in the area she would eventually return to the flower first visited and take off from it on her return flight to the hive. This hypothesis was derived from studies of behaviour upon dishes of syrup which provided an abundant and constant food source (Butler, Jeffree & Kalmus, 1943).

Subsequently Ribbands (1949) provided examples of honeybees which deserted crops which had certainly not failed (they were still attracting new recruits). In addition, he found that although a small proportion of bees usually commenced their trips by alighting on a particular plant in their foraging area, the last visits of foraging trips were scattered at random in that area; Sardar Singh (1950) also found that honeybees neither commenced nor finished their trips upon any particular plant in their foraging area. These results combine with our present knowledge of the mobility of foraging areas to make the 'fixation' hypothesis unacceptable.

The following alternative hypothesis was suggested by Ribbands (1949). He considered that the basic principle of the foraging-behaviour pattern is the continuous exercise of choice. Each honeybee chooses the best portions of the best of any alternative crops with which she becomes acquainted, and appears to compare the crop upon which she is working with any other crop which she remembers. The exercise of choice usually causes the bee to attach herself to a particular area of a crop, because she keeps returning to the part which she remembers was most profitable. Within that area the continuous use of choice and memory enables her to select the most suitable

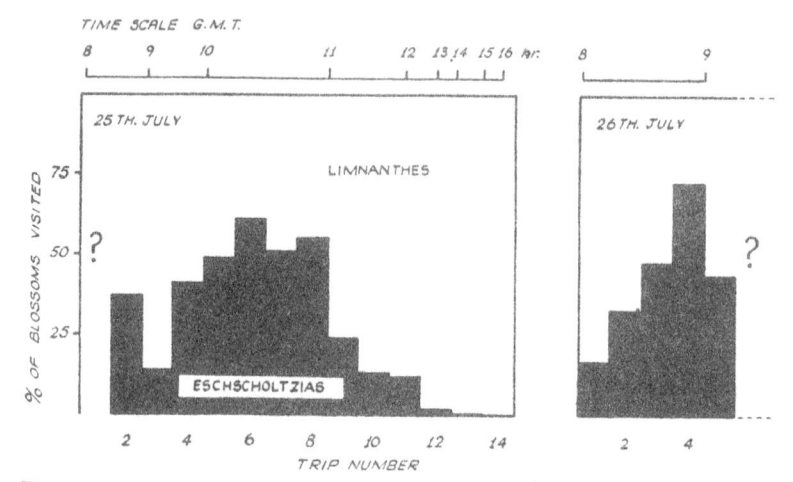

Fig 29 CHANGES IN THE PROPORTION OF *Eschscholtzia* BLOSSOMS VISITED ON SUCCESSIVE TRIPS BY A BEE WORKING ON BOTH *Eschscholtzia* AND *Limnanthes*. The weather was much better on the second day, so the peak proportion was reached more quickly (data from Ribbands, 1949).

blossoms of the crop which she is working; such discrimination was considered to be demonstrated by the great differences in the numbers of visits to neighbouring flowers (one bee working Shirley poppies was observed to make 19, 0, 8, 27 and 28 visits per flower to five neighbouring blossoms).

Continuous exercise of choice can be readily demonstrated among that small proportion of individuals which have become accustomed to working two adjoining crops both at the same time. One bee was watched while she was working flowers of both *Eschscholtzia* and *Limnanthes*, which were growing in adjacent rows and intermingling. It was found that the flowers were not visited at random, but that there were wide and consistent variations in the proportion of the flowers of each species which were visited (Fig. 29); these variations were ascribed to the exercise of choice by the bee, following changes during the day in the relative attractiveness of the flowers.

Many good pollen flowers open early in the day, before nectar flowers have commenced to yield (Parker, 1926; Synge, 1947). Ribbands (1949) found that some bees learned to work both crops regularly, at different times of the day. When collecting from the pollen crop such bees were seen to go to inspect the nectar crop at intervals, and then to return to forage from the former until the latter became satisfactory.

These observations of the activities of marked bees fit in with the experience of Vansell (1934), who found that bees in orchards abandoned pear blossoms at about 10 a.m. each morning, when the nectar concentration in apricot and plum had risen.

Ribbands's experiments indicated that a similar interplay of choice and memory often led to permanent change of crop. As crops deteriorated the bees working them became much more restless and questing; instead of going contentedly from flower to flower, visiting flowers quite close to each other, they had a marked tendency to jump long distances from one portion of their foraging area to another, or to go beyond their previous boundary. This response to less favourable conditions helped bees to concentrate upon the more attractive portions of their foraging area, or to extend or change the foraging area when necessary. The restlessness produced by crop deterioration also induced some bees to inspect other crops growing in the vicinity, and when a superior crop was thus found they changed to the new crop. When a crop was deserted in favour of another the bee usually continued to return, frequently at first and then at increasing intervals, to inspect the original crop. If the original crop improved the bee might change back to it.

About 10% of the individually marked bees which were being observed switched permanently from one crop in the garden to another, and in view of the limited choice available within the garden it was considered likely that there were more switches, not observable, to crops growing outside the garden. It was therefore suggested that change of foraging area and crop is often not an infrequent phenomenon. The world of a bee is a very inconstant one, and she often has to forage in very variable weather, on crops whose nectar abundance and concentration may vary very greatly from day to day and from hour to hour. In such circumstances there may be frequent occasions when her chosen crop temporarily deteriorates, so providing a stimulus for her to move to a better crop if she can find one. One may expect that the proportion of bees changing from crop to crop will vary greatly with local circumstances, such as the disposition of crops and the effect of weather upon the quantity of nectar and pollen available in them; thus one set of observations may indicate a high proportion of changes and another set a low one.

Ribbands & Nancy Speirs (1953b) observed the behaviour of individual bees which were foraging from dishes of syrup, a crop which could be strictly controlled. Each honeybee had its own individuality; in the same circumstances some bees were much more restless than others. If the crop was good no bees were restless, but as it deteriorated all eventually became dissatisfied. The bees were often very adaptable to substantial changes in syrup concentration, so their restlessness was not necessarily visibly increased by considerable reduction in that concentration. Yet bees which had been accustomed to syrups of high concentration might become restive when that concentration was reduced, even though other bees were foraging contentedly from other dishes which had always contained the lower concentration. Long training to a constant source increased the restiveness which occurred when the concentration was eventually reduced, but similar training to a variable source did not reduce it from its initial level.

Although the interplay of choice and memory enables the bee to exploit her environment very adequately, she is not an automaton who invariably chooses the best of the crops with which she has become acquainted. Ribbands & Speirs found that even if a bee had become aware of another dish which contained syrup of a higher concentration she might continue to forage from a dish to which she was more accustomed.

To summarize, it is believed that the continuous exercise of choice and memory (both of which are accurate but not infallible) explains the very wide range of observed behaviour. Bees attached to a particular area of a crop can be considered to be satisfied. The many gradations of behaviour can be explained by supposing that there is no absolute criterion of satisfaction, but a continuous gradation between complete satisfaction and absence of satisfaction. On any crop one may observe a range of behaviour resulting from a similar range of satisfaction, because the state of any individual will depend upon her inherited make-up and her previous experiences as well as upon the state of the crop which she is working.

Chapter 15

POLLEN GATHERING; DIVISION OF LABOUR AMONG FORAGERS

The careful insect 'midst his works I view,
Now from the flowers exhaust the fragrant dew,
With golden treasures load his little thighs,
And steer his distant journey through the skies.

JOHN GAY, *Rural Sports*

Pollen collection and packing

THE HAIRY BODIES of worker honeybees (Plate IX) provide them with a means of trapping pollen from the flowers which they visit. Accidental contamination often plays a part in pollen collection, especially when the process is incidental to nectar gathering, but pollen gatherers tear apart the anthers of some pollen plants with their mandibles (Casteel, 1912*a*) or rapidly run their legs over catkins (Parker, 1926).

Some of the pollen is moistened with honey (Sladen, 1912*b*) and the proportion of honey in the pollen load is substantial—analyses published by Casteel (1912*a*) showed an addition of 10% by weight of sugar. The colour of pollen loads differs from that of pollen in the flower (Reiter, 1947), and different kinds of honey produce different shades of load from the same pollen (Dorothy Hodges, 1952). Part of the admixture may be accidental, but the forelegs can become sticky with regurgitated honey obtained by stropping the protruded tongue, and collect pollen from the body hairs on their sticky surfaces.

The process of packing the pollen load was carefully described by Sladen (1911, 1912*a*), Casteel (1912*a*) and Ingeborg Beling (1931); their observations were extended and summarized by Dorothy Hodges (1952). The pollen is collected from the body hairs by the legs and passed backwards to nine rows of bristles, the pollen combs, which are on the *inside* of the metatarsus of each hind leg (Fig. 30). The hind legs are rapidly rubbed against one another, rising and falling in a pump-like motion. At the downward stroke of each leg its rake (Fig. 30) scrapes off a small quantity of pollen from the opposite leg. This pollen is pressed between the rake and the auricle and squeezed on to

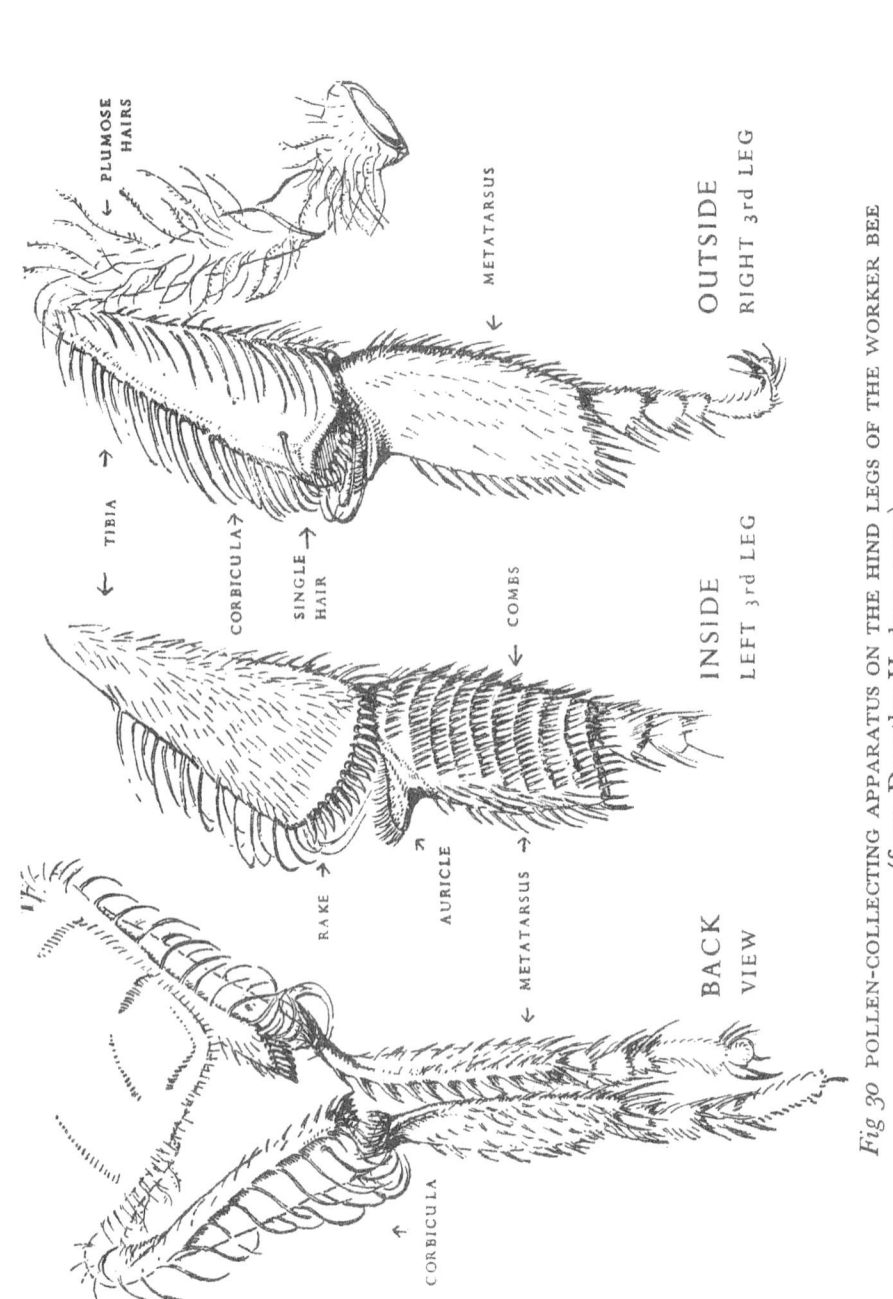

Fig 30 POLLEN-COLLECTING APPARATUS ON THE HIND LEGS OF THE WORKER BEE
(from Dorothy Hodges, 1952).

the *outside* surface of the downward-moving leg. It accumulates there at the bottom and on the outside of the tibia, where there is a smooth flattened area, the pollen-basket or corbicula, which has a sharp spine at its base for holding the load and which is surrounded by a fringe of hairs (Fig. 30). As further supplies of packed pollen are pushed on to the corbiculae, the pollen already there is pushed upwards and the load is shaped by the guiding action of the surrounding fringe of hairs. The middle legs may also pat the load (Parker, 1926), but Casteel found that normally shaped pollen loads could be produced on dead bees if their metatarsi were supplied with moist pollen and moved up and down in the above-described manner. The pollen-packing process usually takes place in the air during passage from flower to flower, but bees may fly up and return again and again to one copiously yielding blossom, or they may carry out some packing movements while they are sitting on a flower.

Pollen and nectar gathering; no usual sequence

Foragers can be divided into groups according to the kind of load they are seeking—nectar, pollen, water or propolis. Rösch (1925) did not find any definite sequence in outside duties similar to that which occurred within the hive, although such a sequence had been postulated by Gerstung.

Ribbands (1949), studying the duration of attachment of individual bees to crops, found that bees preferred nectar crops to pollen ones, and that if they discovered the former they would change to them even when loads took up to twenty times as long to collect. Among 136 bees which were marked on various crops yielding only pollen, the longest recorded attachment was one of 12 days and the average period of attachment was not more than 8 days, but 2 out of 10 of the bees marked on a nectar and pollen crop stayed for 21 days. Bees were often seen to change from pollen crops to nectar crops, but not vice versa. These results, however, showed that there was no deep-seated distinction between pollen and nectar collectors: some individuals were observed to collect pollen each morning from a Shirley poppy or nasturtium crop which yielded only pollen, and to change later each day to collecting either nectar only or nectar and pollen from a *Nemophila* or *Limnanthes* crop which yielded both supplies; they did this even when the earlier pollen crop still contained an abundant supply.

In contrast to this tendency for bees to switch from pollen to nectar crops, Sardar Singh (1950) instanced one marked forager which collected only nectar for five days and then collected pollen (or pollen and nectar?) from the same crop, from which other bees were still collecting nectar.

While some bees collect either pollen or nectar only, others collect both foods from the same flowers. The latter group can sometimes be subdivided; Ribbands (1950b) noted that some bees working on cornflowers collected large loads of pollen with their nectar, as if deliberately, while others on the same crop repeatedly accumulated only very small loads of pollen, as if they were packing into their pollen-baskets only the pollen which they had incidentally acquired. Still other bees on the same crop collected only nectar; although these three different categories usually remained consistent some bees changed from one category to another on successive days.

By daily observation of individually marked bees at their hive entrance, Ribbands (1952b) established that some bees gather pollen throughout their foraging life while others never gather it; that some gather it only at the beginning, some only at the end, and others at various intermediate periods of their foraging career (Fig. 31).

The effect of anaesthesia on pollen-collecting

Ribbands (1950b) showed that honeybees behaved normally after recovery from chloroform anaesthesia, but that when they recovered from anaesthesia with either carbon dioxide or nitrogen their pollen-collecting tendencies were usually inhibited; if such bees had been collecting both pollen and nectar from a crop they returned to it to collect nectar only; if they had been working a crop which yielded only pollen they deserted it. These same treatments applied to young bees made them skip their hive duties (p. 272): these results suggested that the treatments were artificially ageing the bees and so demonstrating a usual foraging sequence; however, this hypothesis was invalidated when it was discovered that the expectation of life of foragers after these treatments was not

Fig 31 THE AGE AT WHICH FORAGERS GATHER POLLEN. Newly emerged bees were individually marked on 6 April, and the *maximum* load of pollen brought in by each bee on every day of her foraging life is recorded in the diagram. Each line is the record of one individual (Ribbands, 1952b).

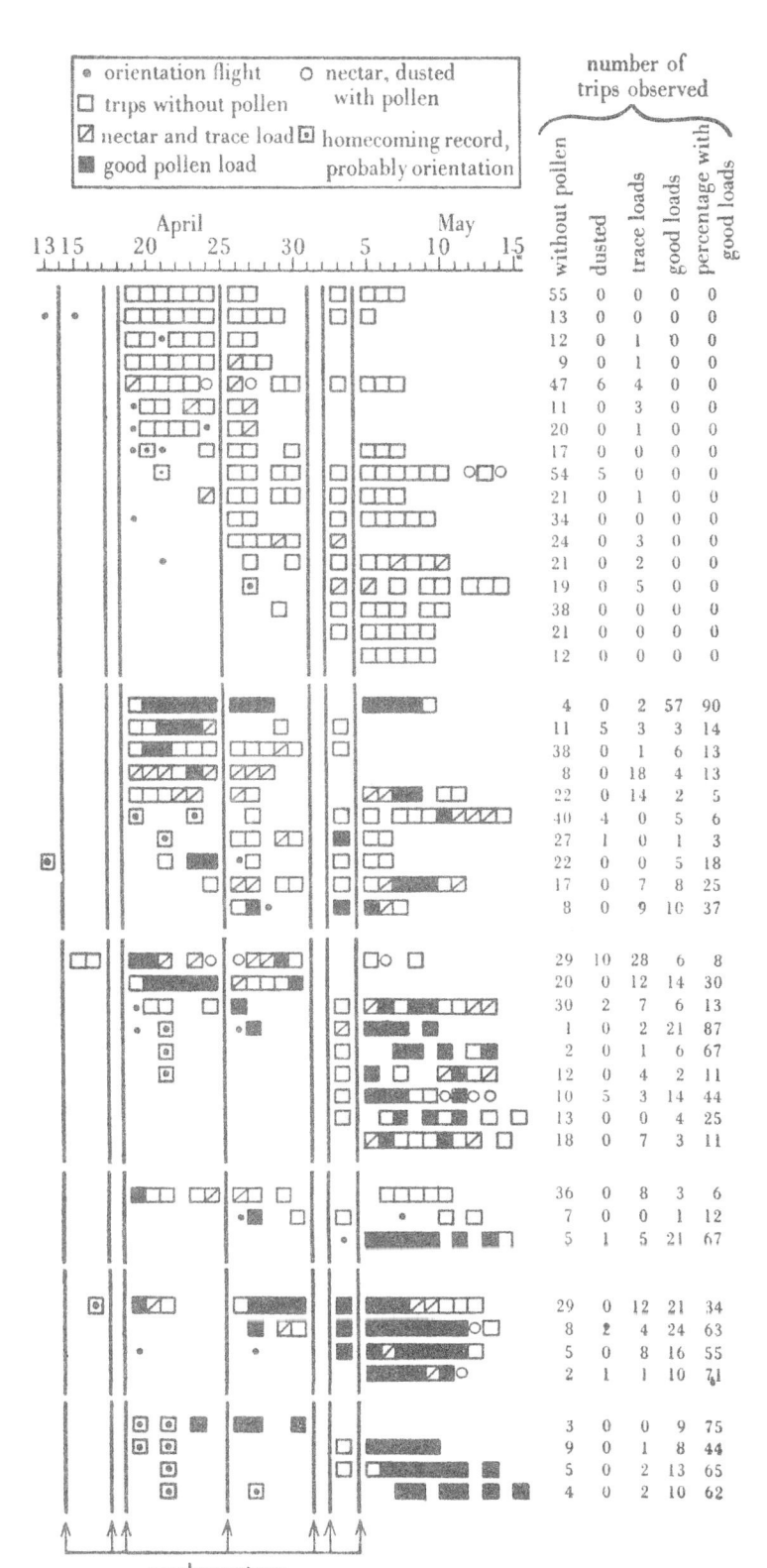

	orientation flight	nectar, dusted with pollen	without pollen	dusted	trace loads	good loads	percentage with good loads
			55	0	0	0	0
			13	0	0	0	0
			12	0	1	0	0
			9	0	1	0	0
			47	6	4	0	0
			11	0	3	0	0
			20	0	1	0	0
			17	0	0	0	0
			54	5	0	0	0
			21	0	1	0	0
			34	0	0	0	0
			24	0	3	0	0
			21	0	2	0	0
			19	0	5	0	0
			38	0	0	0	0
			21	0	0	0	0
			12	0	0	0	0
			4	0	2	57	90
			11	5	3	3	14
			38	0	1	6	13
			8	0	18	4	13
			22	0	14	2	5
			40	4	0	5	6
			27	1	0	1	3
			22	0	0	5	18
			17	0	7	8	25
			8	0	9	10	37
			29	10	28	6	8
			20	0	12	14	30
			30	2	7	6	13
			1	0	2	21	87
			2	0	1	6	67
			12	0	4	2	11
			10	5	3	14	44
			13	0	0	4	25
			18	0	7	3	11
			36	0	8	3	6
			7	0	0	1	12
			5	1	5	21	67
			29	0	12	21	34
			8	2	4	24	63
			5	0	8	16	55
			2	1	1	10	71
			3	0	0	9	75
			9	0	1	8	44
			5	0	2	13	65
			4	0	2	10	62

no observations

significantly different from that of similar untreated bees. The similar results of the treatments with nitrogen and carbon dioxide indicate that the causative mechanism in both cases was extreme lack of oxygen. Simpson (1953) has found that nitrous oxide anaesthesia has the same effects.

One may guess that in normal foragers the type of food collected depends upon at least three factors—availability, the requirements of the colony, and the physiological state of the individual bee—and that usually there is a balance which is easily tipped in either direction, but that after extreme oxygen lack the physiological state of the bee is so altered that the balance is always tipped in the same direction.

Proportion of pollen loads to nectar loads taken into the hive

The proportion of pollen loads to nectar loads varies widely, being largely dependent upon both available forage and colony requirements. In good nectar flows the proportion of pollen gatherers and the total quantity of pollen brought in may not diminish, but the proportion of pollen loads to nectar loads falls off very sharply—because the nectar gatherers complete their trips more quickly. In addition to seasonal and day-to-day variation there is often a daily rhythm. Different plants yield their pollen (Parker, 1926) or their nectar (Dolgova, 1928) at widely different hours, and the bees can only collect what is available. Parker observed a peak of pollen collection between 8 and 9 a.m., a decline towards midday and a secondary peak between 2 and 3 p.m. We now know that several factors contribute to this result. In many plants (e.g. poppies, roses) the anthers burst in the bud and all the pollen becomes available when the flowers open early in the morning, but in some other plants the flowers open in succession (e.g. field bean) or the pollen becomes gradually available (e.g. some Compositae) (Ann Synge, 1947).[1] Many nectar plants, however, do not yield until late in the morning when the temperature has risen. Time perception (p. 133) may enable the individual bee to take advantage of this situation.

Among plants which yield both nectar and pollen the proportions of the two foods collected at any time will partly depend upon their relative availability—if the nectar supply decreases as the day goes on this may induce increased pollen collection without any change in pollen availability.

[1] Further studies were carried out by Mary Percival (1950) and Anna Maurizio (1953).

The proportion of pollen gatherers among the foragers

Filmer (1932) counted the number of frames with brood and the rate of return of all foragers, and also of pollen gatherers,

TABLE 7. The effect of brood rearing on pollen collecting. Results are means of at least 10 one-minute counts of returning bees (Filmer, 1932)

	Six over-wintered colonies (each with c. 3 lb. bees)	Four 3-lb. packages on drawn comb	Four 3-lb. packages on foundation
No. frames of brood	4 4 4 5 6 7	2 3 4 5	2 3 3 4
No. pollen carriers	17 20 35 37 54 75	23 13 30 33	11 16 21 15
No. without pollen	43 10 14 17 4 7	14 27 21 29	32 41 39 57
% pollen carriers	28 66 74 60 95 90	23 13 30 33	25 29 35 20

of 14 colonies of honeybees in a fruit orchard. Eight of the colonies were 3-lb. packages which had been established on drawn combs or foundation 3 weeks previously, and the

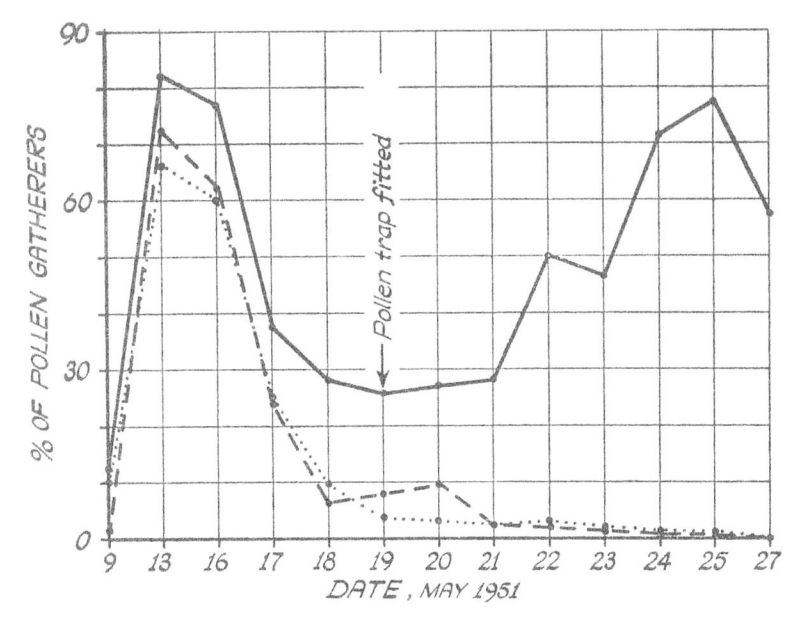

Fig. 32 ADJUSTMENT OF FORAGING DUTIES TO COMMUNITY NEEDS. The proportion of pollen gatherers increased in the colony fitted with a pollen trap, but it did not do so in two neighbouring colonies not so treated (Lindauer, 1952).

other 6 were over-wintered colonies estimated to contain approximately 3 lb. of bees. The results, displayed in Table 7, show clearly that in the circumstances of this experiment the proportion of pollen gatherers increased with the quantity of brood. They also appear to support Filmer's claim that the amount of foraging activity bore a direct relation to the amount of brood rearing, but this claim may have been more apparent than real—a consequence of the more frequent journeys which the pollen gatherers would make (p. 131).

Lindauer (1952) found that the proportion of pollen gatherers increased when a colony was fitted with a pollen trap (Fig. 32). This result indicates that the needs of the colony determined the proportion of pollen gatherers; it would account for Hirschfelder's (1951) conclusion that his colonies were not injured by the presence of pollen traps.

Chapter 16

FORAGING STATISTICS

There is measure in all things; certain limits, beyond and short of which right cannot be found.

HORACE, *Satires*

Speed of flight

VALUABLE OBSERVATIONS OF the speed of flight of honeybees, which took account of the speed of the wind at the same time, were made by Park (1923*a*, 1928*b*). Park, with another observer, timed marked bees during their travels between their hive and a dish of syrup 1/5th mile away. They timed about 25 consecutive trips on each of 4 different days, and they determined the velocity of the wind at the beginning and end of each period with a portable anemometer 'placed at about the height of the bee's flight'.

The bees flew, on different days, with or against or at right angles to the wind. For flight with or against the wind, wind velocity was subtracted from or added to the observed flight velocity. When the bees flew at right angles to the wind they had to hold themselves at an angle to their line of flight, just enough to offset the force of the wind which would otherwise have deflected them, and Park calculated (by triangulation) the correction necessary for the observed 10 m.p.h. cross wind. This correction increases as the observed speed decreases; at observed speeds of 8·8 and 10·7 m.p.h., the equivalent speeds in calm were estimated to be 13·3 and 14·6 m.p.h.

In outward flights the mean speed (corrected for wind) was found to be 12·5 m.p.h., and in homeward flights it increased to 14·9 m.p.h. On both outward and homeward flights the corrected speeds were greatest when the bees were flying against the wind and least when they were flying with it. This possibly happened, as Park suggests, because the bees increased their efforts when they were flying against the wind.

Food supplies taken out from the hive by foragers

Parker (1926) stated that pollen-gathering bees which had just arrived on roses had their honeysacs half-filled, but that those which went to collect nectar from alsike clover arrived with nearly empty honeysacs.

Ruth Beutler (1950) analysed the sugar content of the midguts and honeysacs of bees which were setting out from their hive to forage for either water, pollen or syrup. Foragers going to collect either water or syrup close by the hive carried less than 1 mg. sugar in these organs, bees gathering syrup at 550 yards carried nearly 1½ mg., while those gathering syrup at 1100–2200 yards carried nearly 3 mg. These supplies were equally divided between midgut and honeysac. Pollen gatherers, however, were carrying out nearly 4 mg. sugar, 75% of which was in their honeysacs. The outgoing foragers, therefore, were anticipating the quantities of sugar which they would require during their trips.

Weight of load

Gillette (1897) found that the average weight of a pollen load (both pellets) was 11·2 milligrams. Park (1922) stated that this weight depended upon the species from which the pollen had been gathered, because he found that elm loads weighed 11 mg., maize 14 mg., apple 25 mg. and hard maple 29 mg. He added that the loads from maize looked as large as those from apple or maple although they weighed only half as much. Parker (1926) also found that loads from different species varied in average weight, from 9 mg. for asparagus to 20 mg. for ragweed; the average weight of 233 loads, from 7 species, was 15 mg. Anna Maurizio (1953) found that the mean weights of loads from 35 different species varied between 8·4 mg. (*Erica carnea*) and 22 mg. (horse chestnut). Since the average weight of unladen bees is about 80 mg. (Gillette, 1897; Park, 1922) the pollen load carried is often about one-fifth of the body weight and may be as much as one-third.

Park (1922) found that the maximum nectar-carrying capacity was 70 mg., but that average loads weighed about 40 mg. Gontarski (1935a) found that the maximum quantity of syrup which honeybees could suck up was increased when the syrup was warmer, to a maximum of 74 c.mm. with syrup at 50°C.; he attached a small weight to the abdomen and found that it did not affect the quantity of syrup imbibed, so he

Ben Knutson

PLATE II *Honeybee gathering pollen from columbine*

Lee Jenkins
PLATE III *Honeybee on an apple blossom, using her antennae*

concluded that the bee (which sucked more slowly as her honey-sac filled) stopped when her muscles were unable to pump in any further supply at that temperature.[1]

From these results it appears that the bee can lift a load nearly equal to her own unladen weight, and that the maximum size of load which she carries is more dependent upon her capacity to pack the load (whether pollen or nectar) than upon her ability to fly with it.

Risga (1936) found that the average weight of pollen loads increased with rising temperature (18 mg. at 10–14° C., 20·5 mg. at 14–18° C., 23 mg. at 18–22° C.) and was inversely related to wind strength (25 mg. in calm, 22 mg. at 2–6 m.p.h., 18·5 mg. at 6–14 m.p.h.). Nectar loads showed a similar relation.

Recently Schuà (1952b) has shown that bees trained to a dish at a distance from their colony imbibed slightly greater quantities of sugar solution than did those trained to a dish in the vicinity of the colony. The increased amount corresponded to the larger sugar reserve which is carried out by bees which go to forage at a long distance.

Flower visits and time required per load

The number of flowers which a bee must visit to obtain a load varies very widely. For nectar loads Miller (1902) said that a bee might sometimes obtain a load from a single blossom of tulip poplar, but he stated that Rauschenfels had followed one forager for 640 flower visits on one trip, and held that bees usually visited 50 to 1000 blossoms on each trip. Ribbands (1949) followed one nectar gatherer upon *Limnanthes* for a trip which comprised 1446 flower visits, in 106 minutes; the smallest number of flower visits which he observed during any full nectar trip was 250 (by a pollen and nectar gatherer). It is probable that in very poor summer conditions, when trips lasting 3 or 4 hours are necessary, several thousand flowers are sometimes visited on each trip.

The number of flower visits required in order to gather a pollen load is usually much smaller. Some individual flowers, such as Shirley poppies, contain enough pollen for several loads, and a bee may obtain a whole load from one such blossom although she usually visits several (Ribbands, 1949); on good pollen crops loads were often obtained in less than 30 flower

[1] Kalmus (1938) also attached a weight (50 mg.) to the abdomen of drinking bees; they did not stop drinking and they became too heavy to fly, so he concluded that weight did not play an important part in determining uptake.

visits, but from 66 to 178 flower visits per load were recorded for nasturtiums. Mary Percival (1950) has calculated that white clover flowers contain so little pollen that pollen from 585 untripped flowers (the equivalent of 100 unvisited inflorescences) would be required for a load, yet occasionally bees work this crop for pollen only (Parker, 1926). Weaver, Alex & Thomas (1953) watched a bee gather a full pollen load from 494 flowers of white sweet clover in 12.8 minutes.

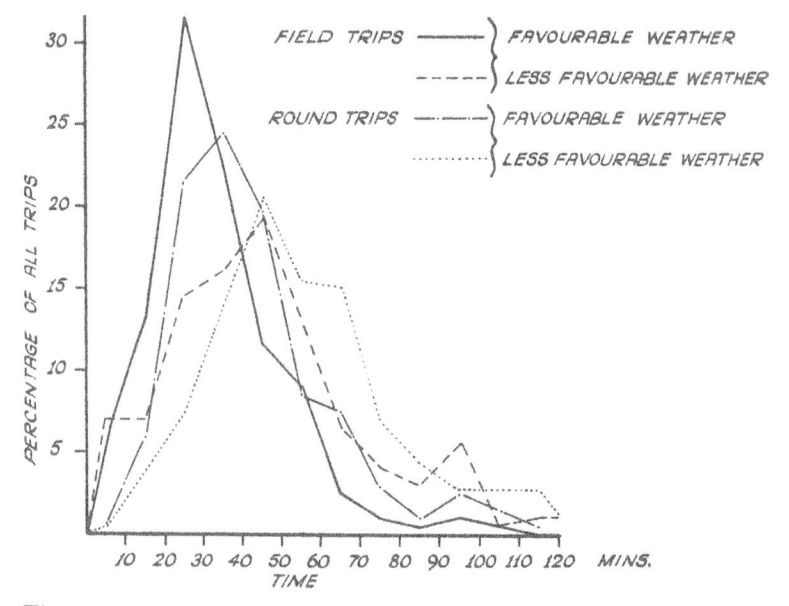

Fig 33 THE DURATION OF TRIPS TO WHITE SWEET CLOVER IN FAVOURABLE AND LESS FAVOURABLE CONDITIONS. Frequency distribution after Park, (1928b).

The time necessary for collecting a load is almost as variable as the number of blossoms which have to be visited. Lunden (1914) marked six individual bees during a good flow from white clover (4 lb. daily hive gain) and found that their trips lasted from 30 minutes to 2 hours and averaged 1 hour. Park (1922, 1928b) investigated the collecting times during nectar flows from sweet clover (Fig. 33). He found that in good conditions (a 5 lb. daily hive gain) the mean time for the field trip (excluding unloading time within the hive, mean value 11½ minutes, mode 4 minutes) was 34 minutes and the mode, or most 'fashionable' time, was 27 minutes. The following year in poorer conditions (1 lb. daily hive gain) the mean time for the field trip was 49 minutes and the mode 45 minutes. Ribbands (1951) found that the mean time for round trips of honeybees (including unloading time within hive) during a heavy lime flow (5 lb. daily hive gain) was 40 minutes. In less favourable conditions Sardar Singh (1950) followed marked bees for 60 individual field trips on golden rod; their mean time was 80 minutes and the longest observed field trip took 169 minutes. Minderhoud (1929a) found that field trips sometimes took up to 3 or 4 hours.

Pollen-collecting trips are usually completed much more quickly than nectar-collecting ones. Park (1922, 1928b) found that when collecting from maize under favourable conditions the mode for field trips was $8\frac{1}{2}$ minutes, but that in less favourable conditions it rose to $15\frac{1}{2}$ minutes. The mode for the period spent in the hive was about $3\frac{1}{2}$ minutes. Ribbands (1949) timed five individual bees during 114 field trips to *Eschscholtzia*, Shirley poppies and nasturtiums; their field-trip times (without journey to and from hive) ranged from 3 to 18 minutes. The mean field-trip times varied on the different crops (from 3·8 to 12·7 minutes) and there was also a marked difference between the mean times of two different bees when they were both working Shirley poppies under similar conditions.

These differences between the mean times of different individuals were similar to those obtained by Butler, Jeffree & Kalmus (1943) for marked bees foraging from dishes of syrup, from which loads were imbibed in less than one minute; water loads are also gathered very quickly—67% of all the field trips which Park (1923d, 1928b) recorded for water carriers were completed in 3 minutes or less; the mode for their round trip was 5 minutes.

Number of trips per day

The number of foraging trips per day depends upon the time required for each. Lunden (1914) found that his marked bees made 10 trips daily during a heavy flow from white clover. During a heavy nectar flow from sweet clover (5 lb. daily hive gain) the mean number of trips per day was 13·5 and the maximum 24, but in poorer conditions the following year (1 lb. daily hive gain) the mean was reduced to 7 and the maximum to 17 (Park, 1928b). Sardar Singh (1950), working at Ithaca, N.Y., reported that the number of nectar trips per day ranged from 3 to 10, depending on the crop; it is probable that in Britain the mean number of nectar trips per day usually falls within a similar range.

Pollen gatherers may make more trips per day than nectar gatherers if the pollen supply is available all day. Park (1928b) found that pollen gatherers working maize made fewer trips per day (mean 8, maximum 20) than nectar gatherers working sweet clover, but this happened because maize pollen is available only in the mornings. Ribbands (1949) observed one individually marked bee while she gathered 47 loads of pollen from *Eschscholtzia* in one day.

The latter value may approach the possible maximum on a good natural crop, but at least 150 trips per day may be made to syrup dishes (Butler, Jeffree & Kalmus, 1943), and a maximum of 114 trips in one day to water have been recorded (Park, 1923d, 1928b). Park considered that the average number of trips per day of water carriers was about half that number; this result agrees with that of Minderhoud (1929b), who found that 4 marked water carriers averaged 56 trips in one day.

Nectar concentration and quantity; flower density

A detailed study of these botanical problems is beyond the scope of this book, but they are material factors in all pollination studies. Nectar concentration and quantity in the same species vary widely with strain, soil conditions, and climate, while hour-to-hour and day-to-day variations are so extensive that any estimates of potential yield must be liable to a wide margin of error.

Recent extensive analyses of nectar concentration and abundance include those of Ruth Beutler & Adele Schöntag (1940) and Boëtius (1948); Beutler's (1949) estimates of the weight of sugar secreted per flower per day in major nectar plants ranged from 0·02 mg. in rosebay willowherb to 7·6 mg. in raspberry. The number of flowers per acre is another important factor, but one which can be more easily estimated. For example, McGregor & Todd (1952) noted that lucerne produced 300 times as many flowers per acre as melons, and could yield 150 times as much honey. Weaver & Ford (1953) estimated that there were 1,900,000 inflorescences per acre of crimson clover, with 95 flowers per inflorescence.

Chapter 17

TIME PERCEPTION

*And, as it works, th' industrious bee
Computes its time as well as we.*
          ANDREW MARVELL, *Thoughts in a garden*

VON BUTTEL-REEPEN (1900) observed that bees working buckwheat commenced to fly early each morning and deserted the flowers at about 10 a.m., when their nectaries dried up; these bees remained in the hive for the rest of the day. This observation might be explained in terms of time perception, as Forel (1906) noted after he had fortuitously discovered that honeybees possess this attribute.

In summer the Forel family took their meals on a terrace in the open air, breakfast from 7.30 a.m., lunch at midday and tea at 4 p.m. One day a honeybee found the preserves at breakfast time, and brought companions. For a day or two bees came whenever the table was laid, but soon they came only at breakfast and tea times—there were no preserves at lunch. On 17 July many bees arrived during breakfast, none at lunch, and a moderate number during tea. Next day the table was laid *without* preserves; bees still came at breakfast time, one bee came during lunch, several at tea time. On 19 July fewer bees came during breakfast, but more arrived at lunch and tea time. Forel remarked that it seemed as if the absence of the morning honey had impelled the bees to continue their search at the incorrect time.

A complementary observation was made by Dobkiewicz (1912) who, having trained bees to gather honey from artificial flowers between 11 a.m. and 1 p.m., put the latter out at a different time and observed that the bees did not visit them.

Detailed experiments were carried out by Ingeborg Beling (1929). At the same time on successive days she fed individually marked bees at a feeding-place, and then recorded their visits to it on a subsequent day during which no food was provided. Beling succeeded in conditioning the bees to come to the same feeding-place not only once each day, but also at two or three different times (Fig. 34). The preciseness of the response increased with repetition of the training, but bees which had been fed for a period of three hours on only one day returned during the same period on the following day (Fig. 35).

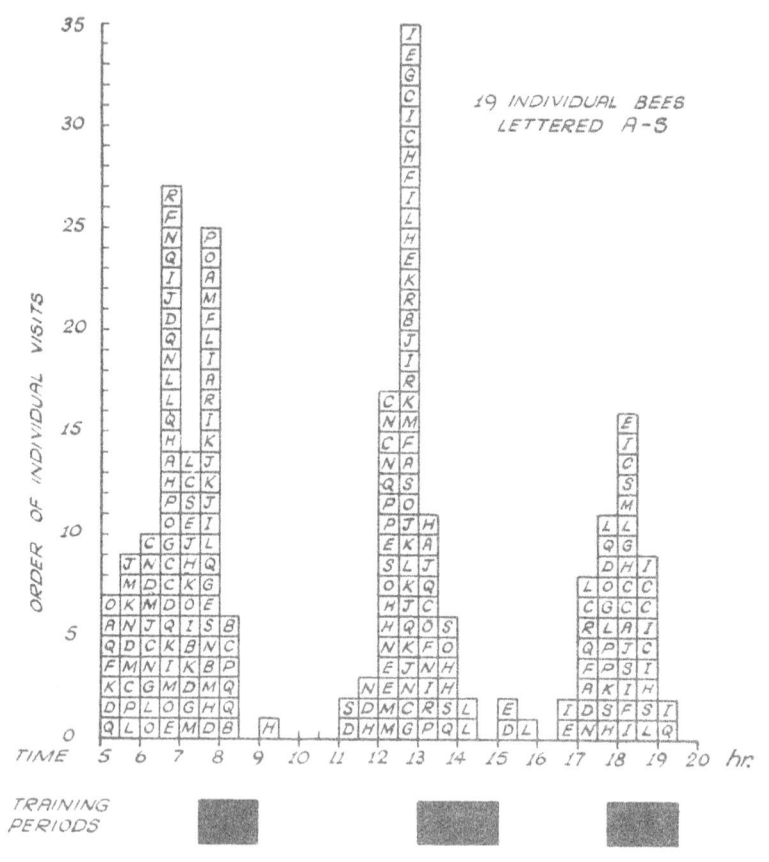

Fig 34 TRAINING OF INDIVIDUALLY MARKED BEES TO SEEK FOOD AT THE SAME PLACE AT THREE DIFFERENT TIMES OF THE DAY. The bees were trained for 6 days and their visits were recorded on the 7th day, when no food was provided (after Beling, 1929).

Beling considered that time perception was tied to a 24-hour rhythm; her attempt to train bees at intervals of 19 hours was not successful. Wahl (1932) was unable to condition bees to a 48-hour interval; they returned every 24 hours. Beling tried to determine whether the 24-hour rhythm was dependent upon some external daily periodicity. She was able to train the bees to any hour even when they were placed in an experimental room in constant light, air humidity and temperature, and the training was not disturbed when the air in this room was irradiated so that artificial ionization would alter the rhythm of electrical conductivity of the atmosphere. Thus she concluded that time perception was dependent either upon some other external factor or upon some organic discharges within the honeybee which recurred at 24-hour intervals; neither mechanism was known.

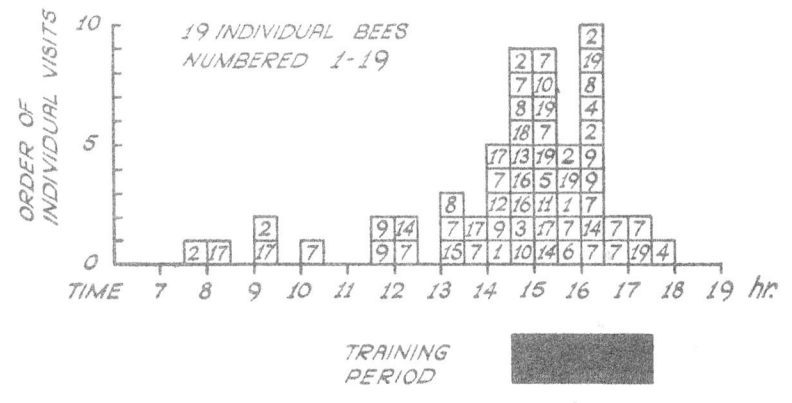

Fig 35 TIME PERCEPTION. RETURN OF BEES AFTER ONE LESSON. Individually marked bees were fed for 3 hours and their visits were recorded on the next day, when no food was provided (after Beling, 1929).

Wahl (1932) re-investigated the problem. He, too, found that honeybees could be trained at several different times, provided that the pauses between individual feedings lasted at least 2 hours. When no food was provided the bees returned punctually during the first period, but less and less punctually during succeeding periods (cf. Forel, 1906). When bees were unable to forage, their time training remained effective for 6-8 days after training had ceased, but trained bees, having been counter-trained to another time interval for 2 days, visited the feeding-place throughout the next day (when no food was provided). Wahl was able to train bees to two different times, at *different* feeding-places; they then sought food at the correct time and place; Kalmus (1934) confirmed that this was possible.

Wahl (1932) trained bees to collect pollen at a definite time also. When fed continuously with a syrup crop which alternated between abundance and scarcity, most of them soon appeared only during the times of abundance. Wahl (1933) subsequently showed that when different *concentrations* of sugar syrup were exposed at different times of the day most of the bees learned to appear only at the times when concentrated syrup was available.

Wahl (1932) took his bees into a salt mine in which all external daily periodicity factors were apparently eliminated, but he was still able to condition them to time. He was able to show that time perception was inborn, by training bees which had hatched into a dark chamber and had never been aware of the alternation of day and night.

Wahl (1932) time-conditioned bees which were kept for 16 days in a dark room at 23° C. He then raised the temperature of both the colony and the room by about 8° C., gave the bees two more days of training at these temperatures, and found that on the third day they were still trained to the same time interval. From this result he concluded that time perception was not dependent upon any internal metabolic rhythm, since such a rhythm would have been accelerated by the increased temperature.

Experiments with ants, which were carried out by Grabensberger (1933), showed that time perception was bound up with food metabolism. Having consulted the textbook of Meyer and Gottlieb (1925), which reported that iodothyroglobulin accelerated food metabolism, Grabensberger conditioned ants to visit a feeding-place at a stated time, added 0·05% of this substance to their food, and found that on the next occasion the ants returned too early. Quinine retards many vital cell processes, in insects as in humans, so Grabensberger then added 0·08% euquinine (a less potent but tasteless product) to the diet of another set of time-conditioned ants; the response of these animals was retarded.

Grabensberger (1934a) then added 0·015% iodothyroglobulin to the food of time-conditioned honeybees. No food was subsequently supplied to them, and on the following three days they returned about 5 hours, 2 hours and 1½ hours too early, successively (Table 8). In the parallel experiment 0·015% euquinine was added, and on the following three days the bees were about 4 hours, 3 hours and 2½ hours late, successively (Table 9). Grabensberger concluded from these results that the effect of either chemical was gradually reversed as it was either digested or excreted.

TABLE 8. The effect of the addition of 0·015% euquinine to the diet of honeybees trained for six days to a 24-hour feeding rhythm (Grabensberger, 1934a)

Number of visits recorded, by 10 marked bees

Date																									
13×33													1	1	2	**12**	5	1							
14×33	×											3	1	**12**	3	2	4	1	1						
15×33	×										1	4	**12**	6	3	3	2		1						
16×33	×									1															
Time	6³⁰	7	7³⁰	8	8³⁰	9	9³⁰	10	10³⁰	11	11³⁰	12	12³⁰	13	13³⁰	14	14³⁰	15	15³⁰	16	16³⁰	17	17³⁰	18

× = Observations commenced or ceased.   ▓ = Treatment and training period.

TABLE 9. The effect of the addition of 0·015% iodothyroglobulin to the diet of honeybees trained for six days to a 24-hour feeding rhythm (Grabensberger, 1934a)

Number of visits recorded, by 50 marked bees

Date																									
1×33							10	**25**	7			1	2										×		
2×33	×										1	1	**12**	3			1	2						×	
3×33	×									1	2	9	7	**15**	5	5	1	1	1					×	
4×33	×																								
Time	6	6³⁰	7	7³⁰	8	8³⁰	9	9³⁰	10	10³⁰	11	11³⁰	12	12³⁰	13	13³⁰	14	14³⁰	15	15³⁰	16	16³⁰	17	17³⁰	18

× = Observations commenced or ceased.   ▓ = Treatment and training period.

Grabensberger (1934b) carried out a further series of experiments with ants, with results which reinforced his previous conclusion. Salicylic acid also has the property of accelerating metabolism, and when this was added to their food the time-conditioned ants again went to the feeding-place too soon. The effects of arsenic depended upon the dose received; with 0·0001% the ants came at the correct time, after 0·0005% they were about $3\frac{1}{2}$ hours late, after 0·00075% they again came at the correct time, after 0·001% they were $3\frac{1}{2}$ hours too early and after 0·002% they came $6\frac{1}{2}$ hours too early. With 0·001% yellow phosphorus they arrived $6\frac{1}{2}$ hours too soon. Grabensberger said that these observations fitted with the pharmacological effects of these poisons upon metabolism.

Further experiments with honeybees were undertaken by Kalmus (1934), who confirmed the effect of euquinine. Kalmus found that ether anaesthesia for several hours, or feeding with desiccated thyroid gland, had no effect upon time perception, but that long and severe chilling (at 5–7°C. for 19 hours) led to delay (Fig. 36), as did narcosis with self-produced carbon dioxide. As the delay produced by the chilling was not of great magnitude it is not surprising that Kalmus, like Wahl (1932), obtained no significant effect with less extreme temperature changes.

A conflicting feature of the experiments on time perception is that Beling did not succeed in her attempt to condition honeybees to a 19-hour feeding rhythm, yet Grabensberger (1933) succeeded in conditioning ants to 3, 5, 21, 22, 26 and 27-hour feeding rhythms, and termites to a 21-hour rhythm. Grabensberger (1933) emphasized that the food metabolism of ants did *not* take place in a 24-hour rhythm, but that if external influences were not materially altered it went on at a fairly constant rate. As ants and bees responded in a similar way to euquinine and iodothyroglobulin, it is likely that the physiological basis of time perception is identical in both groups. Beling's conclusion therefore requires confirmation. Time perception is a development of a property of daily rhythm which is shown in the activities of many insects and which is regarded as dependent upon their rate of metabolism (Wigglesworth, 1950). For example, the emergence rhythms of flour moths (*Ephestia kühniella*) can be induced to fit temperature changes in cycles of 16 or 20 hours, but not in cycles exceeding 24 hours; in a constant-temperature room this inborn rhythm persists for three generations (Scott, 1936).

TIME PERCEPTION

The value of time perception to the honeybee is very clear. Different plants yield their crops at different times of the day and the same bee may gather pollen from an early-morning crop and collect nectar for the rest of the day (p. 117). This

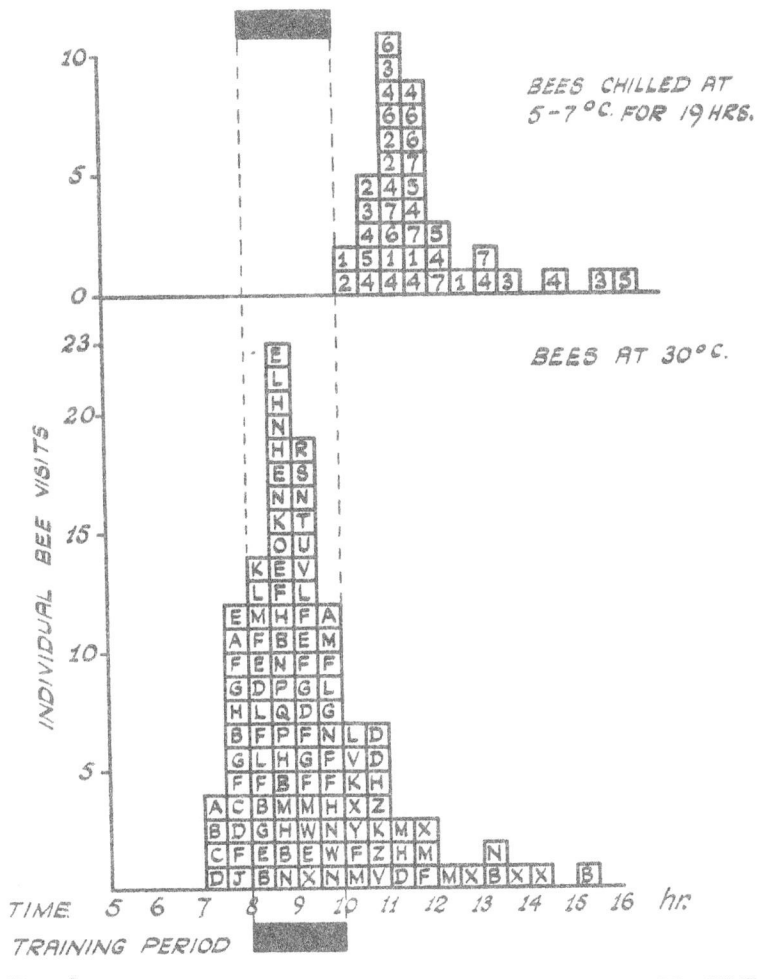

Fig 36 TIME PERCEPTION. DELAYED VISITS BY CHILLED TIME-TRAINED BEES; NORMAL VISITS BY TIME-TRAINED BEES KEPT AT 30° C. No food provided on this day (Kalmus, 1934).

may happen not only if the earlier crop becomes worked out by the time at which the later crop begins to yield, but also if the earlier crops persist but the later crop is superior. Bees working upon nasturtiums, which were often deserted each afternoon although their pollen supply had in no way diminished

(Ribbands, 1949), were probably working two crops in this manner.

Conditioning to time is not carried to unprofitable extremes. Bees which had become attached to a Shirley poppy pollen crop, usually finished by 10 a.m., came to the crop at a later hour when they had been held up by bad weather (Elisabeth Kleber, 1935; Ribbands, 1949). Ilse Körner (1939) observed that bees which had been time-conditioned to syrup dishes usually retired to rest in distant areas of the combs during intervals between feeding; as dances occur on a definite area near the entrance these bees were then unaffected. However, any of these trained bees which happened to be in the dancing area during such an interval, or which were transferred there, did respond to the dances.

A most surprising example of time perception and memory has been recorded by v. Frisch (1950a). Clouds depolarize sunlight, so the dances of bees which are on a horizontal comb lit from a completely cloudy sky are disoriented (p. 157). In nine experiments the dances pointed in a definite direction as soon as such bees were covered by a polaroid sheet. This sheet was only transmitting light polarized in a certain direction, and the bees under it saw a pattern which they would have seen at that time at one point in the sky on a clear day; they were correctly orienting their dances to this pattern. In three other experiments the bees still danced disorientedly; in these the pattern they saw would not have been visible anywhere in a cloudless sky at that time. The polarization pattern of the sky steadily changes with the time of day, and these results were understandable on the supposition that the dancing bees were correctly associating their perception of time with their memory of the pattern which would have appeared in a blue sky at that time, and were orienting their dances in relation to the remembered pattern.

In a further experiment, 29 marked bees were fed at a scented feeding-place 220 yards west of their hive. Overnight the colony was removed to a new site 3 miles away. Feeding-places were arranged 220 yards to the north, south, east and west of the new site, and an observer sat at each place and caught all arriving bees. A southerly breeze would have brought odour to the hive from that direction, but in three hours 5 marked bees were caught on the south feeding-place, 20 on the west one, 1 each at the north and east ones—27 of the 29 trained bees were recovered. During these observations the sun was in the south-east, whereas during the last journeys on the preceding day it had been setting towards the west. Remembered visual landmarks having been eliminated by the overnight move, the bees were apparently searching in relation to the remembered pattern of polarization and adjusting their direction to allow for the time of day at which they were foraging!

Chapter 18
MATING BEHAVIOUR

Rivals none the maiden woo,
So you take her and she takes you.

W. S. GILBERT, *The Bab Ballads*

CHARACTERISTICS SOMETIMES FOUND in the sexual relations of other insects have been fitted into the pattern of honeybee social life. These include the great excess of males, the death of the male during or immediately after copulation, and the storage of sperm from a single mating so that it suffices for the lifetime of the female (Wigglesworth, 1950). In both sexes the absence of certain non-reproductive functions and the individual's inability to lead an independent existence have allowed the reproductive structures and their accessories to reach a higher peak of development than would otherwise have been attainable. Thus the drone is able to produce and store the great quantity of sperm required to fertilize the enormous number of eggs produced by the queen, and the elaborate structure of his accessory reproductive organs enables this bulk of sperm to be rapidly transferred during the mating act.

Mating by capture

Mating has only been observed during flight, and it is probable that the air sacs of the drone can be sufficiently distended only during flight (Cheshire, 1886).

The superior development of the eyes of the drone (Fig. 16) indicate that he plays the active role in the mating quest. This type of difference between the sexes is often correlated with mating by capture on the part of male insects (Richards, 1927). Their large eyes help drones to pursue a flying queen which they have located, but insect vision is so inferior in other respects (pp. 16–22) that these eyes cannot enable the drones to distinguish the virgin queen from other similar insects.

The superior development of drone antennae (p. 65) indicates that recognition of the virgin queen is a matter of smell. Scent organs are of very wide and frequent occurrence in insects, and they often play an essential part in mating (Richards, 1927), but the scent or scent organ which might enable the virgin queen to attract a mate is still unknown. It is likely that it is only operative during flight; the worker's scent gland is not functional in queens (p. 63), but the mandibular glands or the sting glands are possible scent sources.

A few observations, summarized by v. Buttel-Reepen (1923), suggest that drones tend to congregate at selected places in the air, to which the virgin queen is attracted.

Further statements of this kind have been made recently (Müller, 1950). However, there are numerous records of isolated matings, and congregation of drones (which seems to be at least unusual) might be a response to some environmental condition or to the scent from a virgin queen.

Mating position

Eyewitness statements from the United States, reviewed by Shafer (1917) and accepted by Bishop (1920), allege that the mating pair meet and clasp face to face, the procedure taking a few seconds only. European statements, reviewed by Annie Betts (1939b), indicate that the drone mounts on the back of the queen. The latter is the mating pose which has been observed in many bees, wasps and ants (Richards, 1927). Moreover, a face-to-face position is difficult to reconcile with active pursuit of the queen; such accounts are probably fanciful and anthropomorphic.

Physiological details of the mating act have been thoroughly reviewed and discussed by Fyg (1952). Fyg suggests that there is doubt whether the drone mounts the queen or vice versa, but the writer thinks that this doubt is resolved by the evidence of the sex ratio, together with the superior eyes and antennae and stronger wings and more extended flight range of the drone (pp. 63, 146).

Mating age

Janscha (1771) wrote that virgin queens fly out to mate on their third or fourth day, but the mating flight may be delayed by inclement weather.

Huber (1792) noted that virgins which mated when they were over 21 days old became drone-layers. Oertel (1940) summarized various opinions which suggested that virgin queens usually mate between their third and fourteenth day of adult life. Sixty queens which were observed by him at Baton Rouge, Louisiana, mated on the 6th to 13th day after their emergence (6 at 6 days, 10 at 7, 16 at 8, 16 at 9, 7 at 10, 3 at 11, 2 at 13).

Howell & Usinger (1933) marked 100 newly emerged drones and placed a drone trap on the entrance of their hive. The mean age at their first flight was 8·41 days (range 4–15 days). Their maximum length of life was 59 days, but successful matings by overwintered drones have been claimed (Kalnitzki, 1949). Bishop (1920) noted that drones are not sexually mature at the time of emergence, but undergo a growth period of at least 9–12 days; the sexual behaviour of insects is apparently not influenced by the gonads (Wigglesworth, 1950), so it is possible that if opportunity arises drones attempt to mate before they reach maturity.

Mating-flight statistics

Oertel (1940) timed the flights of 56 virgins, and found that mating flights lasted from 5 to 30 minutes and non-mating flights from 2 to 30 minutes. He never saw queens at the hive entrance in the morning, and their successful mating flights occurred between 1.23 and 4.35 p.m. Roberts (1944) timed 114 successful mating flights, which lasted up to 50 minutes; he observed that the mean duration of successful flights shortened significantly as the season advanced (April 19·3 minutes, May 15·3 minutes, June 11·9 minutes), although the duration of unsuccessful ones increased (9·7, 9·9, 11·4 minutes respectively), so he concluded that the shortening was due to an increase in the drone population.

Howell & Usinger (1933), who noted the exits and entries of individually marked drones, observed a mean of 3·1 flights per day on sunny days (maximum 7 flights). One peak of flights was of short duration (less than 6 minutes); 75% of these short flights were first or second trips, so they may have been orientation flights or trials of the weather. Most of the flights were of long duration (mode 27 minutes, maximum 207 minutes) but Howell & Usinger remarked that drones, so far as is known, do not usually alight at all during their flights. The drones were seen to clean their antennae and their eyes very elaborately before commencing flight. Most of the drone flighting occurred between 2 and 4.30 p.m., with a peak at 4 p.m. Howell &

Lee Jenkins

PLATE IV *A pollen gatherer approaching a peach blossom*

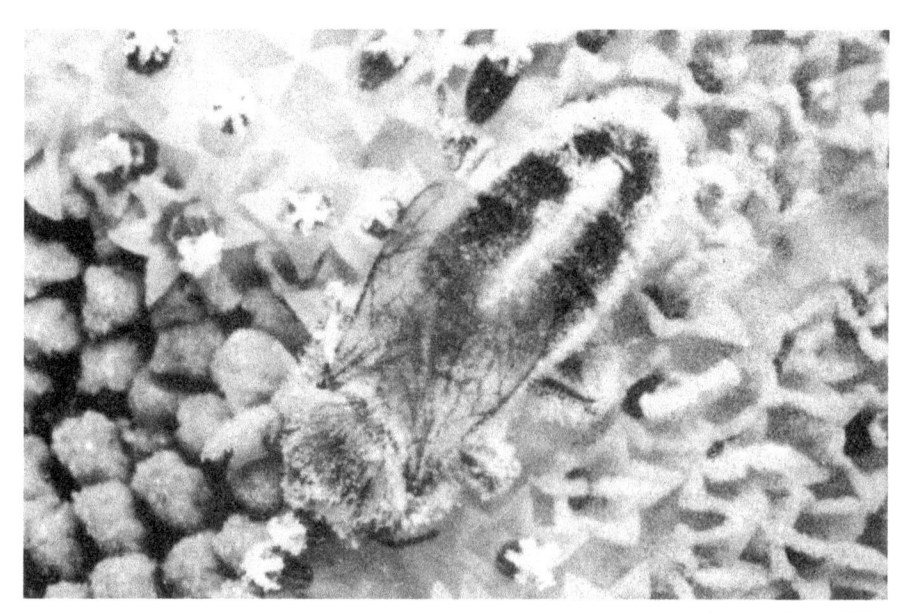

Günter Olberg
PLATE V(a) *A pollen-covered forager, gathering nectar from a sunflower*

Günter Olberg
PLATE V(b) *Honeybee gathering nectar from flax*

Usinger could assign no reason for this; they pointed out that the peak did not fit with the peak of temperature, but it bore some relation to minimum relative humidity.

Single and multiple matings of queens

Risga (1931), who constructed a device which enabled him to trap queens returning to their hives, said that most virgin queens left the hive once only but that some went out again two or three days later for a second trip and a few even went out a third time. Oertel (1940) observed 56 virgin queens, of which 18 mated on the first flight, 25 on the second, 9 on the third and 4 on the fourth flight. Two of these queens also made a short flight on the day after mating, but they showed no sign of a second mating.

A review by Nolan, which cites reports of multiple matings by 35 observers, is quoted by Roberts (1944). There is also a review by Fyg (1944). Roberts observed the flights of 110 marked virgin queens, of three unrelated strains; 43% of their flights produced matings, 55 mated once and 55 mated twice. One of the latter mated twice on the same day (out 2.32 p.m., returned mated 2.42 p.m., out 3.07 p.m., returned mated 3.14 p.m.), 45 mated on successive days, 8 mated again after 2 days, 1 after 3 days. In addition, 14 of the queens which mated only once flew on the following day without mating. Moreover, the entrances of the mating nuclei were covered with queen excluder so that virgin queens could not fly out unobserved; Roberts considered it likely that some desired flights were not permitted, so that his results underestimated the proportion of multiple matings in ordinary conditions.

Mating of laying queens

Experienced beekeepers know that the mating of a laying queen, if it ever occurs, must be a very exceptional event. There have been at least two reports of this occurrence (Kupetz, 1931; Baumgartner, 1948), each instance involving a marked queen which was seen to return with a mating sign.

Nevertheless, one may doubt whether a mature queen could ever mate so that her spermathecae were replenished, because even aged virgins cannot do this (p. 143). There are no reports of matings by old queens which fly out with a swarm, although drones are often abundant in such circumstances.

Mating flight range

The usual flight ranges of queens and drones have not been accurately estimated, but differences in the duration of their flights, and structural differences, provide evidence that drones range more extensively than virgin queens.

V. Buttel-Reepen (1923) reported two incidents which suggested that queens had mated with drones from apiaries 8 and 5 miles away respectively. Roberts (1944) mated 18 yellow-strain virgins in an isolated apiary containing 6 yellow-strain colonies with drone brood. Instrumental insemination of 8 sister virgins demonstrated that the strain bred quite true, and search of the surroundings for other bees revealed only four black colonies 1¾ miles to the north-east and four black colonies 2 miles east of the mating site. Yet every one of the 18 naturally mated queens threw a substantial proportion (minimum 32%) of black offspring. Multiple mating must have contributed to this result.

The flight range of the drone is a factor which must be taken into account in any bee-breeding programme which involves natural mating. We do not know whether the drone searches in widening circles or whether he utilizes his perception of distance and direction (p. 88) in order to travel out and back from the hive in a line. Nor do we know whether searching drones are attracted to the vicinity of other colonies by the scent which comes from them.

PART III COMMUNICATION BETWEEN HONEYBEES

Chapter 19
RECRUITMENT TO CROPS

The bees' behaviour in the beehive is unbelievable.

CHARLES CHAPLIN, *Limelight*

SPITZNER (1788) WROTE that 'When a bee finds a good source of honey somewhere, after her return home she makes this known to the others in a remarkable way. Full of joy, she waltzes around among them in circles, without doubt in order that they shall notice the smell of honey which has attached itself to her; then when she goes out again they soon follow her in crowds. I noticed this in a glass-walled hive when I put some honey on the grass not far away and brought to it only two bees from the hive. Within a few minutes these two made this known to the others in this way, and they came to the place in a crowd.' This observation was forgotten until 1920, when v. Frisch published his first paper on communication between bees by dances.

Dujardin (1852) described an experiment which led him to conclude that honeybees could communicate the presence of good sources of food to their hivemates. Having placed a saucer with syrup in a niche of a wall, he introduced to it a forager from one of two colonies 20 yards away; she fed there and then she flew away, orienting to the niche. Soon about thirty other bees came and searched the locality, apparently seeking the entrance. For several days afterwards many bees from this colony went to the niche, but none went there from the second colony from which no bee had been trained. Moreover, when the syrup had been used up one bee still went to inspect the niche from time to time; when it was empty she did not return, but after the supply was replenished she soon returned, with companions.

Lubbock (1874) was unsuccessful in similar attempts to demonstrate communication, but Emery (1875) believed there to be no doubt that it occurred, and that when a single bee discovered a store of honey she was soon followed by many others. Emery said that the bee-hunter in the United States turned the faculty to good account. 'Going with his box of honey to a field or wood at a distance from tame bees, he gathers up from the flowers and imprisons one or more bees. After they have become sufficiently gorged, he lets them out to return to their home with their easily-gotten load. Waiting patiently a longer or shorter time, according to the distance of the bee-tree, the hunter scarcely ever fails to see the bee or bees return, accompanied by other bees. These are in like manner imprisoned till they in their turn are filled, when one or more are let out at places distant from each other. In each case the direction in which the bee flies is noted, until by a kind of triangulation the position of the bee-tree is approximately ascertained.'

The dancing of bees had often been seen by beekeepers, and variously interpreted, but it was generally believed to be a means of showing that new honey or pollen had been discovered (Root, E., 1908); there were additional observations (Bonnier, 1906) which suggested that successful foragers might be able to inform their companions of the whereabouts of their booty. Then the work of v. Frisch proved that the dances were performed by successful foragers, and that they excited their comrades to fly out to search for the food which the dancers had been collecting. These observations were published in a series of articles (1920–2), then in detail (1923), and then summarized (1924). They proved that the dancers communicate the presence of food, and the odour at its source. Later work (1946a–50) showed that they also communicate the distance and direction of the source from the hive.

V. Frisch (1923) watched individually marked bees in an observation hive, and described two kinds of dance, one performed by nectar collectors and the other by pollen gatherers. He described the former thus (1924): "After she has given her sweet booty to her comrades, she begins a kind of round-dance in which she runs round and round in a circle with quick tripping steps, then suddenly turns round and revolves in the opposite direction, and so on. From 3 to 20 reversals may be made at one place, and the dance may take from a couple of seconds to even a whole minute. It is often repeated on different parts of the comb. Since this round-dance takes place in the midst of a crowd of other bees, the circling dancer comes into close contact with her comrades; these become greatly excited, and turn their heads towards her, trying to keep their antennae touching her abdomen and to trip behind her, so that the dancer draws after her a tail of other bees which accompany her in all the revolutions of the dance. Then the dancer flies alone to the feeding-place, and the newcomers, as though called up by magic out of a trap-door, gather there with her.'

Of the second dance, he said, 'The pollen-collectors also dance if they have found plentiful supplies. However, their dance is different. Especially characteristic of it is a waggling movement of the dancer by which she formally beats her following comrades on their face and antennae with her pollen loads.'

The communication of plant scents

Having shown that recruits would search for essential oils provided with the syrup which had stimulated the bees to dance, v. Frisch (1923) took a colony to the Munich Botanical Gardens, where more than 700 different kinds of plants were in bloom at the time—including a small bed of everlasting flowers (*Helichrysum lanatum*), which are not visited by honeybees. On 14 July, 1921 he trained ten marked bees to a dish of syrup which was surrounded by cut flowers from this plant, and within an hour thirteen honeybees had visited the bed of *Helichrysum* to search in its flowers.

This experiment was repeated, and others gave similar results. V. Frisch concluded that the recruits had perceived the odour of the blossoms which had clung to the bodies of the dancers[1] and he construed the waggling movement of the dance of the pollen gatherers as a device for impressing the scent of their pollen loads upon followers.

V. Frisch (1946b) added that recruits could distinguish not only the scent which adhered to the bodies of the dancers, but also that of the nectar which they brought home and which they distributed to potential recruits before they commenced to dance. At a feeding-place near their hive, he allowed trained bees to sit upon cyclamen flowers and suck up phlox-scented syrup through a slit; more of the recruits then went to a bunch of phlox than to one of cyclamen; reversal of the experiment gave a comparable result. This experiment was repeated with the feeding-place 660 yards from the hive; recruits then went exclusively to the bunch of flowers which carried the odour of the imbibed syrup.

[1] The waxy exudates of the body of the honeybee retain scents very effectively (Wheeler, 1928; Hildtraut Steinhoff, 1948).

Inability to communicate colours

In a parallel experiment to that which showed that honeybees could communicate the scent of the flowers they had obtained syrup from, v. Frisch (1923) fed a group of trained bees upon brilliantly coloured but scentless artificial flowers. Similar artificial flowers were placed in the surrounding meadow, but the recruits did not attempt to visit them.

Subsequently Françon (1939) claimed that honeybees could communicate colour, but his observation lacked an essential control experiment. Ribbands (1953a) confirmed the negative conclusion of v. Frisch.

The communication of distance

Having observed that the recruits did not *follow* the dancer to the feeding-place, v. Frisch (1923) supposed that its direction and distance might have been communicated; failing in an attempt to substantiate this hypothesis, he concluded that the newcomers found the dishes by searching in ever-widening circles from their hive. Park (1923d) also observed bee dances, but his observations disagreed with those of v. Frisch in one important respect; whereas v. Frisch observed that the dances of the bees gathering nectar and pollen were different, Park said that he had seen the waggle dance performed not only by pollen collectors, but also by nectar gatherers, and even by water carriers. However, Park did not appreciate the significance of his observation, and another twenty years were to elapse before v. Frisch accidentally discovered that honeybees could communicate differences in distance. Meanwhile, Henkel (1938) had also observed that nectar gatherers sometimes performed waggle dances; he thought that the round dances were a consequence of the unnatural richness of the syrup sources of v. Frisch's bees.

Having found that feeding bees near their hive did not cause recruits to discover a more distant feeding-place, v. Frisch (1946a) fed eleven marked bees with scented syrup at a dish 330 yards away from their hive and put down two similarly scented but syrupless dishes, one at the same distance from the hive but 110 yards to the right of the feeding-place, and the other 16 yards away from the hive. In two experiments, 118 newcomers visited the former, but only 13 the latter. The two syrupless dishes were now left in the same positions, while a group of marked bees was fed at a dish 11 yards away from the hive but in a different direction; the syrupless dish at 330 yards then attracted only 12 newcomers, but 174 went to the one at 16 yards.

Several similar experiments gave similar results, which could only be readily accounted for by supposing that the dancing bees were communicating the distance of their food source; negative results were, however, obtained from experiments in which the dummy dishes were placed at the same distance from the hive but in a substantially different direction.

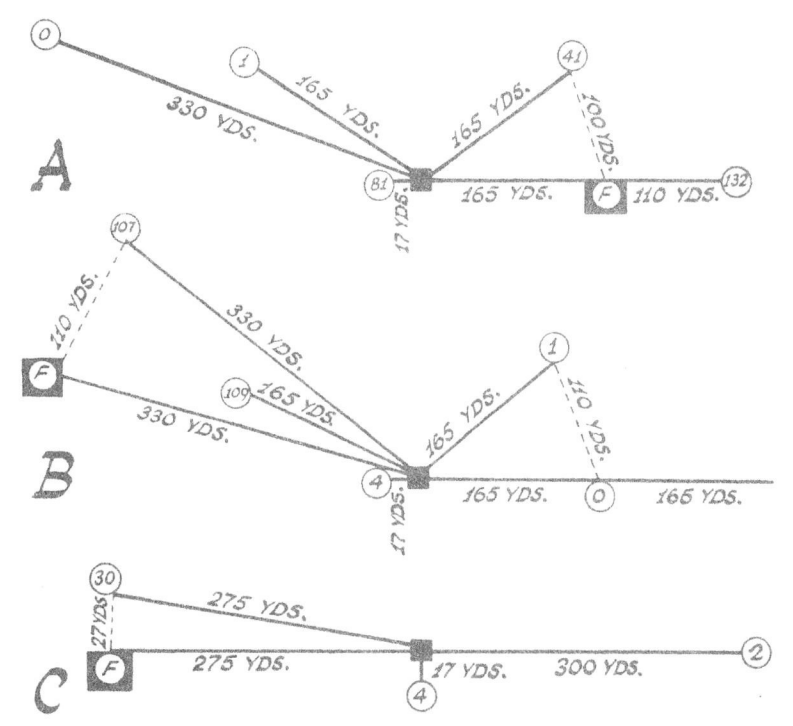

Fig 37 COMMUNICATION OF DISTANCE AND DIRECTION. Three experiments carried out by v. Frisch (1946a). In Experiment C the scent glands of the marked bees were coated with shellac. F=feeding place. Numbers in circles — recruits attracted to syrupless dishes at other sites.

The communication of direction

These results led v. Frisch (1946a) to design a more elaborate experiment, in which marked bees were fed at a dish 165 yards from the hive while dummy dishes were watched at five other sites (Fig. 37a). He found that 132 newcomers went to the dummy sited exactly in the direction of the feeding-place but

110 yards beyond it, but only 41 went to a dummy which was also 165 yards away from the hive, but which was 100 yards to the right of the feeding-place. Several similar experiments were carried out, the result of one of which is depicted in Fig. 37*b*).

V. Frisch was loath to conclude that the dancers were in fact communicating distance and direction. He thought that an odour trial emitted by the scent glands might account for the results, so he designed another experiment (Fig. 37*c*) in which all the scent glands of the trained bees were sealed with a

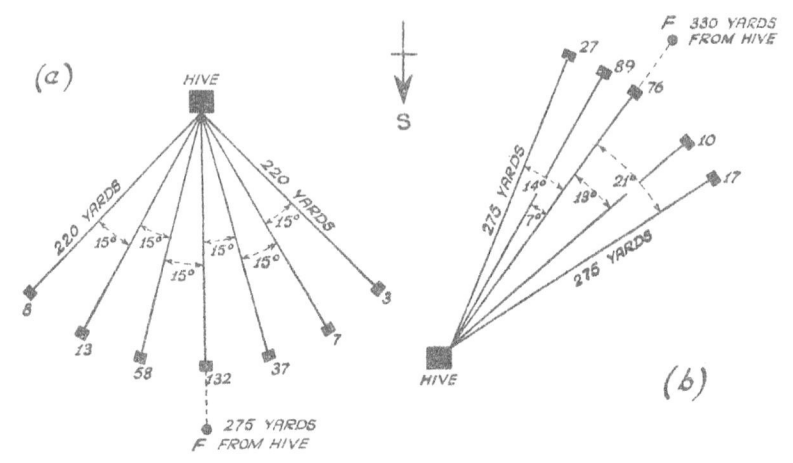

Fig 38 ACCURACY OF COMMUNICATION OF DIRECTION. Marked bees fed at F, and similarly scented empty dishes arranged in a fan, nearer to the hive. The number of recruits attracted to each dish is shown (after v. Frisch, 1948, 1950*b*).

coating of shellac. A strongly positive result was still obtained, so the experiments were not explicable in terms of an odour trail.

V. Frisch (1948) subsequently designed an experiment to determine the distance from the hive at which directed search commenced. Bees were fed at a dish 11 yards east of their hive, and a similarly scented empty dish was placed 16 yards farther away at each of the four points of the compass. The number of newcomers at each dish having been noted, the experiment was repeated with the feeding-place west of the hive. The whole procedure was repeated with the feeding-place 28, 55 and 110 yards from the hive. With bees fed at 11 yards, 63 newcomers went to the dummy in the same direction, while the other three dummies received on the average 50·3 visits; the difference was not significant. With the feeding-place at 28 yards there were 238 newcomers at the dummy in the same direction but a mean of only 34·6 visits to those elsewhere; at this distance the communication of direction was therefore effective. At 110 yards there were 375 visits by newcomers to the dish in the direction of the feeding-place, but the mean number elsewhere was only 6·3.

V. Frisch (1948) also set out to determine the accuracy with which the angle of direction was communicated. In this experiment (Fig. 38a) marked bees were fed at a scented feeding-place 330 yards from their hive, and empty scented dishes were laid out fanwise, in five slightly different directions, 275 yards from the hive. The result indicated that most of the newcomers had appreciated the direction given by the dancers within 10 degrees. A second experiment (v. Frisch, 1950b) gave similar results (Fig. 38b).

The dances in relation to distance of the food source

Having obtained experimental evidence of the communication of distance and direction, v. Frisch (1946a) compared the dances performed by foragers fed 13 yards away from their hive with the dances performed by those fed at 305 yards. The former bees always performed typical 'round dances' while the latter carried out 'waggle dances'. The characteristic of the 'round dance' is that the bee performs a complete circle, whereas the 'waggle dance' is a figure-of-eight. Alterations in the concentration of the syrup fed did not affect the form of either dance. Bees were then trained to various intermediate distances; round dances persisted at 26 yards, but at 55 yards some figure-of-eight movements often occurred; at 110 yards the transition to the waggle dance was almost completed, but circles still occurred occasionally; at 220 yards there were no circles.

The change from the round dance to the waggle dance was studied by Tschumi (1950), who described transitional 'sickle' movements. The diagrams which illustrate his conclusions are given in Fig. 39. V. Frisch observed that the proportion of circling decreased with increasing distance of the feeding-place; Tschumi found that even at 11 yards there were occasional sickle movements among the circles and that as the distance increased these sickle movements gradually expanded into the figure-of-eight of the waggle dance. Hein (1950) reported another variation; bees dancing only 6½ feet from their food supply swung their whole bodies to and fro while making very short runs, and repeated this activity at various places (Fig. 40); these bees usually faced towards the food supply. The first sign of a waggle was observed by v. Frisch at 28 yards, but Tschumi

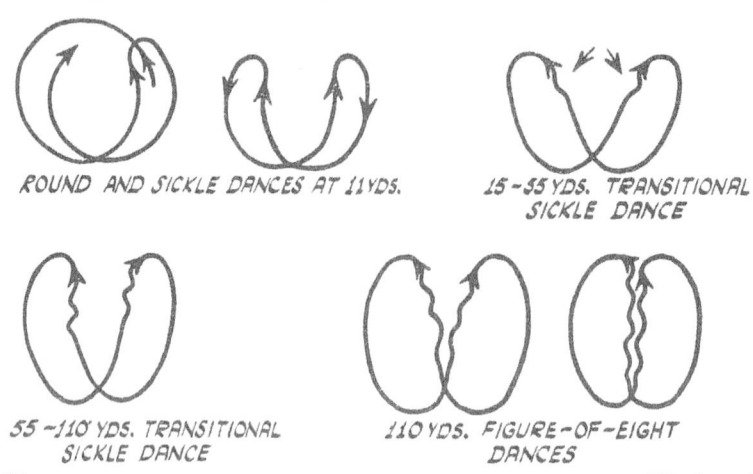

Fig 39 ROUND, SICKLE, AND FIGURE-OF-EIGHT DANCES (Tschumi, 1950).

saw it at 15 yards, Hein at 9 yards. The waggles are performed in the centre portion of the run (Fig. 39); there are more per run as the distance from the feeding-place increases.

These dances seem to vary; they are conducted so quickly that it is difficult or impossible to determine all their details by ordinary observation, but no one has yet analysed them by high-speed cinematography.

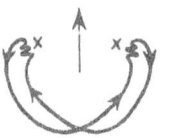

Fig 40 'PULL' DANCES AND A SICKLE DANCE (Hein, 1950).

V. Frisch (1946a, 1948) has shown that with increasing distance the number of waggle runs per minute decreases, whereas both the length of the waggle run and the number of waggling movements per run increases. The length of the waggle run varies irregularly, but the number of waggles per run is more regular (1946a). The latter cannot be counted reliably by the human eye, so v. Frisch paid most attention to the third characteristic of the dances—the number of waggle runs per minute. With the help of a stop-watch, 3112 observations were made, for various distances of the food supply; the results are summarized in Fig. 41, which demonstrates the relation between the tempo of the dance and the distance of the feeding-place. V. Frisch also observed that the bees of some colonies performed their dances more quickly or slowly than the average. At 12,000 yards, the greatest distance at which

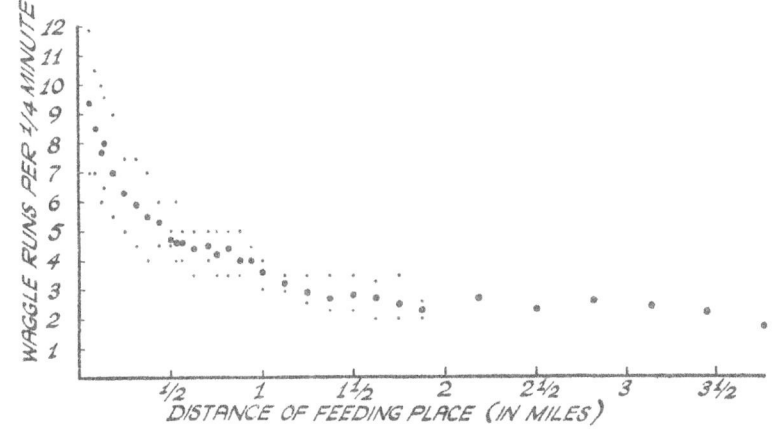

Fig 41 THE RELATION BETWEEN DISTANCE OF FEEDING-PLACE AND DANCE TEMPO. Large dots are means of all results, small dots indicate their range (after v. Frisch, 1951).

dancing was observed, there were only 1·3 waggle runs per ¼ minute (v. Frisch, 1951).

The existence of this relation between tempo and distance does not prove that the recruited honeybee appreciates distance through the tempo of the dance; the speed of the dance is likely to be correlated with the number of waggles per run; honeybees are not equipped with stop-watches, but their antennae may appreciate the number of waggles per run more readily than the human eye can.

The dances in relation to direction of the food source

Watching the dances on the vertical comb of his observation hive, v. Frisch (1946a) observed that in different waggle dances the straight run might be in any direction, but that all the marked bees coming from one feeding-place at the same time were dancing with the straight run in one direction. Yet the marked bees from the same feeding-place altered their direction of dancing in the course of the day; the direction of the waggle run changed in a few hours by approximately the same angle as that through which the sun had moved in the sky. However, the sun moved in a clockwise direction while the waggle runs moved

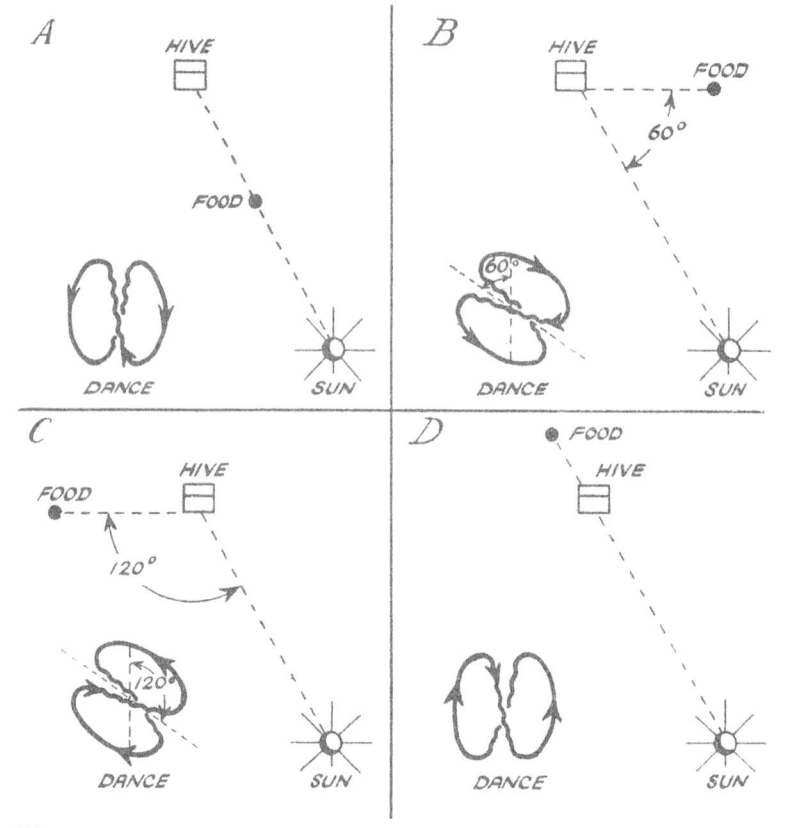

Fig 42 THE RELATION BETWEEN THE DIRECTION OF THE FEEDING-PLACE AND THE DIRECTION OF THE DANCE ON A VERTICAL COMB (v. Frisch, 1946a).

anti-clockwise. Moreover, the waggle run made the same angle with respect to the direction of gravity as the feeding-place did with the sun; for example, waggle runs 60 degrees to the left of vertical indicated that the feeding-place lay 60 degrees to the left of the sun's direction (Fig. 42). Then two feeding dishes were arranged 44 yards from an observation hive but in almost opposite directions from it. Bees were fed at both dishes at the same time, and the directions of the waggle runs of the two groups 'differed fairly accurately' by the very obtuse angle between the two dishes.

Thus the direction of the waggle run indicates the direction of the food source in relation to the position of the sun. Wolf had already shown that foraging honeybees could orient in relation to the sun (p. 84).

So far dances upon a normal (vertical) comb had been observed. V. Frisch now laid the comb flat, so that the bees had to dance in a horizontal plane. The waggle run then pointed directly towards the feeding-place. Moreover, when the comb was rotated in the horizontal plane the dancer adjusted herself 'like a compass needle' so that the waggle run still pointed towards her feeding-place. (The dancers were not disturbed by the approach of a strong magnet, from any direction.) A hut was built so that the horizontal dances could be observed by red light (imperceptible to the bees)—the dances were then completely disoriented (v. Frisch, 1948). Dancers so disoriented would orient in relation to a torch, shone from any direction, as if to the sun. In daylight the dancers oriented to the actual direction of the sun, not to the direction of maximum brightness; if for instance the north side of the hut was opened, the dances were correctly oriented in relation to the sun. The comb was then illuminated only by daylight coming through a tube (6 inches in diameter and 8 inches long) which was inserted in the wall of the hut; when the tube pointed towards clear blue sky the dancers were oriented correctly in relation to the sun (*see* p. 30); if the sky was hazy the orientation was uncertain, and if the patch of sky was cloud-covered the dances were oriented as if that spot of sky was the sun (= torch experiment). Next, a mirror was fixed so that the comb was illuminated by a sector of southern sky, coming as if from the north; the dances then clearly indicated the mirror-image direction.

Professor Kiepenheuer suggested to v. Frisch (1949) that these observations were explicable if the dancers were responding to the plane of polarization of light, since this varies in the blue sky according to its relation to the position of the sun (p. 29), and the degree of polarization is greatly diminished in the presence of clouds. This hypothesis was supported by the observation that the dances were reoriented when a large polarizing screen, placed above a horizontal comb on which bees were dancing, was rotated. The dances retained their original orientation when the polarizing screen was placed so that there was no change in the plane of vibration of the polarized light reflected from the visible sky, but any change in this plane caused the direction of the waggle run to change also. Extreme changes caused the waggle runs to be completely disoriented (v. Frisch, 1949).

Dances upon vertical combs, which provide the usual dancing surface in the hive, have also yielded most important information. V. Frisch (1948) found that although dances on horizontal combs were disoriented in the absence of light, the dances on vertical combs still indicated direction correctly (in relation to the direction of gravity) when they were performed in complete darkness. V. Frisch (1946a) had found that the angle of the dance on a vertical surface was indicated most accurately at noon and least accurately in the early mornings and late afternoons, but that the mistakes, then not interpretable, seemed consistent. A detailed analysis (1948) revealed that the misdirections occurred when the vertical comb was illuminated by sunlight, but disappeared when the dances were carried out in a closed hut or when the sky was completely overcast, so v. Frisch concluded that the errors had arisen because the dancers were orienting to a compromise between the two stimuli of gravity and light. His analysis of dances on a sloping comb (1948) showed that here there were also transitional dances; as the slope of the comb was changed from vertical to horizontal, orientation by light became more and more dominant over orientation by gravity.

The perception of direction under a cloud-covered sky

The preceding experiments have not solved all the problems of direction perception. V. Frisch (1948) moved a colony to a new site on the morning of a day when the sky was closely covered with clouds. The bees were soon trained to a feeding-place (at first 110 yards, later 220 yards from their hive), and their dances were observed. At 110 yards the mean of 20 dances indicated the actual direction with an error of 2 degrees,[1] and at 220 yards the error was 9 degrees. Thus under an overcast sky and in a district strange to the bees the direction-giving was as accurate as in fine weather. This result is not explicable in terms

[1] The actual solar angle was 66 degrees, and the individual dance records were 75, 67, 53, 60, 60, 60, 60, 60, 70, 60, 60, 60, 60, 60, 60, 67, 68, 75, 67, and 75 degrees. These individual errors, not elsewhere recorded by v. Frisch, are of the same order as the errors in the directions taken by recruits (p. 152).

of orientation to the pattern of polarization, which could not be detected in such a sky even with a Nicol prism (Hartridge, 1950); v. Frisch thought that the bees might have been able to perceive heat radiation from the sun, which can penetrate cloud, but one of his pupils was unable to train bees to directed heat radiation. V. Frisch concluded that 'the riddle remains unsolved'.

Flights involving a change of direction

V. Frisch (1948) arranged the hive and the feeding-place 100 yards apart, but separated by a mountain ridge 200 feet

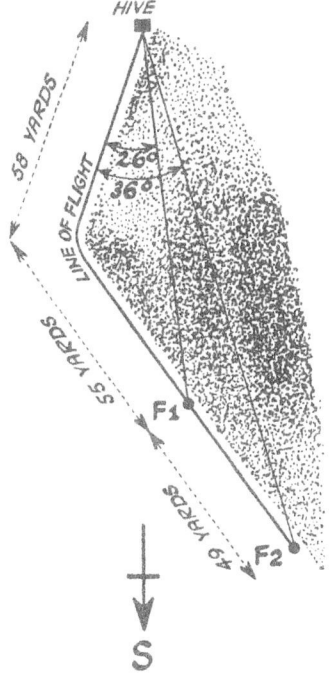

Fig 43 MOUNTAIN EDGE EXPERIMENT (after v. Frisch, 1948). Dancing bees indicated the actual direction of the feeding-place, *not* the direction in which they had flown.

high. The dances then indicated the *total* distance (165 yards) which bees had to fly over the obstacle.

In a companion experiment v. Frisch (1948, 1951) found that dancers which had flown to their feeding-place round two sides of a mountain ridge indicated the actual direction to it —a direction in which they had never flown themselves. Moreover, even when the lengths of the two legs of the journey were unequal (Fig. 43, feeding-place F_2), direction was indicated with similar accuracy; the sensory receptor mechanism had summed the total flying time as well as the two angles.

The flight on which the dancer registers distance and direction

A series of very interesting observations by v. Frisch (1948) showed that if the bees flew *with* the wind from their hive to the feeding-place the rhythm of their dancing was quicker than

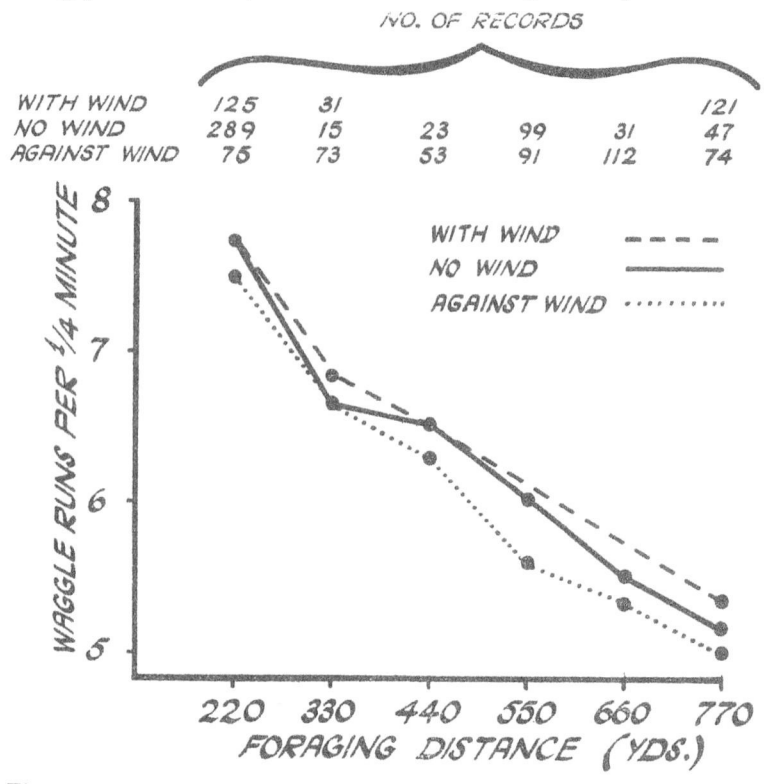

Fig 44 THE RELATION BETWEEN DANCE TEMPO AND THE WIND ON THE WAY *to* THE FEEDING-PLACE (after v. Frisch, 1948).

if they flew *against* the wind (Fig. 44). Thus a contrary wind on the way to the feeding-place had the same effect on the dancing as increase of distance. From this result v. Frisch concluded that (i) distance was recorded during the flight *to* the source, and (ii) the information given by the dancer was related to the energy and time required to reach the goal rather than to the actual distance from hive to goal. In addition, Heran and Wanke (1952) found that honeybees danced more quickly—as if the distance were smaller—when the feeding place was downhill from the hive, but more slowly—as if the distance were greater—when it was uphill.

On the other hand, Khalifman (1950) took marked bees from a feeder due south of their hive, released them at a point due north of the hive, and recorded that their dances indicated the *direction* of their homeward journey.

The circumstances in which dancing occurs

Bonnier (1906) suggested that honeybees which found a new food source invited to it only sufficient companions to exploit it adequately. V. Frisch (1923) fed bees upon two separate dishes of syrup, one dish being abundantly supplied while at the other a thin film of syrup had to be sucked from filter-paper; having observed that bees at the former dish danced more vigorously and attracted many more recruits, he considered that this relation between syrup abundance and recruitment explained how optimum proportions of bees worked the various crops available at any time. Later (1934) he demonstrated that the proportion of returning foragers which danced was related to the concentration of the syrup which was collected.

V. Frisch (1942) found that the proportion of bees which danced was significantly reduced when the $\frac{1}{2}$ molar solution of sucrose to which they had been trained contained also $\frac{1}{2}$ M lactose (tasteless, viscous, without food value)[1]. Dancing was also reduced when 1 M sorbitol (tasteless, viscous) was added, although this substance trebled the food value of the solution.[1] Yet when 1 M sucrose was substituted for $\frac{1}{2}$ M sucrose, so that the sweetness was increased as well as the viscosity and the food value, the proportion of dancing bees increased substantially. The addition of salt ($\frac{1}{8}$ M) also reduced the proportion of dancing. The volume of syrup sucked up at each visit, which increases with increasing concentration of sucrose (v. Frisch, 1934), was not affected by the addition of lactose and was reduced by the addition of salt.

From these results v. Frisch (1942) concluded that neither the degree of filling of the honeysac nor the food value of the solution determined the release of the dances, but that taste was a main operative factor. He added that a psychological factor seemed to be involved, since a sugar concentration which gave rise to active dances when there was no nectar flow might produce no dances when a good crop was available elsewhere.

Further experiments were conducted by Lindauer (1949).

[1] See p. 34.

He showed that the threshold concentration of sugar which induced the dances varied seasonally, and was related to weather and alternative food supplies. The threshold sugar concentration varied from 2 M during the June nectar flow to $\frac{1}{8}$ M late in July (Fig. 45). Park (1923d) had previously reported that water collectors danced in springtime. Lindauer observed that on warm sunny days the proportion of dancing was reduced between 1–3 p.m. each afternoon; this reduction 'did not appear to be related to the daily rhythm of pollen and nectar collection'.

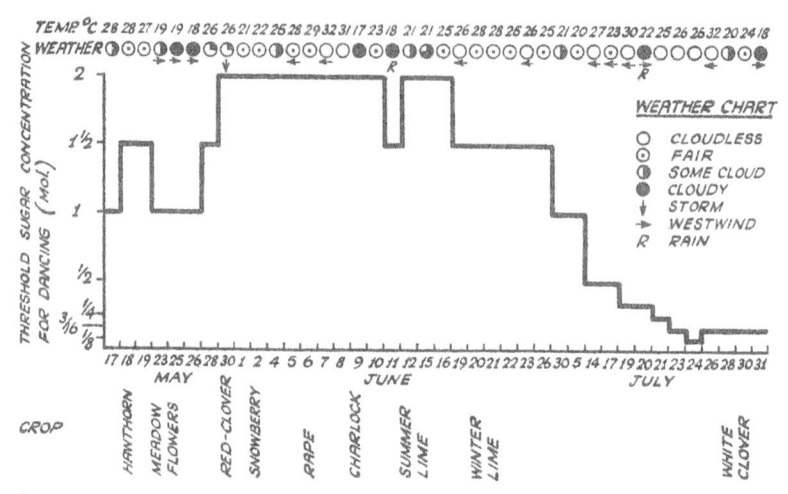

Fig 45 SEASONAL EFFECTS ON THE MINIMUM CONCENTRATION OF SUGAR SYRUP WHICH INCITES COLLECTORS TO DANCE (Lindauer, 1949).

Lindauer studied the effects of adding tastes and scents to a sugar solution of the threshold concentration for dance activities. He confirmed that the dancing was decidedly reduced by the addition of salt, and also by adding either dilute hydrochloric acid or quinine; other unpleasant substances would probably produce the same effect. He obtained reductions following the addition of the foul odours of either skatol or cow parsnip. On the other hand the odours of peppermint oil, phlox, elder, rape, jasmine or lime increased the proportion of dancing bees. This influence, particularly evident when the offered concentration of the sugar solution was just below the threshold for dancing, was much less important than the influence of the tastes.

Lindauer found that bees collecting pollen substitutes would only dance if their colony lacked pollen, and also that the effect of adding attractive scents to these substances was less marked.

Bees did not dance until they had paid several visits to a food source, and a deterioration in the quality of the food might temporarily inhibit dancing by individuals which subsequently danced in response to the deteriorated supply (Lindauer, 1949). The later recruits to a food source danced least readily, seldom before their tenth trip. Lindauer also noticed that different individuals varied in their dancing propensities.

Ribbands (1953d) trained equal numbers of marked bees to three different feeding-places. One was supplied with

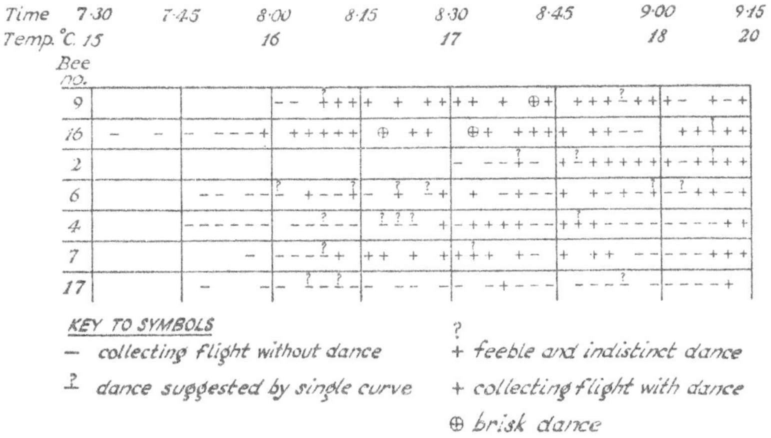

Fig 46 THE DANCE BEHAVIOUR OF SEVEN COLLECTING BEES, FROM THE BEGINNING OF FEEDING WITH SUGAR SYRUP ($1\frac{1}{2}$ mol.). Dancing commenced after several trips, and individuals varied in dancing propensity (after Lindauer, 1949).

concentrated sugar syrup, and two with very dilute syrup. After some time one of the two last was supplied with concentrated syrup, and recruits were attracted to (and killed at) this dish at a significantly greater rate than to the one at which concentrated syrup had been continuously supplied. The experiment was twice repeated. Inferring that the attraction of recruits was related to the amount of dancing, Ribbands concluded that bees danced more readily in improved than in constant conditions.

The origin of the dancing habit

The foragers' dances follow an instinctive pattern, not a learned one; Lindauer (1952) showed that in a nucleus made entirely from newly emerged bees there were successful dances after foraging trips.

The origin of the relation between light and gravity, which

is expressed in the dances, has rightly been a matter for speculation. An important contribution has been made by Vowles (1953a), who has shown that a similar relation between light and gravity exists in ants, in which it has no known adaptive significance. Ants (which do not dance) orient in relation to either light or gravity, and Vowles found that if the stimuli were interchanged the response of the ant indicated that they could be substituted for one another as in the honeybee. Vowles suggested that one nervous mechanism serves for orientation to gravity, light, polarized light, and possibly other stimuli, thus providing an economy of organization and explaining the results of substitution of the stimuli. Perhaps the dances may then be regarded as an extension of the foraging flight, undertaken by excited foragers.

The relation between dancing and scenting

V. Frisch (1923) observed that foragers which had danced often flew around the feeding-place with exposed scent glands on their return; this helped to guide recruits (cf. p. 86). Dancing and scenting are probably both related to the excitement of the foragers concerned, but the correlation between dancing and scenting is not always a close one. On 29 occasions v. Frisch timed the duration of both processes (his Table 14); on three occasions there was neither dancing nor scenting, on five dancing lasted 6–26 seconds and scenting averaged 10·6 seconds; on 10 occasions dancing lasted 35–60 seconds and scenting averaged 53·6 seconds, but on 10 occasions when the bees danced for more than 1 minute their mean scenting time was 43·8 seconds. Bees often dance without subsequently scenting, and Lindauer (1949, his Fig. 6) records a bee which scented without having danced.

The response of crop-attached bees to dancing bees

Most experiments have been designed in order to demonstrate the recruitment of newcomers by dancing bees, but successful foragers also excite resting bees which are already attached to a crop. This was demonstrated in one of v. Frisch's earliest experiments on communication (1920). One morning he marked individually 24 bees which were trained to a dish supplied with sugar syrup. The food supply was removed at 12.15 p.m. and replaced at 2.34 p.m. No bee visited it from that time until 3.05 p.m., when one marked bee (β) arrived, imbibed a load, and left at 3.09 p.m. Between 3.11 and 3.13 p.m.

four other marked bees came to the dish, at 3.13 p.m. β returned, and one minute later another marked bee also arrived. No marked bees had arrived at the dish during the half-hour before β came, but 18 arrived during the following half-hour.

V. Frisch (1923) took a nucleus of bees into a large empty greenhouse in the Botanical Gardens at Munich, and arranged there two large bowls, one containing lime flowers and the other *Robinia* flowers, about three paces apart. One morning three individually marked bees were trained to the lime bowl, six others visited the *Robinia* flowers, while two bees (Nos. 4 and 8) had visited both sources (having learned to collect from lime when the *Robinia* had been taken away).

At 12.42 p.m. both bowls were removed. At 12.58 p.m. the *Robinia* flowers were replaced. No. 8 soon found them, made two trips and danced to three *Robinia* foragers, all of whom went to the *Robinia*. The *Robinia* were covered at 1.40 p.m. and lime flowers were offered at 1.54. Nos. 4 and 8, having failed to visit the *Robinia*, soon went to the limes. They made several trips to the lime flowers, and during their dancing they made a contact with each of the three lime foragers and with four of the six *Robinia* foragers (two contacts with one of these). All three of these lime foragers went quickly to the crop, but the *Robinia* foragers did not leave the hive.

This experiment was repeated with pollen crops (roses and campanulas), similar results being obtained. There were 19 observed contacts, all without result, between dancing pollen collectors and bees attached to other pollen crops, but after all but one of 25 contacts between dancing pollen collectors and bees attached to the same pollen crop the inactive forager arrived on her crop within five minutes.

When composite flowers were arranged, with rose petals and campanula stamens, a bee which collected from them returned to the hive, danced, and aroused only the campanula pollen collectors; the reciprocal experiment also succeeded. V. Frisch found that pollen collectors from two species of rose which had distinct scents (*Rosa moschata* hyb. and *R. rugosa* hyb.) did not arouse each other.

This series of experiments proved that resting bees heeded others which carried the scent of the crop to which they were attached, but ignored those coming from other crops. V. Frisch (1950b) found that two groups of bees, trained to scentless dishes of syrup either close to or far from their hive, responded only to the dances of bees which had been foraging at the appropriate distance.

The biological value of communication about crops

There is no need to stress the advantage to a colony of the ability of successful foragers to communicate the whereabouts of a good crop to their unsuccessful comrades; the value of a signal by which attached foragers can learn that their crop has become ready to harvest is also clear.

The communication of information about crops has several consequences which are important in relation to problems of honeybee behaviour, and these will now be discussed.

Some results of Eckert (1933) can now be interpreted as a consequence of communication of the position of crops. Eckert, noting the direction of flight from twelve apiaries, flying mainly to sweet clover, found that the bees from each apiary went in either one or two main directions, with only occasional ones going elsewhere. The bees from one apiary flew across and beyond a second apiary, while bees from the latter went off in another direction. Eckert found that there was no definite relation between flight direction and prevailing winds; he said the direction evidently depended upon the major source of supply to which the bees oriented themselves when the plants came into bloom, and he added that in only a few instances were the bees of any apiary taking full advantage of their location, as judged by the directions of flight in relation to the fields of sweet clover surrounding them.

At the height of the honey season each colony in an apiary puts many thousands of foragers into the field at once; if there were no communication about crops they would forage as individuals and the proportions of each crop brought into each colony in any one apiary would be very similar. However, striking examples of dissimilarity have been recorded. For instance, Mendleson (1908) recorded that as an exceptional extreme one colony collected 'exclusively' from mustard while 199 others in that apiary were working on sage. Moore Ede (1947) moved seven stocks to a site on a heather moor within reach of a valley which contained white clover. Two of the seven colonies gathered very large crops (one 83 lb.) of 'practically pure clover honey without a trace of heather', while the other five 'had in the supers a pure heather honey'. He recorded other similar instances (*see also* p. 105).

Todd and Bishop (1940) analysed trapped pollens and reported that 'pollen samples from colonies side by side may come from predominantly different sources'; Eckert (1942) reached a similar conclusion. Ann Synge (1947) recorded big differences between the numbers of loads of various pollens collected by two neighbouring colonies in the same apiary (Table 10).

TABLE 10. The number of pollen loads collected from various sources by two neighbouring colonies (Ann Synge, 1947)

Approximate season	Plant	Pollen loads into colony A	Pollen loads into colony B
1 March–20 April	willow	392	3,160
	box	114	809
20 April–20 May	gorse	2,956	701
	aubrietia	483	1,069
	berberis	424	1,653
	beech	0	2,328
	oak	2,587	79
20 May–10 June	buttercup	4,967	1,507
	field bean	4,979	223
10 June–15 July	white clover	67,444	32,415
	Filipendula ulmaria?	8	1,942
	rose	608	2,008
	poppy	1,834	8,066
15 July–30 August	willow herb	1,735	3,576
	onion?	228	1,387
	red clover	29,386	45,148
October	ivy	6	2,610

A consequence of crop communication is that the colony, not the individual, is the foraging unit; thus the pollen trap records for a single colony, or even several colonies, cannot provide an accurate estimate of the relative value of different pollen crops at a site, let alone their value throughout a district.

Thus the proportions of the various constituents of the diet of neighbouring colonies usually differ markedly; this consequence of crop communication makes possible another important system of communication. The different food supply is evenly distributed among the bees of each community by thorough food transmission (Ch. 24), the distinctive odour derived from the waste products of this different food supply is emitted by and can be distinguished by the members of each community, and it is the basis of their mutual recognition (Ch. 21).

Chapter 20

SELECTION OF A HOME

> *Round the fine twigs, like cluster'd grapes, they close*
> *In thickening wreaths, and court a short repose,*
> *While the keen scouts with curious eye explore*
> *The rifted roof, or widely gaping floor*
> *Of some time-shatter'd pile, or hollow'd oak*
> *Proud in decay, or cavern of the rock.*
>
> JOHN EVANS, *The Bees*

Communicating the whereabouts of a home

AN INTERESTING description of the selection of a home was given by Knight (1807), thus: 'Whenever a swarm came (to a cavity in a tree), I constantly observed that about fourteen days previous to their arrival a small number of bees, varying from twenty to fifty, were every day employed in examining, and apparently in keeping possession of, the cavity. A part of the colony which purposed to emigrate appeared in this case to have been delegated to search for a proper habitation; and the individual who succeeded must have apparently had some means of conveying information to others: for it cannot be supposed that fifty bees should each accidentally meet at, and fix upon, the same cavity, at a mile distant from their hive; which I have frequently observed them to do, in a wood where several trees were adapted for their reception. It not infrequently happened that swarms of my own bees took possession of these cavities, and they were observed to deviate in flight very little from the direct line between the one point and the other.'

Knight added that when a swarm issued from the parent hive it generally settled nearby, but this was apparently only in order to collect its members. They had 'generally, and I believe always, another place to which they intend subsequently to go'.

The behaviour of eleven natural swarms was studied by Lindauer (1951*b*). Soon after a swarm had clustered at its first resting place, a few bees began to dance on its surface, and from hour to hour their numbers increased. These bees carried

SELECTION OF A HOME 169

no nectar or pollen and some were red or grey with brick or mortar dust, or black with soot. The dances were similar in form and rhythm to those of foragers, but they often lasted for a much longer time—usually more than five minutes, and occasionally the same marked bee was seen to dance for more than an hour (with appreciable intervals for rest, but without quitting the cluster).

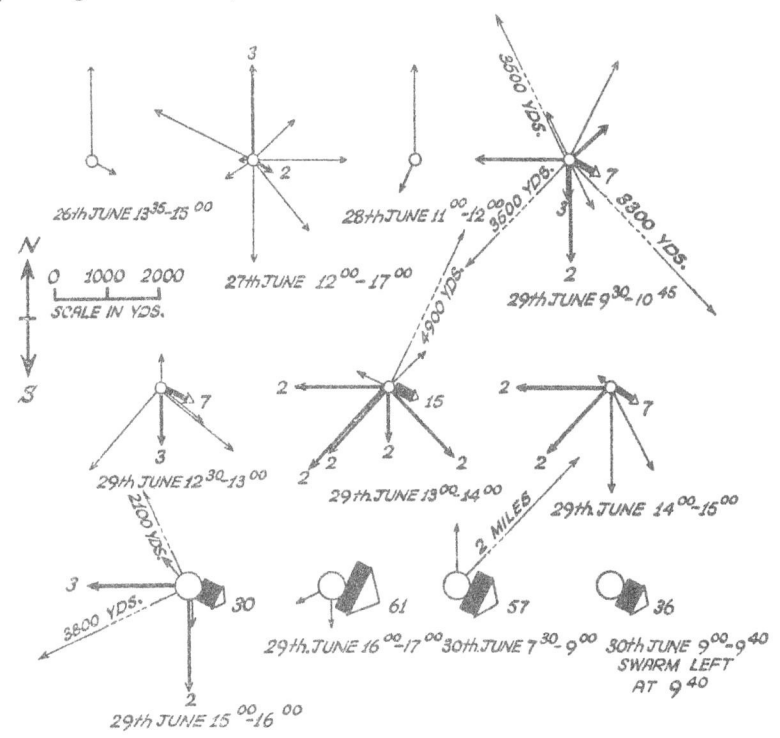

Fig 47 THE DANCES OF BEES ON A SWARM ABOUT TO DEPART FOR A NEW HOME (Lindauer, 1951b). Several sites were indicated at first, but the dancers came to agree on one, and the swarm departed. All bees were marked when they began to dance. The distances and directions indicated by new dancers are shown by the angles and lengths of the arrows, and the numbers observed are given at the points of the arrows.

At first the dancing bees were indicating different directions and distances, but after a few hours one dance pattern had become more common and the others were disappearing. Finally, nearly all the dancers indicated the same direction and distance, and 5 to 10 bees were often dancing at the same time. One set of these observations are recorded in Fig. 47.

Lindauer observed repeatedly that during this process dancers were converted from one dance rhythm to another: only when the dancers were unanimous was the swarm likely to depart. All the dancers of six swarms were marked and counted; in each at least 100 indicated the correct site, from which Lindauer deduced that more than 1000 bees in each swarm had learned its whereabouts.

Nine of the eleven observed swarms departed in the direction indicated by the dancers; two were interesting exceptions. In one, there were two rival groups of dancers which could not agree—that swarm did not depart, but built combs on the branches. In the other, two rival groups of equal strength indicated sites in different directions; the swarm flew off, but tried to divide itself, one half trying to fly W.N.W. and the other N.N.E. For half an hour the groups kept separating and rejoining, and then the swarm settled in a plum tree about 33 yards away from its starting place. Then the dancing was resumed, and the W.N.W. group became dominant.

Lindauer also made an artificial swarm, and watched it move into a decoy hive nearby. The first scout found the hive at 8.55 a.m., flew in and out several times, spent $2\frac{1}{2}$ minutes inside, and then flew back to the cluster without dancing. Eight minutes later she returned to the hive, and spent 9 minutes in exploration. She danced on the cluster only after her third trip, during which she had fanned at the hive entrance. Then she flew backwards and forwards to the hive at 5-15 minute intervals, dancing on the cluster and exposing her scent gland at the hive. New bees attracted to the hive frequently fed each other. By 11.30 a.m. 42 dancers had been marked, and at least as many unmarked bees had also visited the hive. Only one bee had indicated any different site. At 11.32 a.m. the swarm flew off to the hive. Although the natural swarms observed by Lindauer sometimes did not leave their temporary resting place for several days, such swarms often depart within an hour or two. One may suppose that when they depart quickly they make for a site chosen before the swarm issued from the parent colony, as Knight's observation implies.

Competition for sites

Homes may be difficult to find in some localities. Lindauer (1951b) found the site which was being indicated by the dancers on one of his swarms; an hour later a swarm arrived and moved in, but it was not the swarm which he had been watching! All dances on the watched cluster then ceased abruptly, and not until the next day did the swarm agree on a new site and depart to it. Knight (1807) watched a swarm enter a tree cavity already occupied by another one; they joined amicably.

Chapter 21

RECOGNITION OF COMPANIONS

*The daisies in the dell,
Will give out a different smell,
Because pore Jud is underneath the ground.*

RODGERS AND HAMMERSTEIN II, *Oklahoma*

SOCIAL INSECTS OFTEN suffer vigorous competition from their own species, so an effective method of mutual recognition is of great advantage.

Forel (1874) proved that some ants could distinguish their comrades from other members of their own species. He showed that well-established colonies often fought to the death when they were brought into contact, but in difficult circumstances they would form alliances—which once made could not be unmade. McCook (1877) observed that ants which had been immersed in water were attacked by their companions, and he concluded that recognition was a matter of smell. Adele Fielde (1904–05) carried out a series of experiments which led her to conclude that mutual recognition by ants was based on an odour which they inherited from their mother, together with some progressive change in that odour according to their age.

Meanwhile, Bethe (1898) had carried out experiments on the defence of the honeybee colony (p. 179) which led him to conclude that honeybees possessed distinctive nest odours, which were produced by each individual and were inherited. V. Buttel-Reepen (1900) agreed with this suggestion, but gave only general statements in support of it.

The attractiveness of bee scent

A new approach was made by Sladen (1901–02), who investigated the manner in which swarming bees attracted their comrades—by standing still and vibrating their wings so rapidly (fanning) that they set up a peculiar hum. Sladen noticed that a distinct odour was emitted at the same time, and that each fanning bee stood with the apex of her abdomen elevated, and exposing the membrane which connects the fifth and sixth segments of its dorsal surface; he dissected a number of bees and found that when he exposed this membrane the same odour was often, but not invariably, perceptible. Sladen concluded that this scent was the instrument of communication and that the fanning effectively distributed it; he noted that the mechanism was also invoked by tired or young bees which had returned to the entrance of their home, and that it served to direct their comrades towards the hive.

V. Frisch (1919b) found that bees were attracted to boxes which had recently been visited by other bees. A box which had recently received 25 bee visits obtained 33 further visits, while three similar but unvisited boxes received only 7, 6 and 6 respectively. However, the effect was ephemeral; after an interval of 10 minutes it had almost disappeared.

V. Frisch (1923) noted that foragers used the odour from their own scent gland to attract other bees to syrup dishes. They often exposed their scent gland as they flew to and fro over the feeding-place on their return, and also for a time while they were drinking. There were differences between the behaviour of different individuals on the same dish; some might scent at each visit, some irregularly, and some not at all. Two similar groups of marked bees were trained to a pair of dishes close to their hive. Syrup was readily available at one dish but at the other it had to be sucked from moist filter paper; the supplies were reversed from time to time. The trained bees at the plentiful supply were exposing their scent gland, but those at the other dish were not, and the plentiful supply attracted ten times as many newcomers as the other one (291 to 30). Then plenty of syrup was supplied at both dishes, but the scent glands of the foragers at one of the dishes were covered with strong shellac varnish; these bees attracted only one-tenth as many recruits as the normal ones did (12 to 123). From these results v. Frisch concluded that the odour from the scent gland was another important 'word' in the bees' language.

Jacobs (1924), who studied details of the structure of the scent gland, found that a path of fanning and scenting bees could develop between the top of a colony and its feeder; he suggested that what v. Frisch observed was an extension of this behaviour, since no one had seen a bee fanning upon any natural food source. Only v. Frisch & Rösch (1926) have since reported this last occurrence.

Preferential attraction to foragers from the same colony

Until this time experiments had demonstrated that the scent gland served for communication, but it had not been suggested that the odour emitted by it varied in different individuals. Now v. Frisch & Rösch (1926) took a pair of colonies, A and B, which were unrelated but not conspicuously different, and trained a group of marked bees from each colony to separate dishes of syrup, *a* and *b*. The dishes were moved together by stages until they were only $6\frac{1}{2}$ feet apart, but the groups of trained bees did not mix. Then the concentration of the syrup was increased, and the recruits which were then attracted to each dish were differently marked. An observer was stationed at each dish and at each hive, and the hive to which the recruits from each dish returned was noted. In two experiments with the same pair of colonies there were 29 recruits to *a* of which 23 returned to A and 6 to B, and there were 61 recruits to *b*, of which 58 returned to B and only 3 to A. Thus most of the recruits from each colony had been attracted to the dish visited by their own comrades.

These experiments were made with only one pair of colonies, and the results might have been a consequence of a simple genetical difference between the two stocks, in which case one might expect that only a small number of different colony odours would exist among honeybees. Alternatively, it might have been one example of a general situation in which the workers of any colony could attract members of their own colony in preference to all other bees.

Kalmus & Ribbands (1952) reinvestigated the problem, using a slightly different technique. Two small dishes, placed about thirty inches apart, were separated by a small wooden screen in an attempt to reduce intermingling of the scents in their vicinity. An improved form of this apparatus is shown in Fig. 48. Bees from Colony A were trained to collect syrup from one dish, and were marked, while Colony B was not allowed to fly. Next day bees from Colony B were trained to and marked at the second dish, while Colony A was not allowed to fly. Colony A was then released, and dilute syrup was supplied at both dishes. With few exceptions (which were killed), the marked guide bees from each colony visited the dish to which they had been trained; concentrated syrup was then supplied and the recruits to each dish were distinctively marked. The feeding was discontinued when sufficient recruits had arrived, and that evening the combs of both colonies were searched and the recruits were killed and recorded. These results confirmed those of v. Frisch & Rösch.

The work was then carried a step further. A colony was divided into halves, the queenless half requeened, and the recruits from these halves compared on the third day after the division. The recruits again showed a highly significant preference for the dish frequented by their own comrades. Since the bees in both halves were sisters, this preference was not caused by differences in their genetical constitution. Another

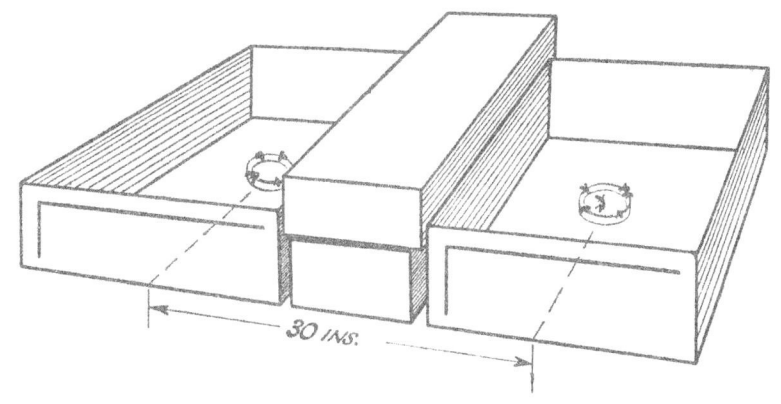

Fig 48 APPARATUS FOR STUDYING ATTRACTION TO COMPANIONS. Bees from two colonies were trained to opposite dishes, and most recruits went to the dish visited by their comrades. The dishes are surrounded by broodchambers to reduce disturbance by wind; the box in the centre contains notebooks, supplies of syrup, paints and brushes.

experiment showed that the preference could be demonstrated in queenless colonies, but no preference was shown after one week by recruits from the queenless halves of a dequeened and split colony.

The role and the persistence of bee scent

In another experiment, using one colony only, Kalmus & Ribbands demonstrated that the differential attraction of the recruits was a consequence of odour differences, and that the odours in question were fairly persistent. Throughout this experiment all recruits to both dishes were killed and recorded on arrival. On a dull calm October day fifty marked bees were visiting a syrup-filled dish, *a*, on one side of a screen, but no bees were trained to a similar syrup-filled dish, *b*, on the other side.

During 70 minutes 24 recruits arrived at *a* and joined the trained bees; only 2 recruits went to *b*. One drop of strong lavender water was then added to the syrup in both dishes. In the next (second) hour 21 recruits arrived at *a*, while one went to *b*. This showed that the presence of this strong extraneous scent did not disturb the preference; the bees were not in this instance searching for the scent contained in the syrup.

The dishes, and small wooden boxes on which they were standing, were then exchanged. The trained bees were jumpy at first, but they all soon settled down to the new dish *b*, and at any one time about twenty marked bees were there. During the next (third) hour the numbers of newcomers killed at each dish in successive five-minute intervals were:

b, with trained bees	0	0	1	4	2	2	1	1	1	6	4	4
a, now without bees	2	1	4	0	2	0	2	0	3	0	2	2

At the end of this hour *b* was removed, together with its stand, and a similarly filled new dish, *c*, with a stand, put in its place; *a* was not altered. During the next (fourth) hour both dishes attracted recruits in about equal numbers. The distribution of the new recruits in successive five-minute intervals was:

c, with trained bees	1	1	0	1	0	2	0	3	2	1	3	2
a, now without bees	1	0	1	0	3	0	1	3	0	2	2	1

During the first twelve minutes after the first change no recruits had arrived at *b*, to which the trained bees were going, but six had arrived at the previously visited dish *a*, which was now without trained bees. Thus the new recruits were not attracted by the sight, sound or smell of trained bees, but by something pertaining to *a*, the dish which these bees had previously visited for more than two hours. This could only be an odour. The results during the fourth hour indicated that the attracting odour persisted at the dish *a* for at least two hours.

Moreover, the trained bees had been returning to their side without mistake, but they were jumpy after each change; during the first ten minutes after the second change of dishes, marked bees had to be driven away from the empty dish *a* on nine occasions, but thereafter they made no mistakes. This result indicates that the odour was used by the trained bees for their own guidance, as well as by the newcomers.

The origin of the distinctive odour which serves for recognition

The crucial experiment of Kalmus & Ribbands was commenced on 20 September 1951, when there was very little natural forage. A colony was dequeened and divided into thirds, D, E, and F, which were placed in a line at 10-yard intervals on a new site. All these nuclei were queenless. Next day a mixture of ½ lb. heather honey and 1 tablespoonful black treacle was poured over the tops of the frames of nucleus F: D and E were not fed. On 29 September, D and F were compared by the use of the above mentioned experimental technique (p. 174); in the fed nucleus 43 correctly marked and 14 incorrectly marked recruits were found, while there were 11 correct and 6 incorrect newcomers in D: this distribution was highly significant. On 5 October, the unfed nuclei D and E were compared; there were 22 correctly and 20 incorrectly marked recruits in D and there were 16 correct and 13 incorrect ones in E: this difference was not significant. On 6 October, E and F were compared; there were 37 correct and 16 incorrect newcomers in F and 35 correct and 5 incorrect ones in E: this distribution was again highly significant. Thus the foragers of the fed nucleus distinguished their comrades from the bees of either unfed nucleus, and vice versa, but no significant distinction was made between the foragers of the two unfed nuclei.

It was considered that the preferential attraction which had been demonstrated between the foragers of the fed and unfed nuclei had arisen in the following way. The food supply of the fed nucleus was evenly distributed among the foraging population by food sharing (Ch. 24), with the result that each forager received more or less the same diet and produced from it the same waste products. Some or all of the aromatic parts of these waste products were collected in, and given out from, the scent gland of each individual, the result being that each one gave out the same odour[1]. The foragers of either unfed nucleus also emitted a common odour, and as they had shared the same food stores their odours were the same, but different from the odour of the fed nucleus. Each individual thus produced a distinctive odour, characteristic of the nucleus and distinguishable by its members.

Colonies foraging normally in an apiary collect from various crops in different proportions (p. 167), so on this hypothesis

[1] Kaltofen (1951), using v. Frisch's scent-perception technique, claimed that a group of trained bees could distinguish their own fanning scent from that of another group of bees from the same colony, but in my opinion his experiment was not adequately controlled.

derivatives of the scents contained in the nectars would be conveyed to the scent gland and the foragers of each colony would give out a different odour. This implies that the differences in olfactory attraction are a consequence of distinguishable differences between mixtures of odours, the components of which may be identical but present in varying proportions. This interpretation is in harmony with the pronounced ability of honeybees to discriminate between different mixtures of the same odours (p. 41); it would lead one to expect that the scent from bees of one colony would often attract bees from other colonies, but to a lesser degree. This did happen in the experiments of Kalmus & Ribbands; sight played but little part in attracting recruits to the experimental dishes (p. 176), so the considerable proportion of newcomers which were attracted to the wrong dish in several experiments demonstrated the potency of the scent from strange bees.

If the above interpretation is correct, it would follow that if a group of colonies were deprived of all their stores, and then only given or allowed to obtain an identical diet, they would come to possess an identical odour and the members of these colonies would be unable to distinguish their companions from other bees in the group. Ribbands (1953d) found that this was so with colonies which had been shaken on to empty combs and left for two months on a heather moor where *Calluna vulgaris* provided the only source of forage.

The uses of mutual recognition

It would seem that the uses of mutual recognition in the field must be very limited, although foragers may at least use their distinctive odour for their own guidance (p. 176). Kalmus (1941) trained foragers from two differently coloured colonies of bees to a dish of sand which was sparsely supplied with sugar syrup; the bees foraged amicably until the syrup supplied had been almost used up; then fighting commenced, but it was never between hivemates. This exceptional circumstance demonstrates a use of mutual recognition in the field, but there can be little doubt that its really important role is at the entrance to the community, where mutual recognition facilitates defence.

Chapter 22

THE DEFENCE OF THE COMMUNITY

> *But O! Too common ill, I brought with me*
> *That which betray'd me to mine enemy,*
> *A loud perfume, which at my entrance cried.*
>
> JOHN DONNE, *Elegy*

DISTINCTION BETWEEN FRIENDS and enemies is a necessary feature of any complex social organization, which enables it to defend itself against rivals of its own species. This chapter is concerned with the defence of the community against robbers. A related aspect, defence against usurping queens, is considered in Chapter 36.

The recognition of enemies

Bethe (1898) divided a colony into two parts and found that a fortnight later bees from either half were received without much hostility when placed in the other, but after a further three weeks old and young bees of either colony were violently attacked when placed on the wrong alighting board. Bethe thought that different inherited odours were the basis of this hostility, though if he had used controls he might have found that the differences shown between the two occasions were linked up with seasonal factors.

The existence of these inherited odours was accepted by v. Buttel-Reepen (1900), but nevertheless he emphasized the importance of untoward behaviour in recognition of the enemy. He said that bees which returned with a full load and blundered into the wrong hive were seldom attacked, but that if they were they tucked in the abdomen, extended the proboscis, and offered nectar to their tormentors. Moreover, bees made queenless and then shaken in front of another colony entered it 'joyfully' and without opposition; bees which had been washed and dried in an attempt to remove their odour could still enter their own hive unmolested because of their quiet and sure manner. On the other hand, the to-and-fro flight of robber bees was 'timid and anxious', and guards often flew at them while they were still in the air.

I do not know of any further experiments on this problem until recently, by which time modern beekeepers had come to consider that the visibly different behaviour of robber bees accounted for their hostile reception (Cale, 1946). Interesting experiments were carried out by Lecomte (1951), who kept queenless groups of 20–30 workers in small cages, and introduced groups of five strange workers to them. Sometimes the strangers were treated with indifference, but at other times they were thoroughly examined; if they remained still their examiners soon lost interest, but combat ensued if they moved. Lecomte made decoys from dead bees, and from wool or cardboard models; the decoys were introduced and either kept still or moved about; in twenty-five experiments with each of several types of decoy there was only one attack on an immobile one (a fresh dead bee), but the moved decoys were readily attacked. Lecomte considered that movement, colour and texture were the stimuli which provoked attack, in that order of importance.

In 1951, experiments which demonstrated the presence and origin of distinctive colony odours (Ch. 21) focused attention upon the role of scent in colony defence, and investigations were commenced immediately by both Butler & Free (1952) and Ribbands (1953b).

Butler & Free confirmed Lecomte's results with jerked decoys and found that foragers made to enter the wrong hive (in summer) were seldom attacked, while robbers trying to enter (in autumn) invariably were. They concluded, 'The attacking of robber bees by guards appears to be released entirely by the characteristic horizontal darting of the robber bee . . . scent is thought to play no part in the recognition of the robber by the guard bee. Although scent apparently plays a part in the recognition of all intruders, except robbers, it is the subsequent behaviour of the suspect, and in robbers exclusively such, which determines the resultant actions taken by the guard bees towards them.'

Ribbands (1953b) used pairs of colonies, one of black and the other of yellow bees, sited within a few yards of one another. Their entrances were restricted to one inch, and after a day or two the hives were moved (overnight) so that they were touching, with their entrances only 2 inches apart (Fig. 49). In these circumstances many of the bees tried to get into the wrong entrance when they returned to the colonies, but all of them were well intentioned. The experiment was repeated several times.

In April, and again during a June nectar flow, there was little or no hostility and the bees were soon mingling indiscriminately. In late August the story was different: most of the bees which mistakenly tried to enter were repelled, and some which landed at the wrong entrance were mauled and pushed away. Some intruders succeeded in entering, but most of these entrants were subsequently hauled out, and a proportion of them stung and

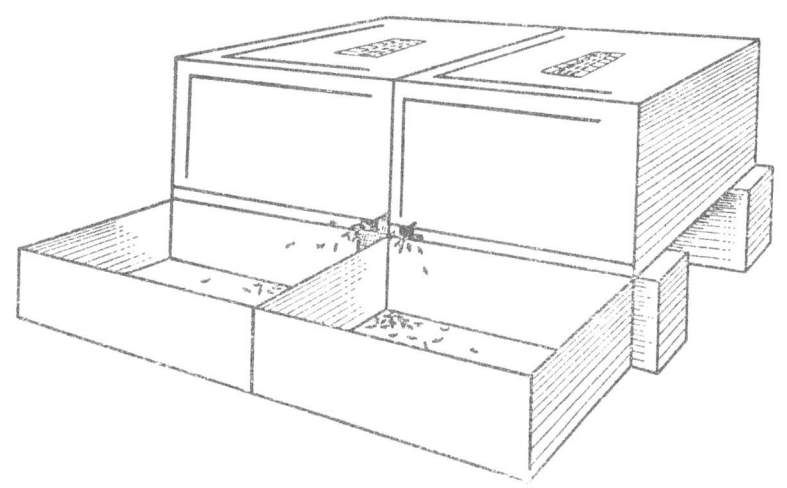

Fig 49 APPARATUS FOR STUDYING COMMUNITY DEFENCE. Hives with differently coloured bees are placed so that their very small entrances are 2 inches apart and separated by a small gauze partition. Many bees orient inaccurately and return to the wrong entrance. Boxes underneath the entrances receive dead or injured bees thrown out by the colonies (Ribbands, 1953*b*).

killed. Nevertheless, some bees did become accepted in the wrong colony, but more corpses were found outside than successful intruders within. It was concluded that when robbing was likely no intruders were willingly admitted, irrespective of their behaviour (some bees gave food to the guards, and were then stung and killed); the intruder's chance of entry depended upon circumstances, for if several tried at the same time some or all of them were more likely to succeed.

Ribbands regarded it as unlikely that robber bees had evolved a pattern of behaviour to their own disadvantage, and suggested that their characteristic flight was the *consequence* of their recognition, not its cause—though it might nevertheless contribute in some degree towards their subsequent recognition and discomfiture. He occasionally observed what he believed to be the genesis of such jerky behaviour, when individuals which had been rebuffed from the wrong entrance several times in succession approached it again; yet if such bees then went in this manner to their own entrance they were admitted.

From these experiments Ribbands concluded that the presence of a scent alien to the distinctive odour of the colony enabled its members to recognize and repel both robbers and other intruders, any other factors being of minor importance. This conclusion can be related to differences in the value of sight and smell to honeybees and humans (Ch. 2, 3).

The circumstances in which intruders are attacked

V. Buttel-Reepen (1900) said that if a colony tolerated robbing, a stimulating food—such as fermenting honey or a mixture of honey and brandy—should be given to it (we now know that this treatment would produce within it a more distinctive odour, p. 177). He also said that if the hive were shaken to arouse the anger of its bees, their irritability would increase and robbers would be more readily repelled. Lecomte (1951) noted that there were marked differences in individual and group irritability, and that aggressiveness was diminished by sparse feeding or long sojourn in his cages; moreover, when he presented his decoys at short intervals, the proportion of attacks on them was steadily raised.

Butler & Free (1952) confirmed this last conclusion, with live bees, and agreed that guards could be alerted by thumping the hive. They also interchanged a pair of hives, in one of which there were a large number of both 18-day-old bees (marked blue), and 28-day-old bees (marked white); they noted that only 7 out of 29 of the younger group entered the hive without stopping when they were met by guards, but 59 out of 75 of the older group did so. The authors suggested that the younger bees were more submissive and the older ones dominant, but it seems to me more likely that the younger bees were less well oriented to the entrance and therefore more likely to hesitate. If successful intrusion is a matter of luck, she who hesitates is most likely to be lost.

For his final experiment, carried out in October with a pair of colonies which had equally repelled intruders in late August,

Ribbands (1953b) trained a large number of foragers, from the black colony only, to collect from a dish containing concentrated sugar syrup. The two hives were then placed together (Fig. 49), and the observer sat at the dish and marked with paint every bee which visited it. No bees from the yellow colony went to the dish. After one of these trials there were 16 marked (i.e. syrup-fed) and 32 unmarked black corpses outside the yellow entrance, but no black bees inside that colony; there were 17 yellow corpses outside the black entrance, but 22 yellow bees had entered and survived in the black colony. Thus no intruders, with or without syrup loads, succeeded in entering the colony in which the foragers were inactive, but a high proportion of intruders were accepted into the black colony, from which several hundred foragers were visiting the syrup dish. The behaviour of the members of the colony—not the behaviour of the intruders—determined whether the latter were likely to effect an entry; this is understandable since guard and foraging duties are interchangeable (p. 304).

Defence against beekeepers

Huber (1814) noted that bees were irritated by the smell of an isolated sting, and stung beekeepers can confirm that the odour from a sting precipitates further attacks; this response of the honeybee facilitates the defence of the community against its larger enemies.

Individual defence by immobility?

Beecken (1934) observed that a worker being groomed by another occasionally assumed a rigid position, with her wings spread widely and her abdomen contracted and lowered. Having observed this rigid immobility[1] in several situations, Grosdanič (1951) was able to produce it experimentally by stroking the rear part of the thorax of a bee with a pointed straw; the immobility lasted for about 30 seconds. He also induced it in queens, but he failed to do so in drones. Butler & Free (1952) noted that the same pose was infrequently assumed by intruders which were being hauled from a hive by guard bees—in which circumstance it might be considered to have a protective function.

[1]'Akinesis.'

Chapter 23

HOW HONEYBEES CAN BE DIRECTED TO PARTICULAR CROPS

> *I schmokes mine pipe und I vatches dose bees,*
> *Und I laughs till mine schtomack goes schplit,*
> *Ven I see dem go schtrait for Hans Brinkerhoff's flow'rs*
> *Und nefer suck Yakob's vone bit.*
>
> EUGENE SECOR, *Songs of Beedom*

Preliminary experiments with scent direction

SCENT-TRAINING EXPERIMENTS (p. 39) and discoveries concerning communication about crops (Ch. 19) contain the implication that one might direct honeybees to particular crops which require pollination.

Gubin & Romashov (1933) fed bees at a feeder inside their hive with sugar syrup which contained mint. Just outside the hive entrance they placed four saucers, with mint and syrup, mint and water, syrup only and water only; in three trials they attracted 318, 105, 47 and 2 bees respectively. In another experiment air was pumped over lilac flowers and then through sugar syrup for 6 hours, by which time the syrup had acquired a noticeable lilac odour. It was then fed inside the hive, and bunches of lilac and of three other kinds of flowers were placed just outside; 13 bees went to the lilac but only 8 visited all the other three bunches together. These results encouraged Gubin & Romashov to study the possibility of directing bees, especially to red clover.

From field studies Gubin (1936) developed a technique which involved (*a*) soaking freshly picked flowers of the target crop in 1:1 syrup, and (*b*) feeding this scented syrup inside the hive, either at night or very early in the morning.

This technique was reported to produce great increases in the number of honeybee visits to red clover and substantial increases in seed set. For instance, Kapustin (1938) recorded a 24-fold increase in honeybee visits and a doubling of the seed crop, and Gubin (1939) a 19-fold increase in the honeybee population on red clover, with a trebling of the seed crop.

Gubin wrote that bees had also been successfully trained to visit vetches, sunflowers and lucerne.

V. Frisch (1943) reviewed these results and pointed out that as the control colonies had not been fed, it was not possible to determine how far the results were due to odour and how far to mere feeding. He noted that there was an important difference between this technique and his own, because the crop was searched for by the fed bees themselves and not by recruits encouraged by their dancing. Experimenting for himself, v. Frisch then fed 85 ml. (0·15 pint) syrup inside a hive in an atmosphere scented with star-anise, and laid out three dishes 33 yards away; one of these dishes was also scented with star-anise, and it attracted 158 bees in 2 hours; in that time the other two dishes, differently scented, attracted only 5 and 8 bees. However, 12 bees were fed outside the hive with only 12 ml. syrup from a dish surrounded by bergamot scent; in 1 hour they caused 206 recruits to be attracted to that dish and 216 to another similarly scented dish, while only 3 and 1 went to two differently scented dishes. From these results v. Frisch concluded that either external or internal feeding might succeed, but that much more syrup was required with the latter.

V. Frisch observed that sometimes an increase in the bee population was not apparent until the hive feeding had finished, so he supposed that the bees at the feeder afterwards went into the field (as they must have done if Gubin's successes are explicable). However, he found that early-morning feeding in the hive, which encouraged recruits as well as fed bees, was considerably more effective than overnight feeding, which presumably sent only fed bees to the crop.

In other experiments foreign odours were added to natural flowers. When syrup, scented with thyme oil, was fed to the bees, larkspur which had been sprayed with thyme oil was visited. Visits to yellow stonecrop and to Kirsten Poulsen roses were similarly induced, but visits to strongly scented flowers (elder and garden jasmine) could not be encouraged in this manner.

V. Frisch also obtained some successful results by mixing the scent with the syrup, as Gubin had recommended, but he emphasized one difficulty; the training scent had to be the same as the target scent, not merely similar. No bees were directed to fresh thyme by feeding with syrup which contained commercial thyme oil, which is obtained by distilling the flowers of this plant. Moreover, from rape flowers v. Frisch was unable to produce any syrup extract which smelled even remotely like the fresh blossoms; he referred to Hesse's (1901–02) record that the quality of flower scents might change after they were picked, and he added that scent extraction with syrup was a rather crude technique. To overcome this difficulty, v. Frisch fed his bees in a feeder garlanded with freshly picked flowers, which were wired in so that the bees could not drag them out of the hive.

Further experiments were reported by v. Frisch (1944). He found that bees fed within the hive danced less actively than those fed outside it, and that a sugar concentration four times as great was required in order to stimulate them to dance; feeding within the hive might therefore produce no dances when crops were good (p. 162). When the same individual bees visited both inside and outside feeders, this relation still held good. V. Frisch estimated that by outside feeding up to fifty times as many bees might be directed with the same quantity of sugar, but he pointed out that internal feeding reduced robbing risks and enabled the colonies to receive equal treatments. This paper also reported the results of further field experiments.

Scent direction to red clover

In the following experiment bees were fed from a feeder within their hive (v. Frisch, 1943). Bees were counted on a 4 × 1-yard strip of $3\frac{3}{4}$ acres of red clover. Between 29 July and 14 August a nearby colony was fed with unscented syrup, and there was an average of 1·3 bees per count on this strip. At noon on 14 August, 150 ml. ($\frac{1}{4}$ pint) of syrup garlanded with red clover were fed to this colony, and next morning there were 16 bees per count. Feeding was repeated each day, and in 51 counts between 15 and 29 August there was an average of 29 bees per count. This was equivalent to a 22-fold increase in visiting.

A repetition of this experiment on another site was a failure. Here the nectar was inaccessible because the corolla tubes of the red clover were longer. V. Frisch concluded that scent-directing to such a crop was 'as pointless as an advertisement for an inn in which there is nothing for sale'.

Summarizing his results, v. Frisch (1943) concluded that training methods could produce quicker finding of a crop, more visits to it and longer hours of honeybee work upon it, but that bees would not continue to visit any crop which offered them nothing. He suggested, as Gubin (1938) had done, that competition from rival nectar sources might be counteracted by directing nuclei of young bees.

In a later article v. Frisch (1946b) reported that in 1943 and 1944 twelve selected pairs of similar fields of red clover, at least one mile apart, had been supplied with 1–2 colonies of bees per acre. All the colonies were fed for 5 weeks with 100 ml. ($\frac{1}{8}$ pint) sugar syrup each day, supplied within the hives; one set received blossom-scented syrup, the other unscented syrup. About three or four times as many honeybees of the former groups visited the fields, and seed production was 40% greater in these fields than in the others.

Further details of these experiments were reported by v. Frisch (1947). In four of the trials the seed harvest per acre was recorded in the directed and the control fields; in the former it had been increased by 39, 43, 73 and 124%. However, in nine trials the seed yield was estimated by determining the percentage seed set in samples of 100 flower heads gathered from the experimental fields; for the same four trials these increases were only 18, 0, 13 and —11%, respectively; for five other trials they were 9, 15, 21, 23 and 26% (mean increase in 9 trials = $12\frac{1}{2}$%). There is a big discrepancy between these two sets of results; v. Frisch considered that the former were the more reliable, suggesting that the flower-head samples had been harvested too late, but I think that the latter set were the more reliable index of the effect of scent direction on seed yield, because irrelevant influences (e.g. number of flowers per acre, harvesting technique) could have affected the yields per acre.

In 28 red clover trials, in 3 successive years, v. Frisch found that the average honey yield from directed colonies was 12, 14 and 11% greater than that from the similarly fed but undirected ones.

V. Frisch reported an increased proportion of pollen loads containing red clover pollen in the directed colonies. Further suggestion of this effect was given by Firssov (1951), who placed 9 colonies of bees near a field of red clover; the following proportions of trapped loads were of red clover pollen:

        3 unfed colonies—10%
        3 colonies fed with unscented syrup—28%
        3 colonies fed with clover-scented syrup—57%.

The divergence between the first two (control) groups illustrates the latitude obtainable between small groups of similar colonies (cf. p. 167).

Scent direction to crops other than red clover

V. Frisch (1947) also reported the result of 47 controlled trials of scent direction, involving 14 different crops. The fields to which bees were directed were more freely visited. Most of the directed bees brought back pollen from the crop to which they had been guided, while the undirected ones returned with more varied loads. The directed bees were also said to commence work earlier and to be less easily deterred by unfavourable weather.

The results were divided into 25 groups, the experiments with each crop in one year being put together; the honey yield from the directed colonies was more than 50% greater in 7 groups and 20–50% greater in 10; it was less than that of the undirected colonies in only 2 groups. On the average the directed colonies stored about 4 lb. more honey than the undirected ones—both having been fed with approximately 4 lb. sugar in syrup. Some experiments with two sets of controls —unfed and fed with unscented syrup—indicated that the latter set obtained slightly more nectar than the former one.

I find these results difficult to understand. One might have supposed that there would be no lack of dancing bees to encourage recruits to crops as good and as abundant as heather (increased 13–34%). At the other extreme, one might have supposed that bees directed to crops which they would otherwise not have worked would gather less nectar than their undirected rivals which were not supplied with misleading information.

V. Frisch paid more attention to honey yields than to seed yields, but favourable effects on seed yield were reported for some crops. For alsike clover, in two experiments the threshed seed yields were increased by 26 and 48%, but in the former of these a comparison of the seed set on 100 flower heads showed an increase of only 6.5%.

For rape, the seed harvest from 4 directed fields exceeded that of the undirected fields by 12, 16, 20 and 33%—in the first of these 4 fields the seed production, based on a count of seed per plant, was only 2.1% greater; the weight of 1,000 seeds was 5.1% greater. For turnip, the seed harvest from 3 directed fields exceeded that from undirected fields by 27, 44 and 52%; there were no counts of the seed yield per plant.

For field bean, flighting to two directed fields increased four-fold and the seed harvest per acre was 8% better; v. Frisch also counted the number of pods developed from 37 flowers on each of 200 plants in both fields, and found that 74·2% had developed in the undirected fields and 79·7% in the directed ones—an increase of 6·7%.

Gubin (1945a) reported that directing increased bee visits to flax seven-fold, thus reducing by 85% the number of colonies required for pollination; he also found (1945b) that bees directed to ridge cucumbers in greenhouses were more than twice as effective as undirected ones.

Economic aspects of scent direction

The results obtained by Gubin and v. Frisch show that honeybees can be successfully directed to crops by feeding them with scented syrup, yet this technique has not been widely imitated. Scent direction usually produces a substantial increase in the number of visits to the crops concerned, but only a small proportion of that increase is reflected in increased seed yield. The latter is the main purpose of the treatment.

The method has limitations. V. Frisch pointed out that bees would not continue to visit a crop which yielded nothing, and most of the difficulties which attend the problem of red clover pollination are said to arise when nectar production is so small that the honeybee cannot reach the supply. If favourable conditions for nectar secretion remove this obstacle the crop becomes very attractive to honeybees; pollination can then be effected without scent directing.

Any tests of the efficiency of scent directing must be oft-repeated, because climatic and soil conditions produce such wide variations in seed yield; v. Frisch's red clover trials satisfy this condition, and indicate that an increased seed yield of at least one-eighth was obtained by scent direction. His work also indicated that increased honey yields were obtained; however, feeding with scented syrup can provoke robbing, and the beekeeper may consider that his potential loss on this score, combined with the labour involved in daily feeding, outweighs in value any potential increase in honey yield.

British farmers seldom take steps to increase the force of bees available for pollinating their crops, except in fruit orchards;

it would usually not be difficult for them to make arrangements for the hire and transport of bees for this purpose, but it might be easier and cheaper to hire a large number of colonies, to be left unattended, rather than some smaller number which would require daily scent directing.

Other methods of directing bees to crops

Pollen collectors are more efficient than nectar gatherers in pollinating lucerne (alfalfa), and red clover is worked for pollen when its nectar is inaccessible. Thus both of these crops might be more efficiently pollinated if the proportion of pollen collectors were increased.

This possibility occurred to Stapel (1934), who supplied two colonies with $1\frac{2}{3}$ lb. sugar each day, in 1:1 syrup, and fitted pollen traps to them. The bees soon learned to avoid the grating of the trap, but Stapel considered that the syrup treatment alone was sufficient; he held that the bees were encouraged to collect pollen and so to visit red clover more frequently. The published data did not establish this proposition, since no untreated colonies were used for comparison. Veprikov (1936) also claimed that the proportion of pollen collectors on red clover was increased by removing combs of pollen from colonies and feeding them with syrup; this work has not been accessible to me.

The results of Filmer and Lindauer (pp. 125–6) indicate that pollen trapping might increase the proportion of pollen gatherers in colonies, with benefit to their pollinating capacity on lucerne and red clover. This procedure might possibly be economic if a cheap and simple trap were used, the pollen being fed back to the colonies in spring (p. 240).

Another method of directing bees to crops was suggested by Gataulin (1945). He thought that colonies in an apiary which did not work as well as others might be attached to plants which were bearing less nectar; if so, a comb of freshly collected nectar from another colony could be put into them, and it would direct them to better supplies.

PART IV LIFE WITHIN THE COMMUNITY

Chapter 24
FOOD SHARING

Here comes Monsieur le Beau
With his mouth full of news,
Which he will put on us, as pigeons feed their young.
Then shall we be news-crammed.

WILLIAM SHAKESPEARE, *As You Like It*

The feeding relations of adults and larvae

JANET (1903) AND DU BUYSSON (1903) both observed that wasp larvae produced a sweet salivary secretion, which adult wasps solicited from them by nibbling at their heads. Roubaud (1916) supposed that this act was the means of social cohesion[1] and explained the attachment of young adult wasps to their nest. This view was extended by Wheeler (1918, 1928), who considered that reciprocal food exchange[2] was the foundation of the social life of termites, ants and wasps, and possibly all social insects. This theory has been criticized by Brian & Brian (1952), who found that samples of larval saliva from the wasp *Vespa sylvestris* were not attractive to adult wasps; they agreed that adults pinch the larvae vigorously in order to obtain this secretion, but they regarded this as a sign of zealous attention to larval welfare rather than selfish greed.

Lineburg (1924b) stated that nurse bees made more than 10,000 visits to larval honeybees during their five days of development, and he suggested that this zeal was circumstantial evidence of food exchange, but King (1928) reported that nurse bees placed food at random in the cells and not into the mouths of the larvae, so reciprocal feeding was impossible.

Lindauer (1952) was never able to see anything pass from larva to nurse, although he watched more than one thousand larval feeds in cells constructed parallel to the glass wall of an observation hive. He described the feeding thus: after an inspection lasting about 10–20 seconds, the nurse bee turned herself so that the points of her mandibles were close to the

[1] 'Oecotrophobiosis'.   [2] 'Trophallaxis'.

larval head; her mandibles were opened and vibrated, and after 1–2 seconds a droplet began to appear between the maxillae; it was left either on the floor or wall of the cell or on the larvae itself. The whole feeding process might occupy from ½ to 3 minutes. Since the larva is always very slowly turning in its cell, it reaches food placed in any position. Lindauer found that nurses seemed to feed the larvae at random—they did not care for any special group of larvae, and seldom fed the same larva a second time. Thus each larva was fed by many different nurse bees.

King and Lindauer agree that there is no reciprocity in the feeding relations between larval and adult honeybees. The high temperature and low humidity of the broodnest (p. 225) would evaporate water from the brood, and the water content of the larval food seems to be adjusted so that there is no additional surplus water for the nurse bee to remove.

Absence of food exchange between larval and adult honeybees fits in with Brian & Brian's observations on wasps, and suggests that maternal instinct and the need to dispose of the surplus of water in the larval diet may fully cover the habits of nursing wasps. Elsewhere in the animal kingdom the nourishment of the young is seldom associated with any material reward.

Feeding relations between adults

An observation on food dissemination among adult honeybees was recorded by Park (1923c). In spring, when bees gather water for brood rearing, Park filled the feeder of a small nucleus (2600 adult bees) with coloured water. The following morning the abdomens of about half of the bees were coloured and distended; most of them contained a mixture of honey and water. In temperate climates honeybees do not store water in combs, so Park considered that they were using their honey-sacs as reservoirs.

Nixon & Ribbands (1952) used radioactive phosphorus, fed in sugar syrup, to trace the extent of food transmission within a honeybee community. Their main experiment was commenced on 21 August, 1951, when little of the radioactive syrup was likely to be stored because colonies were consuming more food than they could collect. Six marked bees from one colony were trained to a dish of sugar syrup, and a dish with 20 ml. (about 1 tablespoonful) of sugar syrup containing a very

Günter Olberg
PLATE VI *Returned bees fanning and scenting at their hive entrance*

Günter Olberg

PLATE VII(a) *A returned forager at the hive entrance, fanning and scenting*

Günter Olberg

PLATE VII(b) *Food transmission*

small quantity of radioactive phosphorus was substituted; the six bees took 3¼ hours (379 loads) to collect this booty.

There were 24,600 bees in their colony, which occupied a National broodchamber and two supers. Samples of bees were collected from the colony 5 hours and 29 hours after the radioactive syrup was supplied to the foragers. The hive entrance was blocked, and a sample of returning foragers was collected; the hive was then opened and the same number of bees was collected at random from each comb; each bee was separately examined for radioactivity. In the earlier sample 62% of foragers, 18% of bees from the broodchamber, 16% of those from the lower super, and 21% of those from the upper super, were radioactive. In the later sample these percentages had increased to 76, 43, 53 and 60 respectively.

Analyses showed that the wings and legs of radioactive bees were not radioactive, so external contamination could not have contributed to the result. The foragers became radioactive most quickly, but a subsidiary experiment showed that any supplies which they might have received from dancing bees (p. 149) were insufficient to account for the quantities of radioactive syrup contained in them. In the second sample the radioactivity of the workers in the broodchamber was significantly less than that of those in the upper super, although the broodchamber contained a mixture of both nurse bees and foragers, the latter with the highest proportion of radioactivity. Of the older unsealed larvae, 85 were examined two days after supplying the treated syrup, and were all positive; so were all cells of unsealed honey. It was concluded that the syrup had been passed rapidly among the foragers and was then transmitted and shared among the whole community.

The lowest proportion of radioactivity (27% in the second sample) was found among the drones (although a sample containing only nurse bees might have yielded a similar result). This suggests that the drones are not fed with nectar, but with bee milk (p. 259).

The process of nectar transfer[1] is described elsewhere (p. 210, and Fig. 1). There is no evidence of any *exchange* of food between individuals, but transfers are so frequent that all members share each item.

[1] The speedy process of nectar transfer can be distinguished from the much slower transfer of bee milk, which takes 1¼–4½ minutes when nurse bees are feeding newly emerged ones, and slightly longer when queens or drones are being fed (Perepelova, 1928d).

The social uses of food sharing

The food sharing, which is much more extensive than would be required merely to prevent individuals from starving when food is available, has important social functions. It serves as a method of communication between the members, and it was for this reason that the phrase 'food transmission' was chosen by Nixon & Ribbands. Transmission of small quantities of food from dancing foragers to recruits stimulates the latter and provides them with information concerning the kind and quality of the crop they should seek (p. 149). Ribbands supposed that food transmission also provides information which is the basis for an effective division of labour (p. 311). Moreover, sharing of food among the foragers ensures that the aromatic waste products of their metabolism are similar; these provide a distinctive odour which is given out from the scent gland and serves for mutual recognition (p. 177), thus facilitating the defence of the community against either robbers (Ch. 22) or social parasites (p. 295).

It is thus suggested that food transmission helps the community to feed itself, to organize itself, and to defend itself, and we know that the special diet provided for queens, drones and larvae helps the community to reproduce itself. The system of food sharing welds the colony into a unit in the struggle for survival, so that the colony, not the individual, either succeeds or starves.

Chapter 25
CLUSTERING

*For your friends are my friends, and my friends are your friends,
And the more we are together the merrier we'll be.*

Clustering position

THE CLUSTERING POSITION of a group of queenless bees was studied by Sendler (1940), who found that they preferred horizontal surfaces to sloping ones, rough ones to smooth ones, and the highest part of the box in which they were placed. In a box one foot high they clustered on the roof immediately above a caged queen, but at a distance of 20 inches the queen ceased to have this influence. The effects of various combinations of these stimuli were investigated; it is a far cry from these experimental conditions to those in nature.

Mutual attraction

A grouping of animals may be a response to some environmental stimulus, but in social animals real mutual attraction also occurs. The attraction between queenless bees, shaken into a darkened box, was studied by Lecomte (1950); he found that 100 or more bees formed a single cluster within 2–3 hours, but if there were less than 50 they remained dispersed singly or in very small groups.

Small gauze cages were added, and the bees would cluster on a cage containing bees and ignore an empty one; they showed no preference for an empty cage in which bees had lived for a long time, or one supplied with a current of air from a colony. The bees did not cluster round a cage of decoy bees suspended one-fifth of an inch above the ground, so Lecomte concluded that there was some vibratory stimulus. They would only cluster round live bees in a hermetically sealed box if dead bees were placed outside this box, but as dead bees alone exerted no attraction Lecomte postulated that vibratory and olfactory stimuli were both required to cause the mutual attraction. The abdomens of dead bees could supply the olfactory stimulus, but the heads or thoraces could not.

Aggregation of foragers

Mutual attraction between foragers was studied by Kalmus (1953). Five syrup-filled Petri dishes were placed on a table, but no bees visited them until one was introduced and marked. During the next half-hour many bees came, but nearly all of them congregated on one of the five dishes; when that was emptied they went to the others, in turn, until those were cleared.

Bees so trained to syrup would gather on one or other of five hitherto unvisited empty dishes if these were put in place of their supply, and by altering conditions at one of the five dishes Kalmus studied the stimuli which induced this aggregation. The bees always alighted on that glass dish under which five bees were confined, and if white papers were placed under all dishes (to increase contrast) two captive bees were sufficiently attractive. Black blobs placed on the white ground were also effective, although more (8) were required. These experiments demonstrated the visual component in the attraction.

The olfactory component was then studied. Twenty-five bees, caged out of sight, ensured clustering, but one fanning bee was just as effective. A dish on which fifteen bees had crawled about for one minute, eighteen minutes earlier, was also attractive.

Efforts to attract the bees to a wide range of sounds (including a tape recording of buzzing bees) were unsuccessful. However, they remained motionless, as if stunned, if they happened to alight on a vibrating portion of the speaker, so Kalmus suggested that a vibratory component might help to keep the bees in cluster.

The results of Lecomte and Kalmus, considered together, indicate that vision, smell and vibration are components of the mutual attractions which cause clustering, their relative importance depending upon circumstances.

(The winter cluster is considered in Chapter 29.)

Chapter 26

WAX PRODUCTION AND MANIPULATION

> *Mark how the little untaught builders square*
> *Their rooms, and in the dark their lodgings rear!*
> *Does not this skill even vie with reason's reach?*
> *Can Euclid more, can more Palladio teach?*
>
> Quoted by WILLIAM CHAMBERS, *The Beehive*

BEESWAX IS AN exudate similar to those which are often produced by coccids and aphids which live on the sugary sap of plants and need to rid themselves of an excess of sugar (Wheeler, 1928); the wax of social bees may therefore have originated as a side product of their carbohydrate diet which they utilized and developed.

The secretion and properties of fresh wax scales

The process of wax secretion was studied by Huber (1814), who proved that the wax was produced from honey. No wax scales or new combs were produced by a swarm which was enclosed and fed on pollen for eight days, but another swarm, enclosed and fed on honey only, produced five new combs in five days. These new combs were removed five times in succession, but each time more were built. In similar circumstances bees which Huber fed on sugar only built even more comb, and this experiment was repeated seven times in succession. Eckert (1927) found that bees fed with sugar produced a wax identical with that produced from honey.

The properties of fresh wax scales differ from those of newly built comb. Huber found that wax scales dropped into turpentine dissolved quickly and completely, without making the liquid turbid, but similarly sized fragments of new white comb were not completely dissolved, many particles remaining suspended in the liquid. The comb fragments disintegrated in ether and fell in powder to the bottom, but wax scales preserved their size and shape and lost only their transparency. The different properties may come from the liquid which is added during manipulation of the scales by the mouthparts of the wax worker.

Comb building

Huber observed how a swarm placed in a glass bell-jar began to build a comb. The cluster hung in festoons from the top, the uppermost ones fixed by the claws of their fore-legs, their companions below clinging with their fore-legs to the hind-legs of the bee above them. The festoons remained motionless. Next day wax scales were visible protruding from underneath the abdomens of the bees. Then one worker detached herself from a festoon, went to the top, and turned round and round until she had cleared a space for herself. She put one of her hind-legs against her abdomen, pulled the scale from its pocket and passed it to the claws of her fore-legs, which carried it to her mouth. The edges of the scale were nibbled and with much tongue-moving the fragments were impregnated with a frothy liquid (neutral or slightly alkaline to litmus, Lineburg, 1924a) and issued from the mouth as a very narrow opaque white ribbon. The jaws cut this ribbon and the now sticky pieces were applied to the roof of the bell-jar.

A detailed account of the manipulation of wax scales was provided by Casteel (1912b), who watched the operation through a binocular microscope. He noted that the scale was rubbed from its wax pocket by the hind tarsus. It became impaled on several of the pollen spines and was quickly carried forward to the mouth, where it was manipulated by the mandibles, which were assisted by the fore-legs if the scale was a large one. The process of mastication and application of one scale lasted nearly four minutes. Casteel emphasized that even in the construction of new combs the reworking of wax was a most characteristic feature, and there were always many active wax workers whose wax glands were inactive.

Huber found that at the commencement of comb construction a ridge of wax was formed and was extended to form the basis of the upper row of cells of the comb, with additions of wax by many bees and much sculpturing by their mouthparts. The antennae were very freely used to determine the position of each addition or sculpturing. Huber described the construction of the comb in detail, and other accounts have been given by Darwin (1859) and Lineburg (1924a).

Darwin said, 'The manner in which the bees build is curious; they always make the first rough wall from ten to twenty times thicker than the excessively thin finished wall of the cell, which will ultimately be left. We shall understand how they work, by supposing masons first to pile up a broad ridge of cement, and then to begin cutting it away equally on both sides near the ground, till a smooth, very thin wall is left in the middle; the masons always piling up the cut away cement, and adding fresh cement on the edge of the ridge. We shall thus have a thin wall steadily growing upward but always crowned by a gigantic coping. From all the cells, both those just commenced and

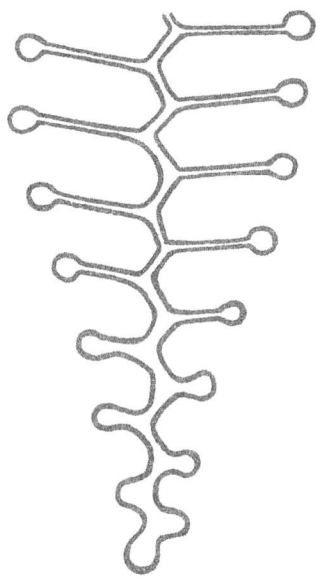

Fig 50 SECTION THROUGH A PIECE OF UNCOMPLETED COMB. The completed cells have thin walls, crowned with a coping; the walls of uncompleted ones are thick (Lineberg, 1924*a*).

those completed, being thus crowned by a strong coping of wax, the bees can crawl over the comb without injuring the delicate walls.' Fig. 50, after Lineburg, illustrates these characteristics of the building.

Mathematicians (quoted by Huber, 1814) have calculated that comb cells are the proper shape to hold the greatest possible amount of honey with the least possible consumption of wax in their construction. Darwin deduced that this perfection had arisen from a few simple instincts, by natural selection. He held that the cells of the Mexican stingless bee (*Melipona domestica*) were intermediate between the combs of the honeybee and the separate rounded cells of bumble bees. The cells of *Melipona* are placed irregularly; they are spherical but, where the spheres would have intersected if they had been completed, flat walls of wax are built between them instead. The intersections save wax and labour; Darwin reflected that if *Melipona* had made its spheres at some constant distance from each other, and of equal size and arranged symmetrically in a double layer, the comb would have been like that of the honeybee.

Darwin supplied honeybees with blocks or flat sheets of wax, and observed that they excavated minute circular pits and extended them until they were shallow basins of about the diameter of a cell. He observed that wherever several honeybees began to excavate basins near together, they began their work at such a distance from each other that the rims of the basins intersected by the time they had reached cell width. When the sheet of wax was thin the bees always stopped excavating to avoid breaking through, and Darwin considered, as Huber had done, that the flexibility of the thinned wax warned them to stop.

Darwin covered the edges of the walls of single partly built cells with red wax, and found that the colour was delicately diffused, particles of red wax having been taken away and worked into the growing edges of the cells around. He concluded that 'The work of construction seems to be a sort of balance struck between many bees, all instinctively standing at the same relative distance from each other, all trying to sweep equal spheres, and then building up, or leaving ungnawed, the planes of intersection between these spheres. It was really curious to note in cases of difficulty, as when two pieces of comb met at an angle, how often the bees would pull down and rebuild in different ways the same cell, sometimes recurring to a shape which they had at first rejected.'

Cell capping

Comb-building activities are spasmodic, but cell capping is a task which is required more constantly and which occupies a substantial place in worker activity. Lindauer (1952) noted that in brood rearing wax fragments were transported and stacked at the edge of the cell from the first larval day onwards, so that most of the wax required was already present when capping commenced. He emphasized that cell-capping bees worked in a very unsystematic manner and often gnawed bits off one partially capped cell and added them to an adjoining cell. Moreover, some individuals worked for only a short while and then went off elsewhere, leaving their work for others to complete; ten different bees helped with the last half of the capping of one particular cell (Fig. 51), and the total time spent on the work was substantial (p. 231).

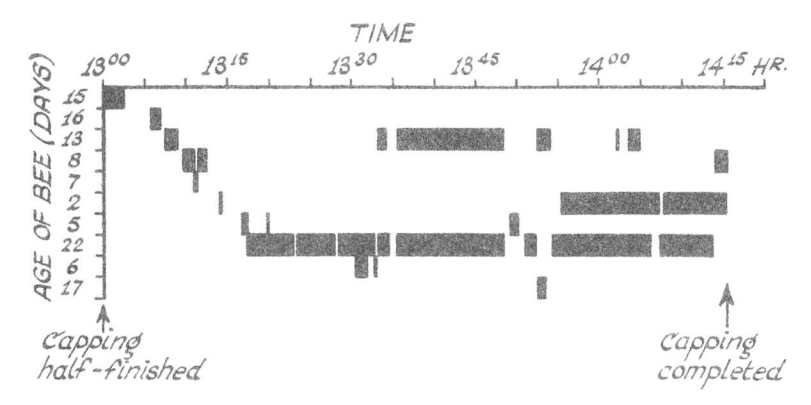

Fig 51 TIME SPENT BY TEN INDIVIDUALLY MARKED BEES COMPLETING THE CAPPING OF ONE PARTICULAR CELL (after Lindauer, 1952).

Cheshire (1886) recorded an exquisite detail of comb construction—the internal sculpturing of the drone cell capping—the building of which has not been described. Anderson (1952) reported that worker honey comb is occasionally capped in this way, and that these cappings are about 50% stronger than the more usual ones.

Dimensions of the combs

The accuracy of comb construction has sometimes been greatly exaggerated by unobservant philosophers, as Darwin and Cheshire emphasized. Cheshire said that it was difficult to find a hexagon with errors of less than 3–4 degrees in its angles; he also pointed out that the bee could not construct angles of less than 100 degrees because it could not insert its head and work its mouthparts in such spaces. It has been claimed that honeybees always build natural comb so that two of the parallel cell walls are vertical, but Thompson (1930) found them horizontal in 123 out of 268 pieces of naturally built comb which he examined. Wedmore (1929) considered that the parallel sides are vertical when the comb is commenced against a horizontal support, and vice versa. Darwin recorded that the finished cell walls averaged $\frac{1}{350}$ inch thick, Cheshire that they ranged from $\frac{1}{280}$ to $\frac{1}{400}$ inch, and Dadant (1926) that they ranged from $\frac{1}{330}$ to $\frac{1}{500}$ inch. The thickness of the bases averages $\frac{1}{230}$ inch (Darwin), is up to $\frac{1}{180}$ inch (Cheshire), or from $\frac{1}{250}$ to $\frac{1}{380}$ inch (Dadant). Crawshaw (1914) found that the walls and bases of drone combs were about 25% thicker than those of workers, and Dadant stated that they were about $\frac{1}{500}$ inch thicker. Dadant's measurements indicated that thin section foundation was drawn out until its base was $\frac{1}{200}$ inch thick (50% heavier than natural comb), while medium brood foundation, initially $\frac{1}{50}$ inch thick, was reduced to about $\frac{1}{180}$ inch.

The cost of comb construction

Cheshire found that 1 lb. wax was built into 35,000 cells, in which 22 lb. honey could be stored, so he calculated that the top cells of a fully filled comb, 1 foot deep, supported 1320 times their own weight. Others have made even greater estimates (but see p. 205). Gwin (1936) calculated that 450,000 wax scales were required to produce 1 lb. wax.

Estimates of the quantity of sugar required in order to produce 1 lb. beeswax have varied widely. It is generally agreed that it is most economically produced during heavy nectar flows. Whitcomb (1946) fed each of four colonies with 6 lb. honey daily for three months (July–September), and provided them with new chambers of foundation, which were removed every 4–5th day. In the last month, when breeding was reduced and when the other chambers of the colonies had become filled with honey, 278 lb. honey was consumed and 55·5 lb. wax was produced—5 lb. honey for each 1 lb. wax. If the conversion could occur without waste of any kind, $2\frac{1}{2}$ lb. sugar would produce 1 lb. wax.

Taranov (1937) provided two sets of ten colonies with empty frames during a nectar flow, and cut out the wax every 3–4 days. In one group, in which the empty frames were interleaved with the brood combs, 6·4 lb. wax and 37·8 lb. honey were produced; in the other group, in which the empty frames were placed outside the broodnest, 1·7 lb. wax and 38 lb. honey were produced. Taranov is said to have concluded that the extra wax was produced without loss of honey. One may more readily assume that the results were obscured by variation between individual colonies, or that wax was less economically produced in cooler conditions beyond the broodnest.

The cost of comb production to the honeybee community is the sum of (a) the sugar used and the energy necessary to convert the sugar into wax and to manipulate the wax during comb building, (b) the wear and tear on the bees, (c) the withdrawal of those bees from other productive work, such as foraging.

Some beekeepers have held that a quantity of new comb can be produced each year, without cost, because a certain amount of wax is necessarily and automatically secreted by the honeybee. Tuenin (1928) appeared to agree with this view, which is complementary to Gerstung's brood food theory (p. 264). However, observation shows that honeybees economize in their wax production, and old wax is used repeatedly. Queen cells incorporate bits cut away from adjoining comb (Cheshire, 1886). Cappings are bitten off by the emerging bee or her helpers, and stuck to the rim of the cell wall, to be used again next time. The wax in them is supplemented with bits of cocoon, traces of propolis and pollen. This explains why cappings of old brood are dark, although those of new combs are light (Lineburg, 1923a; Weippl, 1934). Meyer & Ulrich (1952) state that brood cappings are pared to half thickness, for use elsewhere, before the young bees emerge. Old wax is sometimes employed in comb building; new structures built between old ones may therefore be brown from the beginning, and microscopic examination reveals an abundance of pollen grains and bits of cocoon (Cheshire, 1886). Lineburg (1923b), who noted that bees retrieved wax scales which had been dropped on to the floorboard and used torn-down pieces of old comb and foundation for comb building, supplied a colony with thin flakes of wax (made by spraying hot melted wax on cold water), which they accepted readily.

Perret-Maisonneuve (1927) described an experiment in which thirteen variously coloured 2-c.c. blocks of foreign substances were fixed to the top of a frame of foundation, which the bees built into comb. He found that ruberoid, resins and modelling wax were freely incorporated in the new comb, although pieces of beeswax and of pure paraffin wax were neglected. Roussy (1929) supported this surprising observation. Perret-Maisonneuve concluded that secretion of wax is an economic necessity carried out with parsimony, and that when opportunity arises the honeybee mixes foreign substances with its own wax, as its ancestors did.

Ribbands (1952b) emphasized the relation between wax production and food supply, supposing that when comb space is insufficient the ripening nectar has to be stored in the honeysacs of house bees, which are diverted to this task until their stored loads are assimilated and converted into wax. The wax is used to remedy the lack of comb space, so the need produces the supply. Nevertheless, wax scales are not usually produced during the winter (Tuenin, 1928) although the clustered bees have full honeysacs (p. 219); some other factor, e.g. gland development or temperature, could then exert an inhibitory influence.

Chapter 27

COMB REINFORCEMENT; COMB COLOUR; PROPOLIS

With merry hum the Willow'd copse they scale,
The Fir's dark pyramid, or Poplar pale;
They waft their nut-brown loads exulting home,
That form a fret-work for the future comb,
Caulk every chink where rushing winds may roar,
And seal their circling ramparts to the floor.

JOHN EVANS, *The Bees*

Comb reinforcement

HUBER (1814) OBSERVED that newly completed cells were white and brittle, but within a few days they became yellow, more pliable, stronger and heavier. Their orifices became coated with a reddish varnish, and sometimes there were reddish threads on the inner walls also. Chemical tests indicated that this reddish material was like the propolis which could be taken from the walls of the hive, where it had been used to fill gaps and crannies. Material from both sources was soluble in ether, alcohol or turpentine but was unaffected by nitrous acid, boiling water or caustic alkali (although the last two dissolved wax).

Propolis had long been suspected to be a plant product. Huber placed sticky poplar buds in front of his hives—a bee alighted, tore off threads of the sticky material with her mandibles, took them with one of her middle legs and placed them in the pollen basket of a hind leg. She acquired a load and returned to the hive (cf. Alfonsus, E. C., 1933). Microscopic examination of the residues of poplar buds showed particles of vegetable debris similar to that found in propolis.

Huber then put a swarm to build new combs in an observation hive. For three wet weeks the bees were unable to bring back any propolis, but after warmer weather some bees returned laden with it. These bees went to the top of the cluster, and their companions were seen to bite the propolis off their legs and to carry it away. Some deposited it on the frames, others put it into crannies in the walls of the hive, while others applied it to the inside of the cells. These last deposited the propolis on the glass in the middle of the space between the combs. Other bees entered the cells (against the glass wall), inspected them with their antennae, and cleaned and polished them with their mandibles. One of these then emerged backwards from the cell, approached a heap of propolis, and drew out a thread of it with her mandibles. The thread was broken off by a quick motion of the head and taken in the claws of the fore-feet. The bee re-entered the cell which she had prepared, placed the thread in an angle, used her fore-feet to stretch and fit it and her teeth to embed it. Other bees continued this work and soon all the cell walls were encircled with threads of propolis, which were also put on the orifices.

Huber also noticed another important use of propolis. Shortly after the new combs had been completed the bees began to tear down the walls of the uppermost row of cells and to substitute much heavier pillars composed of a mixture of wax and propolis. The bees used propolis which had been deposited in a lump and had hardened in drying. The bees had some difficulty in removing it from the wall and Huber thought that they impregnated it with frothy matter from their mouth, like that which they used for softening wax. He observed the kneading together of fragments of old wax and propolis. The mixture was used for rebuilding the uppermost cell walls, but economy was entirely set aside and thick pillars were provided. Huber noted that the time at which the structure was so strengthened depended upon circumstances. It is presumably only done as required, and it is more likely to be required when natural comb is built than when modern foundation is provided by the beekeeper (p. 202).

The yellow colour of combs and wax; varnishing

These activities with propolis still left unexplained another attribute of the finished cell—its yellow colour. Huber found that strong sunlight or treatment with boiling nitrous acid would remove this yellow colour, but had no effect upon propolis. On the other hand, propolis was dissolved in alcohol, which did not remove the yellow colour from wax. Thus the yellow colouring matter in the wax had no analogy with propolis. Yet his observations indicated that the tint was not a property of the wax, but of something applied to the cells. Sometimes two or three days were sufficient to turn white combs into yellow ones, but sometimes they remained unaltered for several months of ordinary use. Occasionally he found adjoining cells of yellow and white, or even a single cell with some yellow and some white walls. Combs exposed in the hive, but beyond the reach of bees, remained white for a month.

Huber observed that bees sometimes rubbed the tip of their mandibles against the combs and wood, which then appeared to yellow. In addition, they often swept their proboscis to right and left over the cells, and deposited on them a bright and silvery liquid which was distributed from the end of the proboscis. Huber was uncertain which process was the source of the yellowness. Lindauer (1952) said that the varnish-like coating of the cell walls was produced by belabouring them with the mandibles, but he was never able to see any secretion leave the mouth. He noted that the treatment of cell walls continued after egg-laying in those cells.

Single bees, or whole rows of them, have occasionally been seen performing rhythmic movements on the alighting board. Their head is directed downwards and their tongue half unfolded, and they look as if they were trying to polish or sweep the floor (Morgenthaler, 1931). This activity has been described as 'planing'; Mehring (1866, cited Morgenthaler) said that these bees were then covering the surface with a thin transparent substance, applied with the tongue, and that all surfaces in the hive are covered in this way.

Jaubert (1927) found that 1.3 dioxyflavone is present in brood combs and in the sticky propolis from black poplar but, as Huber showed that cells contain propolis as well as cell varnish, this observation does not lead very far.

Vansell & Bisson (1935), who found that propolis did not impart the characteristic yellow colour of beeswax to white wax but that oil-soluble carotinoid pigments from various pollens were readily absorbed by wax, held that the yellow colour of crude beeswax arose from pollen contamination. They noted that much of the crude wax from Hawaii was practically white. The deep yellow colour of beeswax may be due to this contamination, although Huber pointed out that the colour is evenly spread over the whole of the cell, so that it could not be derived from pollen storage. Simpson (1953) has noted that there are large globules of coloured oil on the surface of the pollen grains of many species, and he suggests that this oil might be transferred to the waxy integument of foraging bees, and thence to the comb surface. Such a process would also explain the rapid staining of all parts of a hive during sainfoin flows. Sainfoin grains are covered with a particularly strongly coloured oil.

Philipp (1928) claimed that there were two kinds of propolis—one gathered from trees and the other produced by the bees themselves. Although he coined a separate name for the latter kind, his identification of it as a sort of propolis—despite the differences noted by Huber—has caused misunderstanding. Philipp found that the varnish was applied to all parts of the hive, as well as to every cell before egg-laying. His microscopic examination of the varnish revealed pollen-shells and little hairs—which could not be found in propolis from outdoors—so he guessed that it was produced from the sticky coat of pollen grains. Philipp suggested that the pollen grains were partially digested in the proventriculus and the indigestible gum was regurgitated, but Evenius (1929) promptly emphasized the weighty objections to this part of his hypothesis. Freudenstein (1932) guessed that the varnish was produced in glands, and merely contaminated with pollen grains. He thought that it might be the same substance as that which was used in the working of wax. McGregor (1952) experimented with the problem; he maintained a colony in a large cage for 9 months, and fed it with 8 lb. pollen pellets and 6 lb. soya cake—it produced no propolis or propolis-like substance but suffered no ill-effects from their absence. Unfortunately, however, his account does not say whether varnish in the cells was looked for.

There is conflicting evidence on yet another issue. Rösch (1925) stated that the polishing of cells was carried out by newly emerged bees, 1–2 days old, but Philipp (1928) maintained that the varnish was produced by nurse bees and house bees. Grosdanič (1931) found that varnishing bees had flown from their colony, which supports the later contention. It would be surprising if the varnish were produced by very young bees, especially as it may well be derived from waste products of the food supply.

If one may derive an opinion from these varied observations and opposite views one could suppose that the cells are belaboured with the mandibles, and perhaps varnished with a glandular secretion. The yellow colour of the combs is probably

Ben Knutson

PLATE VIII *A successful forager, in flight*

Ben Knutson

PLATE IX *A forager on sweet clover*

derived from pollen pigments, but it is uncertain whether it comes via secretion or by contamination spread by the bodies of the bees.

The source of propolis

Chemical analyses of parcels of propolis from the hive have shown that it is extraordinarily variable in composition, and mixed with beeswax (Dieterich, 1911; Rivière & Bailhache, 1921; Shaw, 1924). Shaw analysed some sixteen chemical components contained in it. Others beside Huber and Jaubert have emphasized the importance of poplar buds as one natural source. Alfonsus, E. C. (1933) observed red loads collected from poplar, white loads from pine, and yellow and green loads of unknown origin. His observations of the gathering and unloading of propolis agreed with those of Huber, and he added that temperature plays an important part by softening the material so that it is more easily gathered and handled. He occasionally observed bees on which propolis loads remained overnight; next morning they sunned themselves on the alighting board until their loads were resoftened and could be removed.

Bees are attracted by the physical properties of the propolis which they collect. Knight (1807) watched them collect a mixture of beeswax and turpentine which he had used to seal up cuts in trees, and Dujardin (1852) observed bees gathering white lead in oil, which had been put on another hive as a preliminary to painting; both substances were torn off with the mandibles and placed in the pollen baskets. Freudenstein (1932) recorded that one colony gummed up the spaces between a queen excluder and crownboard with blue propolis, derived from oil-paint. McGregor (1952) observed caged colonies which regularly collected 'tanglefoot', but these bees ignored thick white lead paint, asphalt, or resin, and visited poplar buds only occasionally.

The earlier part of this chapter has been concerned with the use of propolis in comb reinforcement. This important use is not so conspicuous as its more familiar function of filling gaps in order to protect the colony from weather and enemies. It is possible that some of the substances which have been found to serve the latter purpose would not be suitable for the former one.

Chapter 28

NECTAR RIPENING; HIVE HUMIDITY

> *The pedigree of honey*
> *Does not concern the bee,*
> *A clover, any time, to him*
> *Is aristocracy.*
>
> EMILY DICKINSON

SEVERAL IMPORTANT PROBLEMS of honeybee economy are concerned with humidity. In summer, nectar ripening involves the elimination of large quantities of water, and in some circumstances the evaporation of water helps the bees to limit the temperature of their colony (p. 226). In winter the problems are of a different kind. Food combustion provides a large quantity of water which must be eliminated. Sugar combustion produces 60% by weight of water, and honey also contains about 20% water; Phillips (1923) calculated that 8·16 lb. of water comes from each 12 lb. of honey consumed (without allowance for any water which might be added before consumption). This quantity of water may seem large, but in fact honey produces less water than most foods. The dryness of this diet and of the atmosphere within the winter cluster is advantageous to the bee during the winter, when she is less frequently able to evacuate surplus water during flight.

Nectar ripening

Nectar-loaded foragers deliver their booty to other workers soon after they arrive in their hive (Doolittle, 1901); this behaviour has been described in detail by Park (1925). The entire load is sometimes given to one bee but it is usually distributed among three or more. As the bees approach each other, the forager opens her mandibles widely and a drop of nectar appears on the upper surface of the base of her tongue (the whole proboscis remains folded back under the head, as in Fig. 15). The receiving bee stretches out her proboscis to full length and sips the proffered nectar with her tongue (Fig. 1). During the transfer the antennae of each bee are in continual motion, and stroking those of the other, while the receiving bee may stroke the head of the forager with her forefeet. The transfer is accomplished in less time than it takes to describe it.

While in the forager's honeysac, nectar does not become more concentrated (Park, 1927, 1932); instead it is slightly diluted by the addition of digestive juices (p. 61). The surplus water is evaporated in the hive, either during manipulation by the mouthparts of the bees or when the nectar is in the cells.

If there is a heavy nectar flow the receiving bee may deposit her load in a cell immediately (often as a drop suspended from its upper surface), but usually she first manipulates the nectar in her mouthparts (Park, 1925). She goes to an uncrowded part of the hive and rests with her head uppermost. She repeatedly partially unfolds and then refolds her proboscis, exposing to the air a droplet of nectar which appears in the angle between the two portions of the proboscis, and which grows in size each time this unfolds. This process lasts 5-10 seconds. The droplet is then sucked back into the mouth and the process is repeated, with only brief pauses, for about 20 minutes. The times are very variable. The bee then crawls into a cell to deposit her load. The mouthparts and drop of nectar are in position as in a disgorging forager, and the drop is added directly to the honey already present or, if the cell is empty, placed on the upper surface of the cell wall. Park (1933) estimated that within an hour of its arrival in the hive the sugar content of incoming nectar was increased from 45% up to 60% by this activity.

Evaporation of surplus water is completed in the honeycomb. Park (1928a), who placed various concentrations of sugar solution within a colony in cells screened from the bees, found that they soon ripened to more than 80% sugar concentration, equivalent to honey. Ripening depended on the degree to which the cells were filled: when they were three-quarters full it took more than twice as long as when they were one-quarter full. In the latter, 60% sugar solution was fully ripened within 48 hours and 20% sugar solution in 72 hours. Honeybees usually at least half fill their cells, and screened combs filled with nectar by bees took 3 or 4 days to ripen (Park, 1933). Further experiments were carried out by Reinhardt (1939), who found that combs in normally ventilated hives took 1 to 5 days to ripen; in one example additional top ventilation (a screen cover) reduced the time from 5 to 3 days, while reduced ventilation increased it to more than 21 days.

Park's work showed that the concentration of nectar could be satisfactorily accomplished by evaporation from the cells alone, but that it might be speeded up by manipulation in the bees' mouthparts. We do not know the relative importance of the two processes, or whether a sample of nectar may be manipulated several times.

Park and Reinhardt both worked in Iowa in summertime (mean temperatures 25–32° C., relative humidities 40–70%). The rate of evaporation will vary with the drying power or saturation deficiency of the air in the cells, and this in turn will depend upon the temperature and humidity of the atmosphere and the temperatures within the hive.

The effect of fanning on nectar ripening

Evaporation from the cells is not only a consequence of diffusion and convection. During a nectar flow fanning bees expedite matters by directing a current of air between the combs, and on summer evenings beekeepers are pleased to hear the hum which accompanies this activity. The effectiveness of this measure was demonstrated by Jessup (1924a) who sealed all joints and cracks in a large hive (15 lb. bees, Dadant hive with 3 supers) and fitted anemometers to the two small circular openings which were allowed as an entrance. On a hot July day (mean temperature 27° C.) 489–807 cubic feet of air entered the hive each hour.[1] Ventilation was more rapid during the day than at night; it was directed by fanning, and the direction changed at irregular intervals without apparent cause. Unpleasantly high humidity might incite the bees to fan; they also fan in response to high temperature, thereby cooling the hive by evaporation (p. 226). Huber (1814) found that several noxious gases stimulated them to fan, and Hazelhoff (1941) found that fanning commenced within one minute of the introduction of a stream of carbon dioxide into the hive.

The cost of nectar ripening

Evaporation from the combs in summer will utilize surplus heat from brood rearing, so small quantities of nectar can probably be ripened at little cost. The energy needed for ripening large quantities of nectar is appreciable; for instance, Ribbands (1950a) calculated, from a comparison of the amounts

[1] To be compared with a flow of 22 cubic feet per hour, due to the temperature gradient, recorded in a hive in the absence of fanning (Hazelhoff, 1941).

of stores produced by colonies fed with large quantities of concentrated and dilute syrup, that the elimination of each 1 lb. surplus water involved the wastage of 4–5 oz. sugar.

Adult bees in different humidities

Woodrow (1935) kept five groups of 200 bees in cages at different humidities. The bees were kept in the dark and supplied with 50% sugar solution and with distilled water. The relative humidities were 25, 51, 57, 73 and 97%, and the mean lives of the bees were 35, 31, 31, 25 and 8 days. Food consumption was 68, 63, 63, 64 and 38 mg. per bee per day. The bees, unable to fly, suffered from their accumulated faeces which were largely water, so Woodrow concluded that the differences between the groups (small except for the group at very high humidity) were not due to differences in metabolism but to differences in their ability to transpire water. The result indicates that adult honeybees can survive in a wide range of humidities.

Measurements of humidity in the hive

Jessup (1924b) measured the absolute humidity of the air immediately above a colony and deduced therefrom the likely relative humidity within the broodchamber, at 35°C. He calculated that the relative humidity there varied widely, from 20 to 80%. The average for a whole day varied from 30 to 65%, maximum readings occurring during daylight and minimum ones between 3 and 6 a.m., while the humidity tended to rise and fall with that of the atmosphere and in relation to the amount of nectar gathered by the bees. Such calculations can only provide upper limits for broodchamber humidities, since the air is likely to have acquired much water vapour during its passage through the supers with their ripening nectar.

Oertel (1949), who also used a bulky hygrothermograph for humidity determinations, concluded that in summer relative humidity ranged from 40 to 62% in the broodchamber and up to 78% in the supers. Anderson (1948) measured relative humidity in the hive of a wintered colony. It was usually above 55% and higher than in the air outside. The air surrounding a well-insulated colony without top ventilation remained at 75% relative humidity for 10 days during which the humidity of the outside air fluctuated between 15 and 80%.

Study of humidity *among the bees* offers difficult technical problems and has only recently been attempted. Büdel (1948) sucked air out of a beehive through an Assmann aspiration psychrometer in order to determine its humidity. This technique was subsequently criticized by Simpson (1950). In summer Büdel concluded that absolute humidity was constant throughout the hive and somewhat higher than that of the outside air. Büdel's experiments were carried out with a colony without top ventilation, in which the incoming and outgoing air would circulate so that it would be sampled fortuitously. Simpson pointed out that there must be a humidity gradient, but that efficient air circulation might make that gradient very small.

Simpson (1950) devised a miniature dew-point apparatus with which he was able to measure humidity within a winter cluster. The mean saturation deficiency (effective drying power) of the air in the broodless cluster was about 8 mm. Hg. It was substantially above that of the outside air, although the absolute humidity within the cluster was about twice that of the outside air. When brood rearing began the cluster temperature rose substantially, but absolute humidity within the cluster did not rise to a corresponding extent; the saturation deficiency increased to about 17 mm. Hg. and the air became drier. Simpson pointed out that, if the bee is unable to control the rate of evaporation of water from her body, the rate of evaporation per bee at any point in a cluster would be approximately proportional to the saturation deficiency of the atmosphere there. This would have a buffering effect on the saturation deficiency, tending to minimize its variations throughout the cluster and to cause the absolute humidity to rise and fall, both in space and time, with cluster temperature. He observed considerable fluctuations in humidity from day to day, but he was not certain whether they were due to real changes or to movements of the cluster relative to his apparatus.

The humidity of the honeybee colony has not received the detailed attention which has been accorded to its temperature relations (pp. 216–30). Whereas the latter are very important and carefully controlled, I do not think that there is any evidence of careful adjustment of humidity. Since the amount of water vapour in a saturated atmosphere is proportional to its temperature, the drying power of incoming air becomes substantial when it is warmed to the high cluster temperature.

This effect will be greater in winter and will facilitate evaporation from the bodies of bees in the winter cluster. There are probably wide limits to the humidity tolerance of adult honeybees in summer, when any excess water can be readily evacuated.

Since the evaporation of water from any surface takes much heat away from it, the maintenance of a constant broodnest temperature is an objective not readily compatible with maintenance of a constant humidity. Both objectives could perhaps be achieved if necessary by careful regulation of the quantity of nectar ripening there, but there is no evidence of the operation of any such mechanism. If humidities in the broodnest vary considerably, one wonders whether there is any mechanism to compensate for changes in the rate of evaporation from the brood.

Beekeepers are apt to complain about dampness in their hives during winter. The bees give off water vapour into the warm unsaturated air within the cluster, and this air becomes saturated when it leaves the cluster; the water which condenses when it cools may deleteriously affect the combs beyond the cluster, or the hive itself, but it does no direct harm to the bees.

Chapter 29

TEMPERATURE REGULATION; WINTERING

The bees
  Sneeze and wheeze
Eating pollen and honey
  From the lime trees:

But the flies
  Is wise
When the cold weather comes
  They dies.

After L. W. G.

TEMPERATURE AND HUMIDITY regulation are important attributes of the honeybee colony, and so greatly increase its independence of its environment that it has a wider range of climatic adaptability than any other insect. Wherever sufficient food is available in summer, the honeybee can adapt itself and survive, in temperatures ranging from -50 to $+50°$ C. and in extremes of humidity. The metabolic and temperature relations of individual honeybees are similar to those of other insects; regulation of the physical conditions of the colony is brought about by behaviour which is a consequence of social life.

Heat production of bees and brood

The heat production of adult honeybees can be calculated from determinations of either the amount of oxygen used or the quantity of carbon dioxide produced by them; as sugars preponderate in their diet, these two quantities are nearly the same and the respiratory quotient closely approaches unity (Jongbloed & Wiersma, 1934).

Exact measurements of individual metabolism are difficult to carry out and interpret, but those of Kosmin, Alpatov & Resnitchensko (1932) and those of Jongbloed & Wiersma are of the same order (Table 11). Differences in the metabolism of bees carrying out different tasks within the hive have not been investigated. Invert-sugar consumption during flight, based on these results, would range from 9 to 40 mg. per hour, and heat production from 0·036 to 0·158 calories per gm. per hour. The lower of these values is consistent with the results of Ruth Beutler (1936), who measured the flying capacity of bees fed with known amounts of sugar, and calculated their fuel consumption during flight to be 10·4 mg. per hour.

These values are not dissimilar from those which have been found in other insects (Wigglesworth, 1950), and there is no evidence to suggest that the heat production of the honeybee is abnormally high, either in flight or at rest.

TABLE 11. Oxygen uptake of individual bees (c.mm. per hour). Results from Kosmin et al. (1932). Starred results are from Jongbloed & Wiersma (1934)

Temperature °C.	Still	Moving slowly	Moving and cleaning wings still	Flying
35		300	1,500	27,600
21	174*			8,760*
18	54	480	2,160	31,200
11	24	1,200		26,400

Basal metabolism in animals is related to size. A general law cannot be established because each species has its own characteristic basal rate of metabolism, but small animals usually need a higher basal metabolism per unit of body weight than large ones, because they have a relatively greater body surface from which heat is lost. The temperature of the honeybee community can be raised above that of its surroundings because clustering greatly reduces the ratio of exposed body surface to body weight —the cluster is equivalent to a large animal, but its components have the metabolism of small ones.

In breeding colonies the metabolism of the larvae and pupae is not negligible. This matter was investigated by Melampy & Willis (1939). Oxygen consumption per gram of insect tissue fell off rapidly with age (2,900 c. mm./hr. in 2-3 day larvae, 1,063 in 4-5 day larvae, 374 in 7-8 day larvae) but consumption per insect was fairly constant; in larvae more than 3 days old it ranged between 50 and 100 c. mm./hr. and in pupae it gradually rose from 50 c. mm. to more than 100 c. mm.

This rate of metabolism indicates that the heat production of brood is somewhat similar to that of an equal number of resting adult bees, and the metabolism of the brood makes a useful contribution toward maintaining the high temperature of the brood nest. I do not think that the contrary opinion of Himmer (1927a) was a valid deduction from the evidence which he provided.

Temperature of individual bees

Small thermometers had to be used by early investigators of this subject, one of whom recorded that honeybees had a thorax temperature of 35°C. (Ciesielski, 1895); this result and measurements of broodnest temperatures misled him into regarding honeybees as warm-blooded. Thermocouples have facilitated modern approaches to the problem, but even these methods have disadvantages (Krogh, 1948).

Pirsch (1923) made an attempt to measure the body temperature of isolated resting individuals with a thermocouple, and he concluded that they could regulate their temperature within certain limits. A more detailed study was undertaken by Himmer (1925), who found that the body temperature of resting honeybees approached the temperature of their environment, but during muscular activity they generated so much heat that their temperature rose considerably above that of their surroundings. In such circumstances the temperature of the thorax, site of the activity, was higher than that of the abdomen. Similar temperature excesses are found in other insects during activity (Girard, see Wigglesworth, 1950).

Temperature regulation in the broodless winter cluster

The technical difficulties of investigating cluster phenomena should be borne in mind. The presence of combs and the density of the cluster ensure that activities and responses within the cluster are not visible. Early observations of temperatures, which were reviewed by Bachmetjev (1899, 1907) and Armbruster (1923a), involved the use of mercury thermometers. Subsequent investigators used thermocouples, until Himmer (1926) devised a self-registering thermoelectrical method which enabled him to obtain continuous records of temperatures in six different parts of the cluster without any disturbance of the bees.

Observations by Phillips & Demuth (1914) led them to conclude that wintering bees did not form a compact cluster until the hive temperature dropped to 14°C. Below this temperature the cluster had a compact outer shell of quiet bees. The shell might be several layers thick and the bees composing it arranged themselves with their heads inward. The bees inside the shell were more active; in a cluster in an observation hive they walked about, shook their abdomens, or breathed vigorously; their activities generated heat, the outer shell of bees acted as an insulator and the temperature in the centre of the cluster rose. Measurements of individual metabolism demonstrate that most of the heat production and food consumption was carried on by the bees at the centre.

A similar account was written by Wilson & Milum (1927), who noted that the cluster began to form when the hive temperature dropped below 18° C.; it had definitely formed at 13° C. but small groups of bees remained apart till it reached −5 to 0° C. Himmer (1926) considered that the critical temperature at which cluster formation began could not be exactly determined. Wilson & Milum (1927) observed that very active feeding began when clustering commenced (as if the bees were filling their honeysacs with stores) and that bees in the centre of the cluster sometimes moved about without apparent purpose, or stood still, waggling their abdomens or vibrating their wings.

Gates (1914) thought that he had observed a continual interchange between bees on the inside and outside of the cluster, but Himmer (1926) said that there was no interchange in an undisturbed compact cluster, and this conclusion is substantiated by analyses of the glands of clustered bees (p. 270).

The position of a cluster shifts only slightly during the winter (Himmer, 1926; Wilson & Milum, 1927), but Gates (1914) and Himmer (1926) recorded that it expands in warm weather and contracts with cold. Wilson and Milum emphasized these changes (Fig. 52), but Corkins (1930) was the first investigator to point out their significance. The cluster loses heat (by conduction, radiation or convection), and Corkins pointed out that losses by conduction would be negligible, and the rate of loss by radiation or convection (both of similar magnitude; Wedmore, 1947) would be proportional to the surface area of the cluster. Corkins calculated that on one occasion when the outside temperature decreased by 22·5° C. the heat loss in a still cluster would have increased 150%, but the cluster contracted so that the actual heat loss increased by only 18 ± 3·5%.

Movements of the cluster make it difficult to obtain accurate estimates of the temperature at its periphery. This difficulty is lessened in a cellar-wintered colony, in which Himmer (1926) arranged thermometers so that they touched the outer layer of bees. The surface temperature at the top and rear of the cluster ranged from 10 to 12°C., and was usually between 10 and 11°C. The temperature of the undersurface of the cluster was slightly lower, minimum 9°C. Bees chill at 8°C. and would then fall away from the cluster and perish. Himmer noted that many sudden temperature rises commenced when the surface temperature cooled below 10°C., and that the bees on the surface were

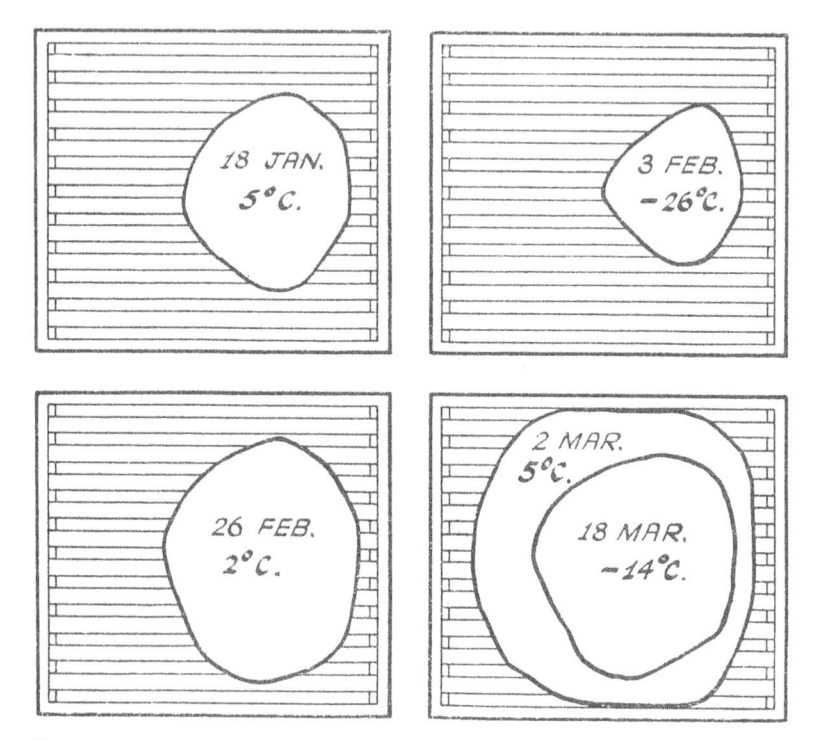

Fig 52 THE RELATION BETWEEN AIR TEMPERATURE AND SIZE OF WINTER CLUSTER (Wilson & Milum, 1927).

also easily irritated by other influences (light, vibration). Therefore, he suggested, the outermost bees perceive and respond to temperature stimuli while the bees within, which do not receive this stimulation, respond to mechanical impulses from the disturbed surface bees. Temperature changes would be perceived through the terminal segments of the antennae (p. 45).

The temperature in the centre of broodless clusters varies within quite wide limits. Gates (1914) found that it was never less than 17°C. and seldom dropped below 20°C.; its usual range was 20–30°C., and it underwent daily oscillations which were usually less than 5° C. Himmer (1926) and Hess (1926) recorded similar observaions; Himmer found that the centre temperature of strong colonies rose higher and varied more, but this result was based on observation of only one pair of colonies.

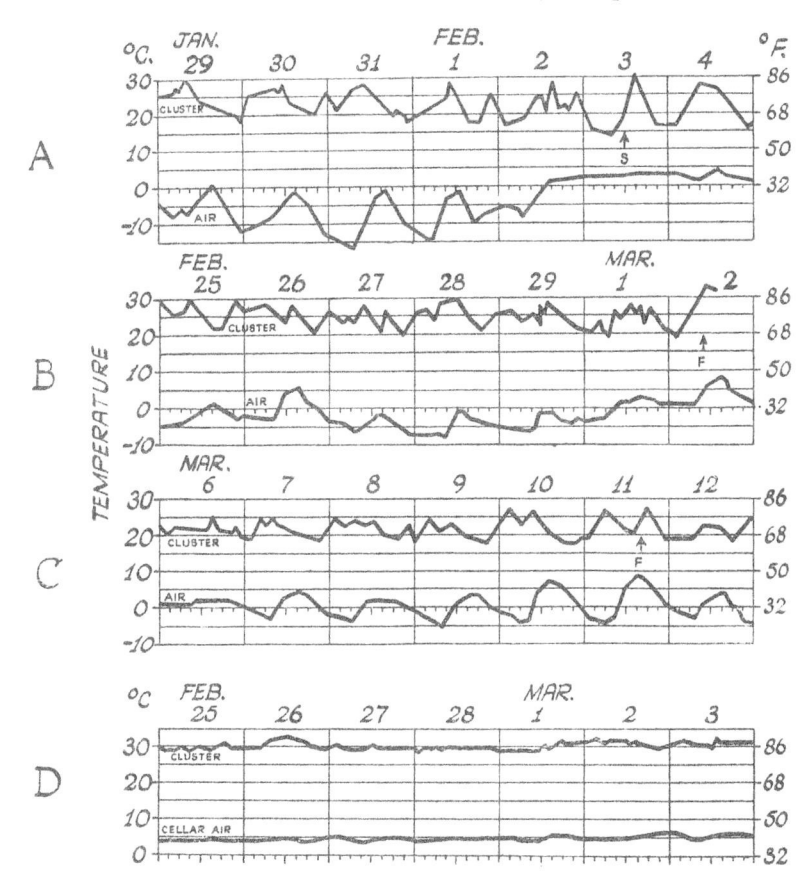

Fig 53 PORTIONS OF A CONTINUOUS RECORD OF AIR AND CLUSTER TEMPERATURES DURING WINTER. A.-C. Records from a normal colony. D. Records from a colony housed in a cellar. Daily fluctuations show an inverse relation, so the cluster temperature is more constant in the cellar. S, disturbance to colony; F, exercise flight (Himmer, 1926).

On one occasion a difference of 59° C. between the temperatures of the environment and the centre of the cluster (−28° C. and 31° C.) was recorded.

The temperature of the cluster centre fell during the early part of the winter but then rose again (Gates, 1914). Himmer (1926) and Wilson & Milum (1927) observed a tendency for the temperature to rise as confinement lengthened, but found that it was lowered again after a good cleansing flight. This fits with the conclusion of Phillips & Demuth (1914) that the accumulation of faeces acts as an irritant and causes a rise in temperature and in the consumption of stores. The intestinal parasite *Nosema apis* has a similar but greater effect (Morgenthaler, 1927).

Any slight disturbance of the winter cluster causes an immediate rise in its temperature (Fig. 53), and this rise may continue until the next day (Gates, 1914); removal of the hive roof may alone produce a rise of 2–3° C. (Wilson & Milum 1927).

Daily oscillations in the temperature of the environment have been found to be inversely related to temperature oscillations at the centre of the broodless cluster (Gates, 1914; Himmer, 1926, Fig. 53A). Gates noted that irregularities in both oscillations were so synchronized that the inverse relation could not have been due to a long time lag in the response of the cluster. The inverse relation usually has a very short time lag. The relation is not always obvious; sometimes Himmer's cluster temperatures showed other oscillations, which were not explained (Fig. 53B). In a cellar-wintered colony, with an almost constant environment temperature, cluster temperature oscillations were markedly reduced (Fig. 53D).

An abundance of stores is a valuable buffer against temperature oscillation in normal colonies. Anderson (1948) measured temperature fluctuations in various kinds of hives, all without bees and heated by a 15-watt lamp; he showed that the addition of 5 combs with 26 lb. honey exercised a considerable damping effect upon daily fluctuations in temperature within the hive (Fig. 54).

Although Gates stressed the inverse relation between *daily* oscillations of temperature in the environment and in the cluster centre, he did not state that this relation held for long-period fluctuations in the outside temperature; his data covered eleven months of observation and were not published in full, but footnotes to his Table IV show that the maximum temperatures of the cluster centre in January and February occurred in frozen conditions, but those in December and March occurred on very warm days. However, Phillips & Demuth (1914), whose published observations covered only three days of experimentation, stressed that cluster-centre and external temperatures were inversely correlated; Wilson & Milum (1927) cautiously agreed that this happens 'up to a variable point'.

Corkins (1930) came to an opposite conclusion. His experiments were conducted on a large scale and in very favourable conditions, with very low humidity and a winter temperature range (−14 to −40°C.) which precluded flying. They extended through four winters, using 3–6 full-strength colonies each fitted with 21 thermocouples which were read once each

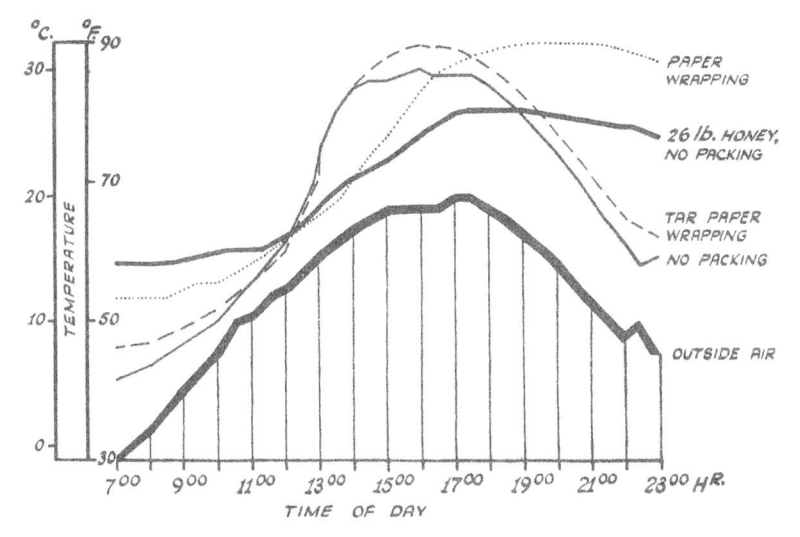

Fig 54 THE DAMPING EFFECT OF HONEY STORES ON TEMPERATURE FLUCTUATIONS. Records from empty hives heated with a 15-watt lamp (Anderson, 1948).

day. Over the whole period there was a slight but not significant positive correlation (0·053±0·026) between temperatures in the cluster centre and in the environment. To confirm these results, Corkins (1932a) carried out experiments with colonies placed in an ice-cream cabinet, in which temperatures were controlled. The bees were exposed to various constant temperatures for long periods, and hourly records were obtained from 42 thermocouples. There was a slight positive correlation between the temperature of the environment and that of the centre of the cluster. Data from one of these experiments are presented in Table 12.

One would expect that any long-term inverse relation which existed would become obvious in any experiments conducted over a wide temperature range, but Corkins's conclusion is supported by the fact that this does not happen (e.g. Himmer, 1926, Fig. 53A–C). The pronounced short-term inverse relation may have caused the long-term one to be assumed by some writers. The amplitude of the daily oscillations of cluster temperature is often more than half that of the oscillations of the external temperature (Fig. 53), and it is obvious that if this was also a long-term relation the cluster temperature would reach an intolerable level when the surroundings became very cold.

TABLE 12. Relations between the maximum temperature in the cluster and the temperature of the environment, in a broodless colony at constant temperature (Corkins, 1932a)

Chronological order of periods	Length of period in hours	Mean cabinet temperature °C.	Mean hive temperature °C.	Mean maximum temperature of cluster °C.
7	48	14·8	18·8	30·0
4	72	14·2	19·6	29·5
1	120	7·5	9·3	29·8
5	48	7·2	9·7	30·3
2	120	−0·2	0·7	29·7
6	48	0·0	0·1	29·8
3	72	−6·5	−5·0	29·2
8	48	−6·2	−4·1	29·2

Armbruster (1923a, b) interpreted some data collected by Lammert in 1895 as evidence for a rhythm of temperature fluctuation in the cluster; he supposed that the cluster became active and generated a large amount of heat whenever it reached a certain critically low temperature, and that this event occurred every 22 hours. Lammert had died and had not left any explanatory notes, so the reason for his 22-hour periodicity is not known; Himmer (1926, Fig. 53) was one of several whose results have demonstrated the untenability of Armbruster's explanation.

Hess's (1926) results led him to conclude that at low air temperatures there was a narrow focus of warmth towards the base of a broodless cluster and that the temperature fell off in all directions from this focus, but particularly in an *upward* direction. He suggested that his results were due to a downwardly directed air flow produced by fanning bees; Bruman (1928) demonstrated such a flow in late March; both experimenters used a padded broodchamber which was only ventilated from the entrance at the bottom.

Temperature regulation in the broodnest

In winter a temporary rise to usual brood-rearing level in the temperature of the centre of the cluster does not necessarily induce the process (Himmer, 1926), yet brood rearing may occur in the winter cluster at lower temperatures: Milum (1930) recorded brood development in a comb at a mean temperature of 31·5° C., and noted that in April 1928 part of the broodnest of one colony fell to 24·6° C. Simpson (1950) noted that the mean temperature at a point within the broodnest during the first four weeks after its commencement on 10th February was 31° C.[1]

Gates (1914) observed that the temperature at the centre of the broodnest remained remarkably constant, and Hess (1926) found that its daily range of fluctuation was about 0·2–0·4° C. Himmer (1927b) obtained a mean daily range of 0·55° C. in a period of four weeks and considered (1927a) that optimum brood development occurred between 34·5 and 35° C. and that the vital limits for brood were between 32 and 36° C.

Dunham (1929a) observed that daily fluctuations in outside temperatures had little if any effect upon the temperature in the centre of the broodnest. Subsequently (1931a) he placed a colony in a controlled-temperature chamber and noted that the mean daily range of temperature in the centre of the broodnest was 34·1–34·5° C. during 5 days at chamber temperatures of 25–27° C.; it dropped only to 33·8–34·1° C. during 7 days at chamber temperatures of 11–19° C. In a second experiment (1933) the broodnest temperature ranged from 34·3–34·6° C. during 3 days in a room at 25–27° C. and increased to 34·6–35·2° C. during 9 days at chamber temperatures of 32–34° C.

Towards the periphery of the broodnest the temperature is lower and less constant than at its centre. On one day with a mean outdoor temperature of 28° C. Dunham (1933) observed mean temperatures of 34·1° C., range 0·8° C., in the centre of the broodnest, of 31·0° C., range 3·5° C., in the outer area of the broodnest, and 30·8° C., range 5·5° C., in the broodless area of the cluster. In the colony exposed to a low temperature (1931a), the temperature of the outer brood area gradually dropped; the decline was associated with absence of egg laying and consequent reduction of brood rearing. Egg laying was also restricted

---

[1] Burtov (1950) found that the eggs of Italian bees withstood temperatures of either 30–31° C. or 37–38° C. for 48 hours; the eggs of Russian bees did not hatch after these treatments, a surprising conclusion which requires confirmation.

by exposure of the colony to high temperatures (Dunham, 1929b).

There is no evidence to suggest that honeybees produce heat specially for the purpose of maintaining the temperature of the broodnest. It seems that the heat is a by-product of the ordinary activities and metabolism of the bees and brood of the colony; in cool conditions the cluster expands and contracts to retain the required amount of heat, and so maintains the required temperature, but in warm weather too much heat is produced and regulation becomes a matter of active ventilation.

Evaporation of water is an extreme resort, but it is likely that in normal circumstances water losses from bees, brood and ripening nectar play a part in temperature regulation. The placing and adjustment of nectar stores might be important, and stores of sealed honey serve as a heat reservoir and buffer.

Honeybees maintain their broodnest at a higher and much more constant temperature than that achieved by social wasps (Himmer, 1927b, 1932), and one may suggest that their larger populations and their supplies of nectar and honey contribute to their greater efficiency.

Ventilation of the colony at high temperatures

Huber (1814) found that bees fanned in portions of an observation hive which were exposed to the sun. Hess (1926) heated a hive to 40°C., but the temperature in the broodnest did not rise above 36°C.

At excessively high temperatures bees bring in water, deposit it in the combs, and fan vigorously so that the broodnest temperature is reduced by the cooling effect of evaporation (Chadwick, 1922). They may even deposit water droplets on foundation which is being worked. Dunham (1931b) observed that when the coolest portion of the broodnest reached 34°C. the bees supplemented these activities by leaving the combs and clustering outside the hive.

During a heavy flow the evaporation of nectar helps to cool the hive, and it is only exceptionally that bees need to bring in water for this purpose. When they do so they deposit it at the tops of the combs. On evaporation it takes up heat from its surroundings, and the cooled air flows downwards through the broodnest (Chadwick, 1931). Chadwick emphasized that during very hot spells the bees bank up at the entrance and so reduce its size and control the intake of hot air; if too much hot air were admitted the bees would succumb. Chadwick described a freak occurence which illustrated the importance of water gathering for temperature regulation. In California, midday air temperature rose to 48°C. one day in June 1916, with little loss from melting combs. It dropped to 29·5°C. at 9 p.m., but at midnight a hot breeze from the desert raised the air temperature to 38°C. The supply of water became exhausted, no more could be collected until daylight, and many combs were melted.

Hazelhoff (1941) produced fanning activity by heating a hive electrically; when a cool air current (c. 15 cubic feet per minute) was sucked through the heated hive all of the 18 bees fanning at the entrance ceased this activity within 50 seconds.

Colony shape

Wadey (1948) stated that he had examined more than 50 wild colonies, situated in all sorts of places. He had noted that wherever the bees were not restricted by lack of space the lateral development of the broodnest had never exceeded 5 combs, and often not more than 3. In one example there were only 3 combs, and the centre one had a patch of brood 4½ feet long by 1½ feet wide. He therefore doubted whether low wide hives formed a natural type of home for bees. Warmth should be somewhat more readily retained and regulated by colonies living on a few long combs, but the extent of this possible advantage has not been determined.

Food consumption of colonies in winter

Modern beekeeping conventions in Britain and America require the over-wintering of ever larger colonies, with ever growing food requirements. Whereas 25 lb. stores were at one time considered an adequate winter supply (Root, 1901), a recent recommendation suggests that 90 lb. is necessary (Farrar, 1952). It would be well for the beekeeper to consider whether all of this expenditure is necessary and ultimately beneficial.

Kleist (1919) concluded from his data that there was no relation between outside winter temperature and winter food consumption, and that the notion that bees eat more in cold weather was erroneous. Örösi-Pal rearranged these results into a curve which showed minimum food consumption at 5°C., with little difference in the range −18 to 8°C. (Annie Betts, 1943). Kleist found that sudden warmth was apt to cause disturbance, with increased store consumption.

Himmer (1926) found that one colony which was kept in the open lost 9·35 lb. between 4 December and 1 April, while a similar colony in a dark cellar at 3–6° C. lost only 4·4 lb. (this included weight loss on its subsequent exercise flight). He found that a lightly insulated colony lost 6·6 lb. between 20 November and 20 February, while a similar well-insulated one lost only 3·75 lb. From these results he concluded that temperature *variations* cause increased food consumption. Wilson & Milum (1927) also found that one cellared and two well-insulated colonies used less stores (12–17 lb.) than two colonies unpacked and left in the open (20–26 lb.),[1] but Corkins (1930) stated that packed colonies usually lost more weight than unpacked ones. Braun & Geiger (1953) reported that in Manitoba 464 cellar-wintered colonies ate more stores per colony than 338 wintered in the open, but they contained fewer bees in the following spring. Packing reduces temperature fluctuations, but it may sometimes increase the mean temperature within the hive, so its effect upon store consumption may well depend upon circumstances. Large reserves of stores act as a temperature buffer, without raising the mean temperature (p. 222).

Corkins (1930) maintained that daily weight losses were greater in warm conditions, and found that at temperatures above −2° C. they were 84% greater than at temperatures below −10° C.

A laboratory technique for the study of store consumption was then introduced by Corkins & Gilbert (1932), who placed a colony in an ice-cream cabinet and measured its output of carbon dioxide during 10-day periods at different temperatures. They found that carbon dioxide output at a hive temperature of 4° C. was 52–62% of the output at 16° C. Then the hive was kept at 8–10° C. for 14 days; the output fell steadily during the first week and then reached a fairly constant level, so Corkins & Gilbert concluded that the higher metabolism during the settling process was probably typical of that during changes in the cluster under natural conditions.

The results quoted indicate that warmth and temperature variation both adversely affect winter store consumption; the latter factor appears to be the more important one. If this interpretation is correct, one may query whether some

---

[1] Walker (1945) found that bees in insulated hives took fewer cleansing flights during winter than those in unpacked single-walled ones.

recommendations for wintering practice have been pushed to undesirable extremes. Restricted ventilation is certainly conducive to dampness, mouldy combs and fermented stores, which have obvious disadvantages; on the other hand, excessive ventilation would produce much wider temperature fluctuations, with increased cluster temperatures and store consumption. Recommendations may need adjustment to different local conditions (winter length, temperature changes, humidity, etc.).

TABLE 13. Relation between winter food consumption and yield during the following season (Farrar, 1952)

Year	Consumption	Number of colonies	Average consumption (lb.) October–April	Average yield (lb.) October–October
1947	Below mean	23	43·0	159·1
	Above mean	28	57·4	208·5
	Difference		14·4	+49·4
1949	Below mean	29	56·7	147·5
	Above mean	28	76·3	150·0
	Difference		19·6	+2·5
1949 (another apiary)	Below mean	83	45·1	85·5
	Above mean	80	56·3	83·5
	Difference		11·2	−2·0
1950	Below mean	83	43·9	172·6
	Above mean	63	60·6	172·2
	Difference		16·7	−0·4

A small part of the winter food supplies is consumed in order to maintain life and temperature, but a much larger part is consumed in winter and early spring brood rearing. Here again modern conventions are opposite to those of an earlier generation. Our elders assumed that winter brood rearing was unwelcome (Phillips, 1915), but in the United States it is now regarded as advantageous (Farrar, 1952). The latter contention seems to derive in part from the opinion that in springtime newly emerged bees will be of much greater value than old and senile ones, but a little evidence even suggests the contrary (Böttcher, 1950), and in addition one might suppose that high winter temperatures would encourage *Nosema* and acarine diseases.

Honeybee populations build up so rapidly in spring that winter brood rearing may not increase the number of foragers present during the nectar flow. Data published by Farrar (1952) seem to confirm this view (although Farrar himself drew the opposite conclusion). Farrar compared the average yields of colonies which had consumed more or less than the average quantity of stores during the preceding winter. His results (Table 13) showed that in the first year the greater winter food consumption was associated with increased honey yield during the following season, but in 2 subsequent years it was of no advantage. In experimental data of this kind it is impossible to determine to what extent the increased store consumption was related to the initial size of the colony and to what extent it was attributable to increased brood rearing.

Data collected in a ten-year period by Braun & Geiger (1953) indicated that increased winter food consumption did not significantly increase subsequent honey yield. They divided their colonies into four groups on the basis of their winter losses, thus:

181 colonies, mean winter loss 27 lb., mean honey yield 184 lb.,
241 colonies, mean winter loss 38 lb., mean honey yield 187 lb.,
231 colonies, mean winter loss 47 lb., mean honey yield 200 lb.,
149 colonies, mean winter loss 61 lb., mean honey yield 197 lb.

These results should not be taken to imply that large colonies were of no advantage. On the contrary, the mean summer honey yield from 288 large colonies (covering 55 combs or more in autumn) was 280 lb., but that of 124 smaller ones (bees on 30 combs) was only 126 lb.; the respective mean winter losses were 43 lb. and 37 lb. One set of Braun & Geiger's data indicated that the larger colonies were more valuable in one year, but not in the next one.

The optimum size of colony for over-wintering should depend upon type of spring and time of nectar flow; even in the same district it may therefore vary from year to year, unpredictably.

Chapter 30
BROOD REARING

> *Vast numbers throng'd the fruitful hive;*
> *Yet those vast numbers made 'em thrive;*
> *Millions endeavouring to supply*
> *Each other's lust and vanity.*
> BERNARD MANDEVILLE, *The Fable of the Bees*

THIS CHAPTER WILL be primarily concerned with the extent and method of rearing worker brood and the effect of brood rearing upon other activities; the diet supplied to the brood is considered in the following chapter (p. 245); the feeding is described on pp. 191–2.

The amount of time required for rearing larvae

The time spent on cell preparation *before* egg laying was recorded by Perepelova (1928b), who organized a continuous watch on thirty cells from the time of emergence of their previous occupants; each cell was prepared by from 16 to 29 different bees, in an average time of $40\frac{3}{4}$ minutes (removing remains of capping, 13 min., smoothing the cell edge, $6\frac{1}{2}$ min., cleaning the walls, 8 min., cleaning the bottom, $13\frac{1}{4}$ min.).

Lineburg (1924b) reported that nurse bees made more than 10,000 visits to each larva during its development, but this high value was probably a consequence of exceptional conditions within his observation hive.

A record of the time spent in rearing one particular larva was made by Lindauer (1952). He found that 59 bee visits, occupying 54 minutes 26 seconds, were concerned with cell-cleaning *after* egg laying. There were 1,926 bee visits of larval inspection, occupying 72 minutes 1 second, but there were only 143 feeding visits, which took 109 minutes 37 seconds. Most of this time was spent during later larval life (Fig. 55). 657 bee visits were concerned with capping the cell or preparing it for capping; this occupied 6 hours 20 minutes 4 seconds. Thus the whole process required 2,785 bee visits, occupying 10 hours 16 minutes 8 seconds. This calculation takes no account of the time spent in food preparation.

The duration of immaturity

The constant development time of the eggs and the larvae reflects the even temperature at which the broodnest is usually maintained (p. 225), but in cold weather, when the temperature of the broodnest is lower than the optimum, the time for development is extended.

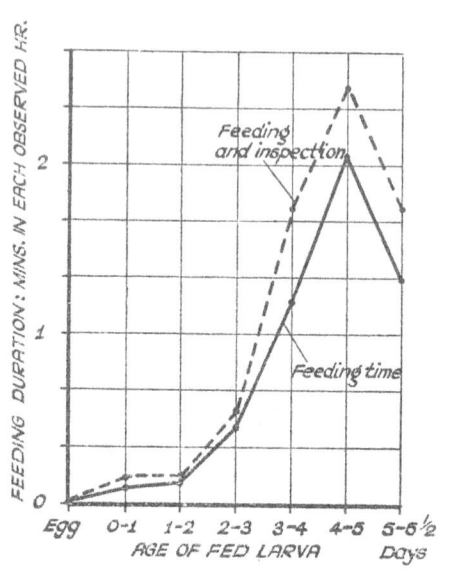

Fig 55 TIME SPENT IN FEEDING LARVAE OF DIFFERENT AGES (Lindauer, 1952).

This effect was studied by Milum (1930), who confined queens on empty combs for 6 or 24 hours and then transferred the laid-in combs to different parts of 5 broodnests, the temperatures in which were carefully measured. The time of capping of the brood varied from less than 8 to more than 11 days, but most of the larvae were capped between the eighth and ninth day after laying. Some workers emerged in less than $19\frac{7}{8}$ days, while others required more than 24 days for complete development from egg to adult. The longer times were for a comb at a mean temperature of only 31°C. Within the temperature range 33·5–34·5°C., 93·5% of the bees had completed their development in less than 21 days (in a mean time of $20\frac{1}{2}$ days).

The effect of food supply on development

Food supplies may also affect the rate of development. Nelson & Sturtevant (1924) showed that more nectar was fed to worker larvae during a nectar flow; in consequence they developed more quickly and might be capped 18–24 hours sooner, and the resulting pupae might weigh up to 8% more than those produced in less favourable circumstances. This nectar flow may or may not have been associated with an increase in pollen supplies.

Levin & Haydak (1951) weighed samples of newly emerged bees from 4 colonies at 7–10-day intervals throughout a season. Their results were grouped and divided into 6 periods; the mean weights of the whole bees in successive periods were 17·6, 16·9, 18·1, 17·5, 18·8 and 17·0 mg. There were similar variations when the heads, thoraces, abdomens or intestines were separately weighed. They found that 'these peaks in weight corresponded fairly closely to the fluctuation in the amount of pollen brought in by a colony with a pollen trap, in the same apiary, during the periods in which it was calculated the larvae from which these bees developed were being fed'. During that season there was a very light nectar flow which showed no correlation with these changes in weight.

De Groot (1950) recorded that honeybees which emerged in September and October had a mean dry weight and nitrogen content of 17·2 mg. and 2·06 mg. respectively, compared with averages for newly emerged bees in summer of only 15·2 mg. and 1·78 mg. These autumn increases (which were statistically significant) may possibly be associated with the small number of bees reared then, and with the abundance of bee milk then available from the well-developed glands of the winter bees. Kuwabara (1947) found that worker larvae were fed more frequently in queenless than in queenright colonies; in consequence they were capped more quickly (111 hr. instead of 130 hr.), and the pupae had three times as many ovarioles.

The amount of food required by larvae

Pollen which had been trapped during summer months was moistened with honey and supplied to 4 colonies (on fresh combs) during the following winter by Alfonsus, A. (1933), who carefully weighed the quantities consumed and counted the amounts of brood reared. Alfonsus found that colonies of different size which were fed the same amount of pollen, reared about the same number of bees, and that the mean consumption per bee reared was 145 mg. (=4·67 mg. N_2 or 29·2 mg. protein). Rosov (1944) compared the food consumption of a colony which reared brood with that of a colony from which all eggs were removed every third day, and calculated that 125 mg. pollen and 142 mg. honey were consumed per bee reared.

Haydak (1935) transferred a confined colony to pollen-free combs and fed it only on sugar syrup. Brood rearing soon ceased, but in the interim the amount of nitrogen lost by adult bees was 3·21 mg. per bee reared; each emerging bee contained 1·73 mg., so 1·48 mg. had been destroyed. These lower values agree with the results of Alfonsus and Rosov, since Haydak had commenced with a material which was already processed and incorporated in the tissues of the bees. Haydak (1949) calculated that 4·36 mg. N_2 were used for each larva reared on a pollen substitute made from soyabean flour + 20% yeast.

There are wide variations in the food value of different pollens (p. 251), the weight of pollen loads (p. 128), and the number of bees reared during one season (p. 235), so calculations of the amount of pollen required for each bee reared or for each year can only be very approximate. Using Alfonsus's result as a yardstick it would seem that about 10 journeys might produce sufficient pollen to rear 1 bee, and that a colony might require about 64 lb. pollen in a season in order to rear 200,000 bees (p. 235). Young bees also require pollen. Records of pollen trapped at the hive entrance have their disadvantages,[1] some of which were pointed out by Eckert (1942) when he published a composite record indicating that a colony could collect 122 lb. pollen in one season; Hirschfelder (1951) estimated that his colonies with pollen traps collected 34–63 lb. pollen in a season; these results indicate that the foregoing calculation is reasonable.

SEASONAL FLUCTUATIONS IN THE RATE OF BROOD REARING

The maximum rate

Berlepsch (1860) counted 38,619 cells of brood in one colony at one time; this is equivalent to a mean production rate of 1,855 per day. Berlepsch also confined the queen to an empty comb for 24 hours, and found that she laid 3,021 eggs during that time. The highest mean which Dufour (1901) obtained for a 21-day period was 1,627 per day; Brünnich (1922–23) counted colonies of Swiss blacks and found that their maximum rate ranged from 970 to 1,750 per day; Merrill (1924a) found that the maximum rate for Italians averaged 1,720 per day and

[1] See also p. 126.

that the highest maximum was 2,030. Counts of *sealed* brood by Nolan (1925), for 48 colonies of Italian bees, showed that their maximum daily means over a 21-day period ranged from 757 to 1,587.

However, Wilson & Milum (1927) reported that one colony in Wisconsin contained 42,120 cells of brood at one time, which represents a mean of 2,005 for 21 days; they added that as this colony had contained only 18,819 cells of brood 19 days previously the rate of egg laying towards the end of the period had probably been at least 3,000 per day.

The lower counts of Nolan were counts of sealed brood, an index of brood actually reared, while the other counts included eggs and young larvae; Nolan pointed out that eggs are often eaten by worker bees. Merrill (1924*b*) published data to illustrate the discrepancy between totals of eggs and sealed brood; sealed-brood totals were not more than 83% of the equivalent unsealed-brood totals, and in unfavourable weather they were sometimes less than half. Myser (1952) observed the loss of 103 out of 117 newly laid eggs from one piece of comb in 12 hours.

The yearly total

From brood counts at frequent intervals Brünnich (1922–23) estimated that 90,000 to 150,000 bees were reared each year by his colonies; Merrill (1924*a*) recorded an average of 177,000 bees reared from March to October inclusive; Nolan (1925, his Table 17) found that 195,000 bees were reared by a good colony in one season. Bodenheimer & Nerya (1937) estimated that one Palestinian colony had produced 220,000 eggs in twelve months.

FACTORS WHICH DETERMINE THE AMOUNT OF BROOD REARED

The relation between brood rearing and size of colony

There must be a relation between the amount of brood reared and the size of the colony. From population studies in other fields one would suppose that the relation might be expressed in sigmoid curves, but the limits of these curves have not yet been explored. In small colonies, or in large colonies early in the season, the rate of brood rearing might be directly proportional to colony size, but at the height of the season one would expect that the rate in large colonies would not be greatly affected by colony size.

The influence of the queen on the rate of brood rearing

In strong colonies at the height of the season the egg-laying capacity of the queen helps to determine the rate of brood rearing. The results of Nolan (1925) related to 16 colonies, half of which were headed by queens of poor quality; in early spring

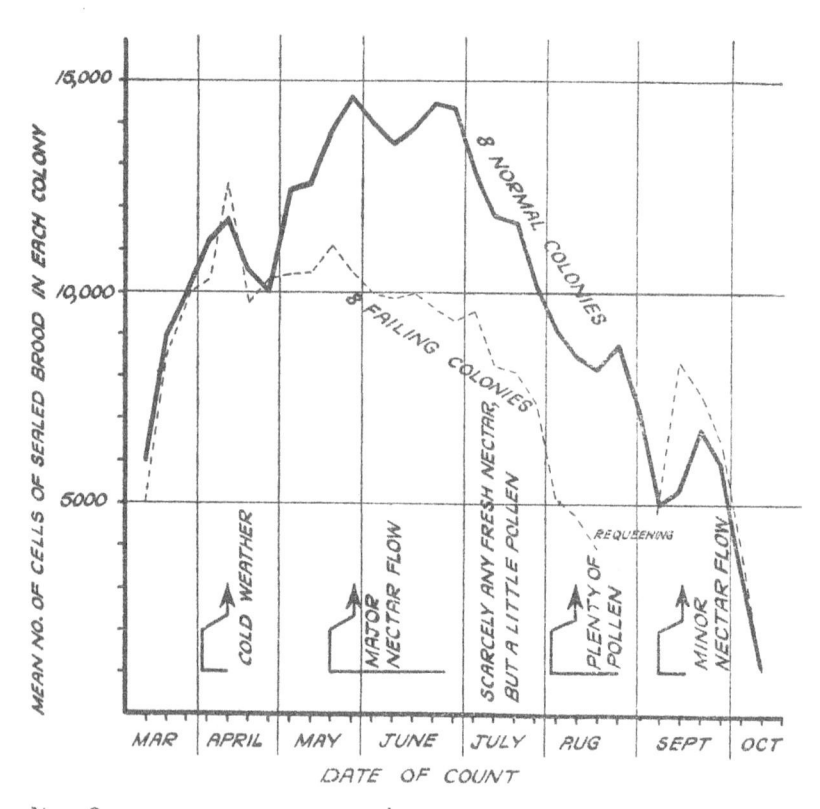

Fig 56 COUNTS OF SEALED BROOD IN TWO GROUPS OF COLONIES, MADE AT WEEKLY INTERVALS DURING A SEASON. To show the difference between colonies with good and bad queens, and the effects of major changes in food supplies (data obtained from Nolan, 1925).

both halves built up at the same rate, but later on there was a marked difference in the average quantity of brood reared (Fig. 56).

Nolan's results suggested that one-year-old queens were more prolific than two-year-old ones, but Brünnich (1922) reported that over 13 years in Switzerland he had obtained the largest honey crops from colonies headed by two-year-old queens, and that the maximum rate of brood rearing was attained in such colonies. In Manitoba Braun (1942) found that in five successive years 189 colonies with one-year-old queens produced more honey (mean yield 189 lb.) than 168 colonies with two-year-old ones (mean yield 171 lb.).

The influence of the queen on brood rearing has been challenged by Merrill (1925a), but an unpublished statistical analysis, carried out by my colleague Mr. M. J. R. Healy, shows that the size of Merrill's experiment was inadequate for the support of any firm conclusions.

The influence of temperature on the rate of brood rearing

Often one cannot clearly distinguish between the direct effects of temperature changes and the indirect effects exerted through changes in food supplies. Nolan's results (Fig. 56), for instance, show an April fall in brood rearing, in which cold weather was associated with reduced foraging.

Dunham (1930) fitted an observation hive with thermocouples and found that in spring, in an expanding broodnest, the queen laid eggs in a temperature range of 23–34·5°C., but most actively in the range 31–33°C. The commencement of brood rearing may quite possibly be determined by the warmth of the cluster. Temperature regulation in the winter cluster is considered elsewhere (pp. 218–24), but one may note that Merrill (1924a) held that brood rearing commenced when a temperature high enough for flight was followed by a sharp drop; this was considered to cause increased store consumption and then a rise in the temperature of the centre of the cluster.

Perepelova's results (p. 267) provide an instance of temporary acceleration of spring brood rearing by a measure designed to raise the temperature within the broodnest. Merrill's (1923) data, which showed that in spring there were fewer bees in unprotected colonies than in those which had been protected by a windbreak during the winter, and more bees in packed than in unpacked colonies, were probably a consequence of temperature differences.

It has often been suggested that in springtime brood rearing reaches a peak level, from which it inevitably declines, and some results (e.g. Figs. 56–58) do suggest that egg laying increases most vigorously after a period during which it has been severely restricted. Low temperatures help to produce such a restriction. There is no certain evidence that temperature fluctuations in summer conditions exert any important *direct* effect on brood rearing. Taranov (1951) reported that Italian and Caucasian bees reacted differently to a cold spell in early June; 10 colonies of Italians reduced brood rearing by 39%, but 2 groups of 7 Caucasian colonies increased their brood rearing by 14 and 31%; this difference might be related to other habits (p. 240).

The relation between brood rearing and pollen income

Pollen provides the honeybee with its only source of protein, so any shortage of pollen will act as a limiting factor which

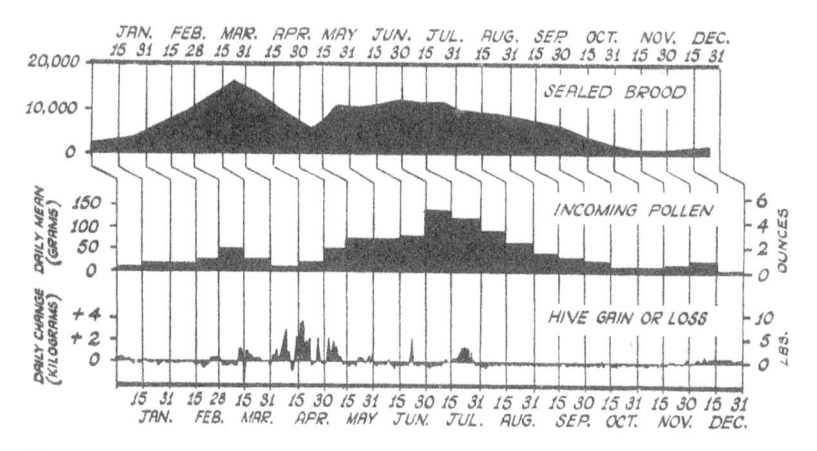

Fig 57 THE RELATION BETWEEN BROOD REARING, POLLEN SUPPLY AND CHANGES IN COLONY WEIGHT, FOR A COLONY IN A CALIFORNIAN ORANGE GROVE (after Todd & Bishop, 1941).

restricts brood rearing. A good example of this effect, on wintering colonies, was provided by Farrar (1934); he over-wintered 8 colonies, dissimilar only in respect of their pollen reserves. Between October and mid-April 4 colonies without pollen reserves reared no new bees, their populations dropped 55–88% and they ate 21–35 lb. stores; 2 colonies with 160–170 sq. in. reserve pollen reduced their population by 40–51% and consumed 38–47 lb. stores; 2 colonies provided with 615–626 sq. in. reserve pollen contained only 5–8% fewer bees in mid-April, having eaten 42–56 lb. stores. More data of this kind were provided by Farrar (1936).

The effects of pollen abundance, in summer, are marked but less spectacular. Nolan (1925) obtained an upswing in August brood rearing which could be associated with a plentiful supply of pollen (Fig. 56). Todd & Bishop (1941) found that the trend of brood rearing in California was parallel to pollen income, and that the springtime egg-laying peak coincided with a peak of pollen collection (Fig. 57); in summer larger quantities of pollen seemed to produce less effect than smaller quantities had achieved in spring. V. Frisch (1947) directed honeybees (Ch. 23) to visit poppies, a nectar-free pollen crop, and recorded that in one experiment brood increased by 5 combs in directed colonies but by only 1 comb in similar undirected ones.

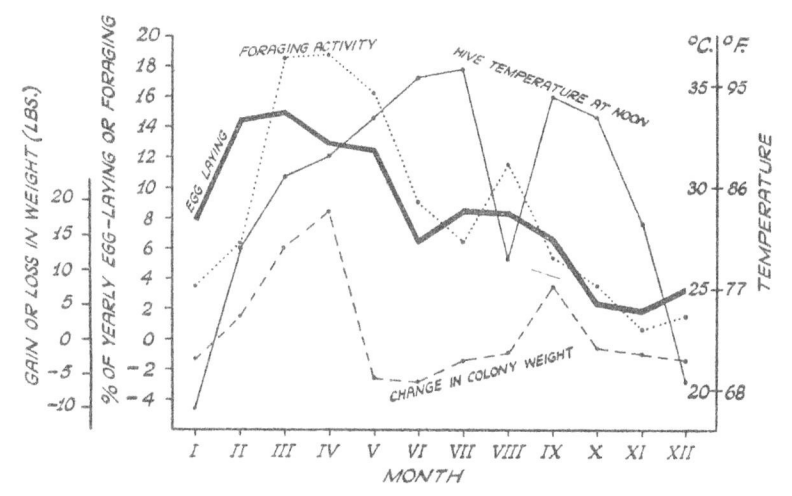

Fig 58 MONTHLY CHANGES IN EGG LAYING, FORAGING ACTIVITY, COLONY WEIGHT, AND MIDDAY TEMPERATURE WITHIN THE HIVE, FOR A COLONY IN THE HILLS OF PALESTINE (after Bodenheimer & Nerya, 1937).

Brood rearing in relation to nectar flows

The effect of nectar supply on brood rearing is more controversial. Merrill (1925b) found that in springtime colonies with plentiful stores (20 lb. minimum) reared 50% more brood than those supplied with only 6 lb. stores (although the latter colonies at no time had less than 2 lb.). Nolan's results (Fig. 56) do not indicate that the initial expansion of brood rearing was associated with a nectar flow, but brood rearing was well maintained during the subsequent main flow and fell off from the end of it; a minor peak of brood rearing in September was associated with a minor nectar flow.

However, Merrill (1924a) found that colonies measured by him reduced their brood rearing during nectar flows, and that in colonies well supplied with stores the biggest peak of brood rearing actually occurred during the last week in May, when bad weather hindered foraging. Many beekeepers consider that heavy nectar flows may cause the bees to block part of their broodnest with nectar, as would happen in colonies provided with insufficient empty comb. Taranov (1951) wrote that there were strain differences in this character; in nectar flows Italian bees increased brood rearing, but bees of a Caucasian strain filled the broodchamber with nectar and reduced the laying of the queen.

Todd & Bishop (1941) drew attention to the important fact that nectar flows are not necessarily related to pollen increases; nectar flows from orange were associated with a reduction in the pollen supply, and in brood rearing (Fig. 57), but sage and sweet clover crops were good pollen sources, so that results might then be different. Currie (1932) had previously recorded that in nectar flows from *Eucalyptus meliodora*, which produces sticky pollen not available to honeybees, Australian colonies could dwindle and die out in dry years when pollen was not available from other sources.

The effects of pollen supplements on brood rearing

Farrar's results (p. 238), coupled with his observation that natural pollen reserves at Wyoming in October varied from 24 to 224 sq. in., provide justification for an attempt to supplement pollen reserves when large numbers of bees are required in early spring for pollination purposes. Moreover, Farrar (1936) was able to show that in Wyoming conditions the additional winter stores expended in rearing extra bees were recouped with heavy interest during spring—although data which he subsequently published (Table 13, p. 229) suggest that this result may not be consistently true, even in Wyoming; in districts without an early nectar flow it may not hold at all.

Soudek (1929) fed 15 groups of newly hatched bees with sugar syrup and different kinds of protein food, and found that with two of these foods only—fresh white of egg and dried yeast—the bees' pharyngeal glands reached full development. When no natural pollen was available, Currie (1932–35) removed the brood and pollen from colonies and fed them on 7 different diets, or on sugar syrup only; the most promising foods were dried milk and white of egg, both of which caused gland development and brood production.

Meanwhile Haydak (1933) reported better gland development

and brood rearing with dried yeast than with other substances tested, and found (1937) that brood was reared by colonies fed with soya flour. He found (1940a) that a colony fed with a 4:1 mixture of soya flour and dried skim milk actually reared more brood than a control colony fed with pollen, but Haydak & Tanquary (1943) obtained only equivocal results by feeding this mixture to package bees in springtime. Other results (Haydak, 1945, 1949) showed that a vitamin deficiency, which could be remedied by the addition of dried yeast, was the chief defect of soya flour. These and other results led to recommendations of a 4:1 mixture of soya flour and dried yeast, made into a paste with sugar syrup and placed immediately over the combs of the colony in early spring.

Palmer-Jones (1947) obtained substantial increases in both brood and honey production by bees fed with this mixture in New Zealand. Walstrom (1950), working in Nebraska, reported that colonies fed in spring with the same mixture produced more honey than control colonies, and added that distiller's corn solubles (the residue left after the distillation of alcohol from maize) were cheaper and more efficient than soya flour. Wolfe (1950) found that in four successive years colonies in Nebraska supplied with cakes of soya and yeast had produced about 50% more honey than unfed colonies; even better results were secured by addition of 12% natural pollen, which Schaefer & Farrar (1946) had found beneficial.

The significant increases obtained in honey production should be attributed to increased brood rearing caused by the supplements. Wolfe emphasized, as Haydak & Tanquary had done, that there may be areas where the supplements would have no value whatever; he doubted whether their use could be justified in areas where weather in spring is generally favourable and natural pollen constantly available.

Vinogradova (1951) claimed that the honey yield from a group of 5 colonies fed daily with 33 grams baker's yeast, in sugar syrup, was twice that of 32 unfed controls. This does not fit in with the foregoing and less surprising results obtained by many other workers who have fed colonies with yeast.

The effect on brood rearing of feeding with sugar syrup

The effects of feeding with sugar syrup are as contrary as the effects of nectar flows on brood rearing. Ribbands (1950a) reported that autumn feeding with concentrated ($66\frac{2}{3}$% by wt.)

sugar syrup not only produced the maximum quantity of stores but also produced a significant increase in the amount of brood rearing; feeding with a similar weight of sugar in dilute (38% by wt.) syrup produced no visible increase in brood rearing. However, Butler (1946) had shown that springtime feeding with concentrated syrup produced a significant *reduction* in the quantity of brood then reared—as well as the disappearance of 12 lb. sugar per colony with no corresponding addition to stores; he obtained a smaller (not statistically significant) reduction when the same quantity of sugar was supplied in dilute syrup.

A new angle to this problem was provided by Eva Crane (1950) who reported springtime feeding experiments which revealed a significant increase in brood rearing when dilute syrup was supplied to small colonies, but a smaller or non-existent increase when the same syrup was fed to large colonies. She also noted that the brood rearing of colonies sited within 50 yards of an available water supply increased less than that of those sited beyond this distance, and that only the latter colonies (which also consumed more stores) showed a significant increase in brood rearing.

Can these diverse results be fitted into one pattern? Butler had experimented with strong colonies, close to a water supply; Ribbands's experiments were carried out with both large and small colonies near to water. My own guess is that the differences are the consequence of effects upon the foraging of the colonies concerned; in bad spring weather provision of a sweetened water supply within the colony removes a handicap which might otherwise restrict brood rearing, and this supply may be too dilute to interfere with foraging; a more concentrated supply may fulfil this function less adequately because it may itself be more concentrated than the sugar content of the larval food (p. 249), while it will tend to produce a state of dissatisfaction (p. 118) with any natural crops which are available but less attractive than the syrup—thus it might discourage foraging on the pollen-and-nectar crops which provide much of the honeybees' spring fare and which are usually available if weather permits them to be reached; autumn feeding is postponed until there is no natural nectar source with which the bees can become dissatisfied, but the excitement which attends the feeding may encourage more of the bees to leave the colony and to work pollen-only sources (e.g. sowthistle) which at that time provide the only available forage. V. Frisch has found that concentrated syrup fed within the hive may encourage dancing and reconnaissance (p. 185).

If this interpretation is correct one may suppose that in districts where conditions differ from those outlined above (e.g. in a heather district if autumn feeding were carried out while the heather was still available) the results would also be different.

THE EFFECT OF BROOD REARING ON FORAGING ACTIVITY

The effect of brood rearing on the ratio of pollen gatherers to nectar gatherers is discussed on pp. 124–6.

Effect on the proportion of foragers, gauged in terms of honey production

In large colonies, in summer, the proportion of brood to bees is likely to be in inverse relation to the size of the colony. Changes in the proportion of brood have a two-fold effect; in the first place the proportion of foragers may be expected to vary inversely with the proportion of brood to bees, and secondly the brood itself consumes a substantial quantity of food. These two factors taken together would explain, or help to explain, why honey production rises steeply as colony size increases. On one occasion Schachinger (cited Miller, 1901) found that a colony of 20,000 bees stored ½ lb. honey, while 30,000 stored 1½ lb. and 40,000 stored 4 lb. Data were collected and subjected to complex statistical analysis by Farrar (1937), who concluded that 1 colony with 30,000 bees had produced 1·36 times as much honey as 2 colonies each with 15,000, 1 colony with 45,000 bees 1·48 times as much as 3 colonies each with 15,000, 1 colony with 60,000 bees 1·54 times as much as 4 colonies each with 15,000. Sharma & Sharma (1950) recorded that the honey yields from 8 colonies of *Apis indica* showed a similar trend; 2 colonies with 11,000 bees in each together produced 14 lb. 1 oz. (10·1 oz. per 1,000 bees), while 2 colonies with 24,000 bees produced 58 lb. 2 oz. (19·2 oz. per 1,000 bees).

One cannot tell whether the whole of the increase can be attributed to a reduced proportion of brood rearing in the larger colonies, but it is possible that there are also minor advantages, e.g. greater efficiency in crop finding, which would make some contribution towards these results.

The relative honey-storing capacity of large and small

colonies will vary widely according to circumstances. Brood rearing itself fluctuates through the season; in addition, if two groups of colonies of varying sizes were placed in different circumstances, so that one group foraged in a major nectar flow while the other worked in poor conditions, the relative proportions of honey stored in the colonies of the two groups would be different. In poor conditions differences between the proportion of honey stored by large and small colonies are amplified (pp. 74–5).

Chapter 31

QUEEN AND WORKER DIFFERENTIATION

> *Both of the babes were strong and stout,*
> *And, considering all things, clever,*
> *Of* that *there is no manner of doubt—*
> *No probable, possible shadow of doubt—*
> *No possible doubt whatever.*
>
> W. S. GILBERT, *The Gondoliers*

IT WAS KNOWN many years ago that new colonies could be produced by division, and that the queenless portions could raise a queen if they contained young worker brood (Jacob, 1568). Then Schirach (1770) showed that such queenless bees constructed queen cells upon, and reared queens from, young worker larvae; both queens and workers were therefore derived from one kind of egg. Schirach's researches caused Huber (1814) to encourage Mlle Jurine to dissect worker bees and discover rudimentary ovaries in them, thus proving that workers were not neuters but females.

Some other insects occur naturally in more than one form, with differences determined by their environment. This polymorphism is most strongly developed in the social insects; the parasitic Hymenoptera provide less extreme examples—some species of the latter are large and fully winged when reared on large hosts, but small, wingless and with differences in legs, antennae and ocelli when reared on small hosts (Thompson, 1923; Salt, 1937).

The time at which differentiation occurs

Klein (1904) placed worker larvae of different ages in queen cells and obtained transition types between workers and queens, in accordance with the age of the larvae when grafted. His studies were extended by Becker (1925), who sectioned the pupae developed from grafted larvae of known age and compared them with queen pupae raised in swarm cells. His results are summarized in Table 14.

While the swarm-produced queens and the earliest graft were possibly slightly superior, there were no appreciable differences

TABLE 14. The effect of grafting age upon queen differentiation (Becker, 1925)

Age at grafting time	Number of queens examined	Number of egg tubes in one ovary	Spermatheca, width in mm.	Mandibular glands, relative size	Pharyngeal glands, relative size
swarm-cell queens	9	154	1·06	1·6	0·07
½–1	7	157	0·97	1·6	?
1	4	160	0·77	1·7	0·03
1½	5	163	0·73	1·5	0
2	8	161	0·89	1·5	0·08
2½	6	148	0·95	1·7	0·02
3	8	166	0·80	1·5	0·04
3½ (5 grafts)	24	7–156	0·16–0·76	0·5–1·08	1·0–2·5
4	3	5	0·18	0·3	4
workers	5	4	0·01	0·6	4

between larvae grafted at 1 and 3 days old. The 3½-day grafts were very variable and very different from the earlier ones, and the pupae which arose from 4-day grafts were not significantly different from workers, so Becker concluded that the striking differences between the organization of queens and workers arise when their larvae are between 3 and 3½ days old.

Zander (1925), in parallel experiments, weighed numbers of queens of different origin, and their weights were:

	Average (mg.)	Range (mg.)
50 swarm-raised queens	187	145–234
53 supersedure queens	185	126–231
14 early-grafted	208	178–233
6 grafted at 2 days	206	189–228
6 grafted at 2½ days	177	98–211
4 grafted at 3 days	170	142–189
4 grafted at 3½ days	170	156–202

Comparing these results with those of Becker, Zander concluded that there were no significant differences between swarm-raised queens, supersedure queens, and those satisfactorily reared by grafting from young larvae, and that larvae used in queen-raising grafts should not be more than 2-days old.

Differences between the average weights of 2-day and later grafts are not conclusive evidence that earlier grafted larvae are superior, since his grafts were made at different times. Zander's conclusion, however, was supported by the results of v. Rhein

(1933), who took 20 mg. (2¾ day) queen larvae from their cells, fed them on the food of older worker larvae and observed that in their ovary development (100 ovarioles per ovary) they came to resemble queens rather than workers.

The biochemical study of Melampy & Willis (1939) also supports this view. They found that the oxygen consumption per gram of the tissue of queen larvae 2 to 3 days old was already 50% greater than that of worker larvae of the same age and that the respiratory quotient of these queen larvae was 1·14 while that of the worker larvae was 1·42. Thus the metabolism of these queen larvae was already much more rapid than that of the worker larvae, and they had grown faster, having attained a live weight of 10·55 mg., compared with 6·08 mg. for worker larvae. Melampy, Willis & McGregor (1940) also showed that in queen larvae 2 to 3 days old the percentage of total lipid was twice as great as in worker larvae of the same age; their analyses of other constituents showed smaller differences. From these results it seems likely that the food of worker larvae is altered or restricted by the time they are 2 days old.

Quantitative differences in food supply

Queen larvae are supplied with so much food that a surplus is usually left uneaten when they pupate. Worker cells are very much smaller than queen cells, and worker larvae receive a much smaller quantity of food; they, too, are surrounded with an excess for nearly three days, but the quantity of their food is then restricted (Nelson & Sturtevant, 1924). This very marked difference in quantity led Spencer (1867) to consider that the lower feeding of the worker larvae resulted in dwarfing and arrested ovary development.

No queen cells have ever been found in colonies of *Apis dorsata* (Roepke, 1930); it is likely that queens and drones are reared in the same kind of cell. This would be a more primitive condition, in keeping with the relative status of *dorsata* and *mellifica*.

Qualitative differences in food supply

Important differences in the quality of the food supplied to worker and queen larvae are clearly recognized. Queen larvae and young worker larvae are entirely fed upon bee milk, but the older worker larvae receive a less nutritious diet which presumably also contains nectar (v. Planta, 1888–89; Adrienne Köhler, 1922; Nelson & Sturtevant, 1924). The carbohydrate content of this food exceeds that of bee milk, its protein content is less and its fat content only half (v. Planta; Haydak, 1943). The food of older worker larvae usually contains pollen grains (Köhler; Nelson & Sturtevant; v. Rhein, 1933). The variation in the amount of pollen is considerable, so Haydak (1943) suggested that the pollen admixture is accidental. Simpson (1953) counted pollen grains in the guts of mature larvae, and found that the quantity ranged from nil (in one winter sample) to a number which probably supplied not less than one-fiftieth of the total larval nitrogen requirement; he suggested that the pollen had been added with the nectar, incidentally. Traces of pollen have been found in the food of queen larvae (Elser, 1929; Haydak, 1943).

It was thought (v. Planta, 1888–89) that the quality of the food of the worker larvae was not changed until their fourth day, but Nelson & Sturtevant (1924) found that it occurred during the third day, at the time the quantity of food was restricted. V. Rhein (1933) thought that the food was not changed until the larvae weighed 35 mg. and were $3\frac{1}{4}$ days old.

Subtle differences between the quality of the bee milk[1] provided for queens and workers of various ages have been postulated. Elser (1929), who investigated the food of queen larvae, using material supplied to him by beekeepers, concluded that its fat and protein content decreased with the age of the larvae and that its acidity steadily increased. A similar investigation by Haydak (1943) produced opposite results; Haydak found that the fat and protein content increased with age, and that the acidity was unchanged. V. Planta (1888–89) and Haydak (1943) recorded the chemical composition of the bee milk supplied to both queen and young worker larvae (Table 15). Gontarski (1949b) reported small differences in the amounts of sugar and water in the foods of queen, worker and drone larvae of different ages, and v. Rhein (1951b) reported that young and old queen larvae received different diets.

Some of the differences between the results obtained by these investigators may have been due to sampling difficulties; in the absence of statistical presentation it is not possible to assess the significance of any of them. Any differences which exist are unlikely to be important in relation to the differentiation of queen, worker and drone types; v. Rhein (1933) reared worker larvae upon royal jelly[1], and later (1951) reared drone larvae

[1] See footnotes, p. 56.

upon the food of young worker larvae; Adrienne Köhler (1922) has pointed out that it is unlikely that the composition of a glandular secretion can be voluntarily changed, and Haydak's (1943) results led him to agree with Langer (1929) and Elser (1929) that the composition of bee milk is variable; Lindauer (1952) observed that each nurse bee feeds larvae of all ages (p. 301).

TABLE 15. Composition of the food of queen and worker larvae. (Vivino, cited Haydak, 1943)

	Age in days	% moisture	pH	Protein ($N \times 6.25$)	Fat	Ash	Carbohydrate, etc.
queen larvae	1	65	4·1	40·4	7·6	3·3	48·6
	2	69	4·1	48·8	5·6	2·9	42·6
	3	70	4·2	50·6	16·1	2·6	30·6
	4	70	4·1	46·2	18·7	2·3	32·7
	5	68	4·2	49·7	15·2	2·3	32·7
	sealed	68	4·2	58·0	12·6	2·4	27·0
worker larvae	1–2	74	4·0	78·3	17·7	4·0	0*
	3–5	65	3·9	50·4	5·9	1·6	42·1

*This surprising result follows from the unusually high value for protein; only one sample was analysed.

The vitamin content of larval food

Melampy & Mason (1936), Schoorl (1936), Evans et al. (1937), and Haydak & Palmer (1938) showed that royal jelly lacks vitamin E. Melampy & Breese Jones (1939) found that royal jelly contains no vitamin A and little, if any, vitamin C, but that vitamin B is present. Results of various determinations, presented in Table 16, indicate that the members of the vitamin B complex occur in variable amounts. These vitamins are probably derived from plant sources: although only present in insignificant amounts in honey—Kitzes et al. (1943) suggested that even these small quantities are contained in the pollen grains suspended in the honey—pollen is rich in all of them; they are probably principally contained in its fatty ether-soluble portions.[1]

The vitamin content of pollens from different species of plants has not been analysed, but Haydak & Palmer (1942)

[1] Todd & Bretherick (1942) reviewed the literature on the fat content of various pollens, and found that this ranged from 1·5 to 23·6%. They themselves analysed samples of pollen from 34 species of plants and found that the ether-soluble proportions ranged from 0·9 to 17·5%.

TABLE 16. Vitamin B analyses of larvae and their foods

Quantity in micrograms per gram of fresh material

Source	Moisture content %	Thiamine (b_1)	Riboflavin	Pyridoxine	Nicotinic acid	Pantothenic acid	Biotin	Inositol	Folic acid
ROYAL JELLY									
Melampy & Breese Jones (1939)	66	3–4.5							
Haydak & Palmer (1940)		3.1							
Pearson & Burgin (1941)						183			
Kocher (1942)		4.5–5.2	6.5	0	80–100	[0.75]			
Haydak & Palmer (1942)		1.5–2.9							
Haydak & Palmer (1942) biological analysis				50					
Cheldelin & Williams (1942)		5.6–7.4	6.6–10.0	2.2–2.5	47–73	65–110	1.6–1.8	78–150	0.16–0.22
Kitzes, Schuette & Elvehjem (1943)		18.0	28.0	10.2	111	320	4.1		0.5
Haydak & Vivino (1950)	68	1.2–1.3	5.3–8.3		91–149	172–200			
Gontarski (1949b)							0.9–3.7		
WORKER LARVAL FOOD									
Haydak & Vivino (1950) 1–2 day larvae	73	1.3–3.2	11.7		81–100	128–147			
Haydak & Vivino (1950) 3–5 day larvae	65	1.2–3.3	10.8		45–52	19.5–20			
POLLEN									
Kitzes, Schuette & Elvehjem (1943)		6.0	16.7	9.0	100	27.0	0.25		
Haydak & Palmer (1942)*		3.8–5.0	15–34.2		60–1250	[0.17–0.45]			
Vivino & Palmer (1944)†		6.3–10.8	16.3–19.2		132–210	16.0–27.6			
WORKER LARVAE									
Haydak & Vivino (1943)		4.6	32.2	14.6	77.2	180.2			
HONEY									
Kitzes, Schuette & Elvehjem (1943)		0.04	0.3	0.1	1.1	0.5	0.0007		0.03

* Pollen from combs.   † Pollen-trapped samples.

analysed three different samples of stored pollen obtained from combs, and Vivino & Palmer (1944) analysed four samples of pollen which had been obtained at different seasons from pollen traps. These results, and those of Kitzes et al. (1943) are summarized in Table 16. The only startling difference between these three sets of analyses is in their pantothenic acid content; a technical fault may have vitiated the analysis of this substance by Haydak & Palmer (1942). A much wider range of variation might be revealed if pollens from separate species were analysed.

Svoboda (1940) and Anna Maurizio (1950) both showed that the nutritional value of different pollens, measured by the amount of activation which they produce in the pharyngeal glands of caged bees, varies widely; Svoboda suggested that sometimes the determining factor is the proportion of digestible protein contained in the pollens. Haydak (1949) demonstrated that the deficiencies of soyabean flour, for brood rearing, could be completely remedied by the addition of 20% dried brewers' yeast, or partly remedied by the addition of nicotinic acid and riboflavin. Maurizio (1950) found that soyabean flour alone had only a slight activating effect upon the pharyngeal glands; her results indicate that the vitamin deficiency in soyabean flour inhibits bee-milk production; but if such deficiencies were less acute a bee milk deficient in vitamins might possibly be produced.

Although bee milk is derived from pollens it does not necessarily contain all the vitamins which are present in pollens. Manuilova (1938) found that the vitamin C content of stored pollen ranged from 50 to 342 mmg./g., and Haydak & Palmer (1942) obtained values from 68 to 118 mmg./g., but the latter confirmed the statement of Melampy & Breese Jones (1939) that this vitamin is not noticeably present in bee milk. Vitamin K, which is not present in fresh pollen (Vivino & Palmer, 1944), is found in pollen which has been stored in the hive (probably because it has been synthesized there by fermentation bacteria), but it is absent in royal jelly (Haydak & Vivino, 1950).

Conversely, bee milk may contain some vitamins in much higher proportions than pollen does. Both Kitzes et al. (1943) and Haydak & Vivino (1950) found that the thiamine, riboflavin and nicotinic acid content of royal jelly was little different from that of pollen, but Kitzes et al. found that the pantothenic acid and the biotin content of royal jelly was 12 to 16 times that of pollen; Haydak & Vivino (who did not analyse the biotin) confirmed this result for pantothenic acid. Since the moisture content of royal jelly is substantially higher than that of pollen, the analyses in Table 16 underestimate the real concentration of these substances.

Haydak & Vivino's analyses showed that, as a consequence of this concentration, the food of young worker larvae was $9\frac{1}{2}$ times as rich in pantothenic acid as was the food of older worker larvae. They also concluded that young worker larvae have an ability to accumulate thiamine, riboflavin, and pantothenic acid, since the proportion of these substances was from 2 to 4 times as great in the dry weight of young worker larvae as in that of their food.

Pearson & Burgin (1941), who showed that royal jelly was the richest known source of pantothenic acid, suggested that this might be the determining factor in queen-worker differentiation. Kitzes *et al.* (1943) concluded that biotin and pantothenic acid might both play a significant part in the metabolism of the young bee, and Gontarski (1949*b*) emphasized the importance of biotin in this role.

Gontarski found that the food supplied to queen larvae 2 days old contained 3·7 mmg./g. biotin dry weight, but by the fifth day this was reduced to 0·9 mmg./g. The food supplied to drone larvae 2 days old contained only 0·7 mmg./g. biotin dry weight. Gontarski concluded that it is not the quantity of larval food, but its quality, and especially its content of substances such as biotin, which determines the differentiation between queens and workers. He supposed that teams of nursing bees are present, each team feeding larvae of only one age and sex. This supposition does not fit easily with the fact that dequeened colonies are quickly able to nourish several dozen queen cells, if so many larvae are grafted into them, it is contrary to the recent observations of Lindauer (p. 305), and it is not necessary in order to account for Gontarski's results. He also found evidence (as did v. Rhein, 1933) that bee milk underwent chemical changes when exposed in the open air; as vitamins are readily destroyed, his biotin analyses (if significant) might be accounted for by a hypothesis that all the bee milk was similar in composition when secreted, but that its biotin content was quickly reduced when it came into contact with the atmosphere; the rapidity of such a process would depend upon the relative surface area exposed and thus it might occur less rapidly in queen cells, more rapidly in drone and worker cells.

Considerable difficulties appertain to these vitamin analyses; the sampling of minute quantities of worker larval food without additional exposure to the atmosphere presents a major problem. Moreover, the activities of the different components of the vitamin B complex are so inter-related that no adequate understanding of them is likely unless they can all be separately analysed and the results compared. This point is sufficiently illustrated by Haydak's (1949) experiments on the addition of vitamins to soyabean flour; for inducing brood rearing riboflavin was ineffective, nicotinic acid partly effective, and nicotinic acid + riboflavin more than twice as effective as nicotinic acid alone. As this work of Haydak has demonstrated the importance of vitamin B in honeybee reproduction, the idea of a vitamin-controlled differentiation based on biotin, pantothenic acid, or some combination of vitamins is not fantastic. But there is no *proof* that these substances control the differentiation of queens and workers, although v. Rhein (1951*b*) found that the addition of pantothenic acid to the food of worker larvae increased their ovary development.

The mechanism of differentiation

All investigators consider that the basis of the differentiation is nutritional, but the food supply of worker larvae differs from that of queen larvae in both quantity and quality and the relative importance of these two attributes has not been agreed.

Marchal (1896) studied the differentiation of queen and worker wasps, and concluded that worker wasps were produced by nutritional castration; the food supplied to them differed in quantity, but not in quality, from that provided for their queens. In both wasps and bumble bees there are only slight differences between queens and workers apart from ovary development, but in honeybees the differences are greater. Furthermore, in honeybees there are clear differences in both the quantity and the quality of the food provided; differences in food quality consist of differences in the quantities of various constituents and also in the balance between those constituents. Haydak (1943) emphasized the importance of the quantitative differences and suggested that extra nutriment assimilated by queen larvae stimulates the development of their ovaries, which then produce a secretion which circulates in the blood stream and inhibits the development of worker characters.

The experiments of Altmann (p. 281), suggest to me that the

available quantity of fresh bee milk, in particular, may stimulate ovary development and determine the differentiation into worker or queen. This conclusion could account for the development of laying workers in broodless colonies (Ch. 35) and the prodigious egg production of laying queens, which are fed on the same diet (pp. 56–8).

Queen-rearing problems

Methods of queen rearing are beyond the scope of this book, but relevant conclusions may be derived from the scientific experiments. *Properly* grafted larvae produce queens which are as satisfactory as those reared naturally, providing the grafted larvae are not more than two days old (p. 247). They should be placed in a queenless colony staffed with plenty of nurse bees, preferably during a pollen harvest (p. 233).

No special category of nurse bee is required for feeding queen larvae, and removal of unsealed worker brood will eliminate rival requirements for bee milk. Goetze (1926) reared successive grafts in the same queenless colony at 14-day intervals and found that the number of queens raised fell off sharply ($42 \rightarrow 23 \rightarrow 12 \rightarrow 2$) and their mean weight deteriorated ($0 \cdot 197 \rightarrow 0 \cdot 181 \rightarrow 0 \cdot 178 \rightarrow 0 \cdot 150$ g.).

Queen development is accomplished more quickly than worker development. The whole process from egg laying to emergence occupies 15–16 days. Thus if queens are produced from grafts of larvae 1 day old they should emerge in 11–12 days. Sometimes they emerge sooner; this may happen because older larvae are used, but other variables must also be taken into account (p. 232). Queens reared from very old larvae may develop more slowly (Becker, 1925).

Chapter 32
DRONE PRODUCTION

*From the Lime's leaf no amber drops they steal,
Nor bear their grooveless thighs the foodful meal;
On other's toils, in pamper'd leisure, thrive
The lazy Fathers of th' industrious hive.*

JOHN EVANS, *The Bees*

MAN'S EXPLOITATION OF the honeybee is based on his ability to thwart its reproductive instincts, or to bend them to serve his ends. His success depends on control of swarming and restriction of drone production, to encourage the maximum surplus of honey instead of the greatest quantity of honeybee colonies.

The production of drones greatly exceeds the production of swarms, or queens, so only a minute proportion of the drones exercise any useful function. The excessive production has played an essential role in the evolution of the honeybee community (Haviland, 1882). It costs the colony far less to rear a drone than to rear a queen with her accompanying swarm; thus if similar numbers of drones and queens were ordinarily produced, as in most solitary bees, so that both stood an equal chance of mating, the queen of any colony which produced more drones and fewer queens would have more grandchildren. Such excessive drone-producing colonies would be naturally selected, and this process would continue until a balance had been achieved in which the cost of producing a swarm was equivalent to the cost of producing a drone which successfully mated.

The intelligent use of wax foundation, which was invented by Mehring in 1857, now enables beekeepers to control the drone-rearing propensities of their stocks. However, the production of *some* drones is still of vital importance. The problem of drone production may be divided into two separate parts (*a*) the building of drone comb, and (*b*) the rearing of drone brood.

Construction of drone comb

Wild honeybees often build large-celled comb for storage as

well as for drone production. Langstroth (1890) said that 'No two colonies in the same apiary will show the same number of drone cells. You will find a colony whose comb will consist of two-thirds worker and one-third drone, the adjacent colony will have but one-sixth drone comb, another a few square inches only.' He gave as generally accepted facts concerning swarms hived on empty frames, (*a*) they always begin with worker comb construction, (*b*) if the queen of the swarm is very prolific her bees will build very little drone comb, but if she is deficient they build a great deal, (*c*) if the queen of a swarm is removed or dies, drone comb only will be built during her absence. He supposed that workers build worker cells when the queen is in the vicinity and anxious to lay in them, but that in other circumstances they build drone comb; this hypothesis does not seem consistent with the fact that drone comb is readily built in the broodnest in spring, when breeding is expanding rapidly.

I have found no records of any experimental approach to this problem. If Darwin's conclusion is correct (p. 200) that worker comb construction is due to a balance between many bees all instinctively working at the same distance from each other, some simple change, e.g. in temperature, might cause them to cluster less closely. Cyprian bees are reputed to build a smaller proportion of drone comb than bees of other strains.

Rearing of drone brood

Drone rearing and swarming are complementary acts of reproduction, but the relationship between them has not been clearly established. Huber (1792) said that queens usually laid only 50–60 drone eggs during their first season, but they produced 1000–2000 in their eleventh month, i.e. with the advent of the next swarming season. He suggested that large-scale laying of drone eggs commenced about three weeks before queen cells were founded. There is a conflict of opinion, but no clear evidence, concerning the role of drone production in the swarming cycle (p. 266).

Drone rearing is seasonal and is dependent upon the condition of the colony. In dearth conditions rearing may cease and adult drones may be thrown out, even in June, but drones may be bred again that season if food supplies increase (Langstroth, 1890). Mathis (1947) instanced two queens raised in May 1941 in strong colonies, which laid drone eggs within a month; similar queens in small nuclei laid no drone eggs; a queen which laid drone eggs when she was in a strong colony laid only worker eggs when she was placed in a nucleus.

Rosser (1934), who considered that a plentiful supply of pollen was necessary for large-scale drone production, emphasized the variation which occurs between different colonies. As extreme examples he cited (a) two nuclei, the progeny of one queen, which both filled half a frame with drone brood before they had six frames of worker brood, and (b) a queen which reared no drones and would not use drone comb even when it was placed in the centre of the broodnest. This characteristic was transmitted to her daughters and grand-daughters.

Huber (1792) placed a colony on drone comb only and observed that the bees did not cluster properly and the queen dropped her eggs haphazardly rather than lay them in drone cells. Root (1883) observed that queens might eventually lay worker eggs in drone cells if the bees had contracted the mouth of those cells with wax. Jordan (1928) found that in the absence of other comb queens would lay worker eggs in drone cells which had not been so contracted. Langstroth pointed out that the queen would behave satisfactorily if all drone comb was removed from a colony, and he held that she preferred to lay in worker cells. On the other hand, Simpson (1953) has noted that in colonies short of drone comb queens will often lay in small portions of such comb well beyond the limits of the broodnest proper. Moreover, when a queen is confined below an excluder her bees will often keep drone cells above the excluder cleaned and ready for her to lay in after adjoining worker cells have been filled with honey.

The relative importance of the queen and the workers in the production of drone instead of worker brood in normal colonies has not yet been clearly separated. When queens become 'drone layers' this is due to a change not in behaviour but in physiology —absence of mating or insufficiency of sperm. In such abnormal colonies the behaviour of the workers is altered, and drones will be nourished and preserved throughout the winter, as they are in queenless colonies (Huber, 1792).

Laying of drone eggs

When Dzierzon (1845) set forth the theory that drones developed from unfertilized worker eggs and queens and workers from fertilized ones, he suggested that the normal queen could lay either kind of egg 'at her pleasure'. Dzierzon's theory of sex determination was easily confirmed, but his suggestion concerning egg laying has been less readily accepted.

Attempts to explain the selective egg laying in automatic terms have not been successful. It is difficult to explain in this manner the laying of the correct eggs in partly built cells, and Jordan (1928) provided other criticisms, including the fact of the initial perplexity of queens which were placed on drone combs.

Jordan measured the dimensions of many drone and worker cells, in both old and new combs, and he noted that these dimensions overlapped. From corner to corner, worker cells measured 5·27–6·32 mm. and drone cells 6·20–7·02 mm. The dimensions of queens were more variable, so Jordan pointed out that the abdomens of large queens would press against drone cells more than those of small queens would against worker cells. Thus any such mechanical influence could not produce the accuracy of egg laying which normally exists.

Jordan possessed one comb which contained a large patch of drone cells and a large patch of worker cells, separated by a 2–3 cm. band of transitional cells gradually decreasing in size. It was placed in the colony on 1 June. The first two batches of brood contained workers in all the transitional cells, but in the third batch drones were reared up to the regular worker cells. He suggested that the queen had been deceived by the gradual transition.

Gontarski (1935*b*) said that in the spring when a queen first commenced to lay in drone cells she tended to lay only in the smallest drone cells, but subsequently she did not discriminate in this way. Carrying out experiments with stretched foundation from which the bees produced wider cells, he reported an instance in which normal drone cells (internal diameter 6·2–6·4 mm.) occurred in the middle of an area of these extra large worker cells (diameter 6·1 mm.), and the queen continued to lay worker eggs even in the drone cells. He concluded from this observation that it was not the absolute size of drone cells which caused the queen to lay drone eggs in them, but the difference in their size. However, the experiment was carried out in July and the size and state of the colony were not given, nor was there any evidence that that particular queen was willing to lay any drone eggs at that time.

Other Hymenoptera have a comparable method of sex regulation. Chewyreuv (1913) observed that the sex ratios of the progeny of three species of Ichneumons varied with the size of the host in which they had been reared, females having come predominantly from the larger hosts. Brunson (1938) studied this phenomenon in the parasite *Tiphia popilliavora*, which attacks larvae of the beetle *Popillia japonica*; 94% of the emergents from second-instar larvae were males, but 67% of those from the larger third-instar larvae were females; transfers of eggs proved that this result was not a consequence of differential mortality and made it clear that the mother regulated the sex of her offspring to suit the size of the host.

The physiological basis for the queen's discrimination is still a matter for speculation (e.g. Flanders, 1950).

The food of adult drones

Perepelova (1928d) noted that drones rarely move about in the colony except when soliciting food. She did not see drones take any food from combs; all the marked bees which fed them were of nursing age (30 records, range 8–14 days), and the time spent on feeding ranged from $1\frac{1}{2}$ to $4\frac{1}{4}$ minutes. Alpatov & Saf'yanova (1950) found that caged drones were fed by nurse bees but were neglected by foragers. Nixon & Ribbands (1952) found a low proportion of radioactivity among drones in a hive treated with radioactive syrup (p. 193). These results suggest that drones are fed with bee milk, but one may guess that they obtain nectar or honey before they go on their long mating flights (p. 144).

Chapter 33
SWARMING AND SUPERSEDURE

> *The hive contained one obstinate bee*
>   *(His name was Peter), and thus spake he—*
> *'Though every bee has shown white feather,*
>   *To bow to tyranny I'm not prone—*
> *Why should a hive swarm all together?*
>   *Surely a bee can swarm alone?'*
>
> *Upside down and inside out,*
>   *Backwards, forwards, round about,*
> *Twirling here and twisting there,*
>   *Topsily turvily everywhere—*
> *Pitiful sight it was to see*
>   *Respectable elderly high-class bee,*
> *Who kicked the beam at sixteen stone,*
>   *Trying his best to swarm alone.*
>
> <div align="right">W. S. GILBERT, <i>The Bab Ballads</i></div>

IN THE DAYS of skeps and sulphuring, beekeepers were interested in the production of swarms, but nowadays their concern is with swarm prevention. Beekeeping journals contain many dissertations on this subject, but they include surprisingly few records of accurate observations or experiments.

Although the cause of swarming is a matter for discussion, many characteristics of the process are generally agreed. For instance, Demuth (1921) and Morland (1930), supporting rival theories of the cause of swarming, differed little in their assessment of the numerous factors which aggravate or reduce the tendency to swarm. There can be no non-swarming strain of wild honeybees, but swarming propensities vary widely in different races and strains (e.g. Taranov, 1951).

Behaviour of a colony preparing to swarm

A description of the behaviour of the bees in a colony preparing to swarm was given by Huber (1792). Sixteen swarm cells were started by bees in an observation hive, and ten were enlarged (unequally, as they possessed larvae of different ages). Eight days after the cells had been started the queen still laid a few eggs, but her abdomen was much reduced in size and she was agitated. None of the workers were then feeding her, and sometimes they would advance brusquely to her, strike her with their heads and climb on her back. This treatment excited the queen, who ran all round the hive agitating the bees, which then excited each other. They all rushed to the entrance and the queen left the hive with them.

Similar observations were made by Taranov & Ivanova (1946). They said that when queen cells were commenced the queen's retinue increased from 10-12 to 22 nurse bees, which were perpetually trying to feed the queen, who could find few empty cells in which to lay eggs and refused to take so much food. Queen cells were started and the bees pursued the queen until she laid in them. When queen larvae hatched, the bees ceased to feed the queen, and struck her and jumped upon her, as Huber had stated. Taranov & Ivanova said that during swarming the queen was pushed out of the hive. The queen does not lead the swarm. Scrive (1948) captured a queen as she left a colony from which most of the bees had already departed; the swarm settled in the top of a nearby tree, not around the cage in which he placed the queen; then it broke up and returned to the hive.

Several swarms often issue from a swarming stock. The prime swarm, with the old queen (Janscha, 1771), usually departs soon after the first queen cell has been sealed, while the first afterswarm or cast goes about 8 days later, when the first virgin emerges. Any later casts issue with intervals of a day or two between them. Huber gives a detailed account of the repeated emergence of casts from a colony in an observation hive. He noted that the bees thinned the tips of the cells when the virgins were about to emerge, and that after the first virgin emerged her rivals were confined in their cells by the bees, sometimes for two or three days; the bees fed these imprisoned virgins and chivvied the emerged one away from them. In the excitement produced by the egress of a cast several virgins were allowed to escape from their cells at once; so Huber found two virgins in one cast and three in another.

Huber pointed out that the bees guarded swarm cells to protect the first virgin from damaging them, but that in a dequeened stock the replacement cells were not guarded and the bees appeared rather to excite the combatants. He said that although they kept the virgins from rival swarm cells they did not restrict the old queen. She was allowed to approach the cells and to destroy them if necessary; an old queen was averse to queen cells which contained pupae or virgins, and if the egress of the prime swarm was prevented by several days of wet weather she destroyed all the cells, so there was no swarm.

This does not necessarily happen; if it did queen clipping would usually prevent swarming, not merely delay it. Clipped queens often leave the hive and get lost when their colonies make an abortive attempt to swarm, but I have observed one colony which swarmed with a virgin, the marked and clipped queen still remaining in the hive (which was at ground level; she may have left and crawled back again); this queen had not disposed of her rival. The amount of congestion of bees in the hive may possibly determine whether the queen cells will be efficiently protected or not.

Taranov (1947) held that the sequence of events which culminates in swarming is (a) the presence of an excess of nurse bees in relation to the brood to be fed, (b) mobbing of the queen, which causes her to reduce her egg laying, (c) clustering of the excess nurse bees, and their conversion into inactive 'swarm bees'.

The inactivity of the swarm bees plays an essential part in the swarming cycle. It is associated with full development of the pharyngeal glands (Maurizio, 1950) and enlargement of the ovaries (Perepelova, 1928b); these organs serve as protein stores, which will facilitate the establishment of the swarm in its new abode. At the same time, reduced brood rearing facilitates the departure of the swarm. Perepelova (1928b) showed that combs of young brood placed in a colony ready to swarm might reduce the enlarged ovaries and cause swarm cells to be torn down by the bees. Taranov (1947) observed the proportion of the colony which departed with 16 different swarms and concluded that the quantity of bees which remained in the hive afterwards was directly related to the quantity of brood in it.

The inactivity of the swarm bees is obtained at the expense of both foraging and brood rearing. Ribbands (1951) found that during a nectar flow the mean gain in weight of each of six normal colonies was $32 \cdot 7 \pm 3 \cdot 6$ lb., while that of three similar colonies with queen cells was only $13 \cdot 3 \pm 3 \cdot 8$ lb. It is probable that the foraging population dwindles because its potential recruits are diverted to become swarm bees.

The ages of bees in a swarm

An experiment of Rösch (1930) led him to conclude that swarms consist of bees of all ages in about the same proportions as in the parent colony. This problem was investigated in much greater detail by Morland, and his work was completed and published by Butler (1940a). Twice each week large numbers of newly emerged bees were marked and added to colonies about to swarm. Eight swarms emerged, the swarms and the parent colonies were searched, and the marked bees were counted. The swarms contained some foragers, but were mainly composed of young and middle-aged bees.[1]

Swarming and supersedure

Supersedure, in which a young queen is reared to replace the old one but the colony does not divide, has features in common with swarming; the bees construct 1 to about 4 queen cells and the virgin which emerges first kills her rivals, and subsequently mates and begins to lay. The old queen and her successor may live amicably together and both lay eggs in the same broodnest (p. 287). Swarming colonies usually construct more cells than this, and the old queen usually departs with the swarm soon after the first queen cell is capped.

The close relationship between the two procedures was emphasized by Cale (1946), who stated that more swarms are the direct result of supersedure than of any other factor. Alfonsus (1932a) recorded three instances in which swarms emerged from colonies which were, in his opinion, superseding. One occurred in spring, when two queen cells were raised in a 6-comb nucleus but a spell of hot weather produced a swarm. Two others occurred in August, when a very warm day caused two colonies, with two and four queen cells respectively, to swarm—both swarms superseded their old queens in September. Conversely, he observed numerous queen cells in two colonies in June, but swarming was inhibited by cold rainy weather from the time of sealing onward—when warm weather returned no swarms issued and examination showed that only one virgin queen had emerged in each colony and that the other queen

[1] The number of marked bees of various ages, expressed as a percentage of the total number of marked bees of that age present in the colony from which the swarm emerged, was: under 3 days, 4–6%; 4 to 15 days, 54–82%; 16 to 19 days, 38–73%; 20 to 23 days, 35–57%; 24 to 27 days, 22–38%; 28 to 31 days, 17–30%; 32 to 35 days, 10–19%; 36 to 39 days, 5–12%; 40 to 43 days, 0–2%. Taranov (1947) erroneously concluded, from a few observations using young marked bees only, that bees more than 22 days old should not be in the swarm.

cells had been torn down. This last observation is consistent with Huber's statement that bees only swarm when the weather is fine, and that, if the sky becomes overcast when they are agitated in preparation for the issue of a swarm, calm is restored.

THE CAUSE OF SWARMING

The 'brood food theory'

The theory that a surplus of bee milk causes swarming was put forward by Gerstung (1891–1926). Gerstung supposed that (1) there is an interchange of food material between all the bees in a colony, (2) there is a rigid division of labour according to age, and the young bees get rid of their surplus protein as bee milk, (3) if insufficient larvae are present the bee milk is fed to the queen and increases her egg laying, (4) each nurse bee 'can and must' produce bee milk for 5–10 larvae, so when the queen reaches her maximum output of eggs an enormous surplus of bee milk occurs, and (5) drone production provides an outlet for this surplus at first, but when it becomes too great another outlet is found in the rearing of queens, which leads to swarming. Gerstung realized that this theory would not explain all the phenomena of swarming, and he suggested that some superimposed nervous stimulus caused the suppression of egg laying, comb building and other activities.

Morland (1930) reviewed Gerstung's theory, and he considered that swarming was aggravated by (*a*) restriction of the broodnest, which causes a sudden drop in egg laying when it is filled, (*b*) very rapid spring build-up, which subsequently produces a more acute surplus of nurse bees, (*c*) old queens, which are less likely to breed steadily, (*d*) lack of ventilation, which might act either directly or by stimulating the activity of the pharyngeal glands; it might be diminished by (*a*) nectar flows, (*b*) increased drone production, (*c*) compelling bees to build comb, (*d*) removing bees and brood. Morland pointed out that the effects of nectar flows and comb building could be related to reductions in bee milk, since Rösch (1925) had found that division of labour in the colony was flexible; he thought that the other factors concerned also exerted their effects through changes in the quantity of bee milk.

Gerstung believed that the surplus of bee milk arose when the egg-laying activities of the queen reached their maximum, and Morland that it arose when these activities were beginning to decline. Koch, R. (1934), who also accepted an excess of nurse bees as the cause of swarming, held that this excess resulted from a reduction in egg laying which was caused by the temporary exhaustion of the queen: Lehnart (1935) pointed out that Koch's views could not be reconciled with the fact that small hives aggravate swarming. Taranov (1947) proved that exhaustion of the queen plays no part in the swarming cycle; he transferred the queens and young bees from six cell-producing colonies into new hives and found that the rate of egg laying slightly increased during the following three weeks (the rate of six similar queens, not transferred, was reduced by 72%).

The 'brood food theory' is consistent with a proportion of the observed facts, but there are weighty objections to it. Confinement or infertility of the queen would be likely to produce the maximum possible surplus of bee milk. Yet Huber (1814) observed that queen cells were not produced when the reigning queen was confined in a small gauze cage, providing that the cage was made so that she could touch the other bees in the colony with her antennae (see p. 285); Lehnart (1935) agreed with this observation; Southwick (1947) reported the acceptance of an infertile queen by the bees of two successive colonies, without cell production. Todd & Bishop (1941) reported that the swarming period in California coincides with a sharp curtailment of breeding due to a decline in the amount of pollen available.

Demuth (1922) reported that colonies gave up swarming when shaken on to empty combs, although there would be no larvae to feed for 3 days, and the bee-milk surplus would thus increase. He also stated that he had obtained several prime swarms from colonies made entirely from large numbers of bees 'old enough to work in the fields', which should therefore have been deficient in bee milk. Work on winter bees shows that pharyngeal gland development reaches a maximum during the winter (p. 269); it has been calculated that the proportion of nurse bees to brood increases as the summer advances, and reaches a maximum in autumn (Bodenheimer, 1937).

Despite these difficulties the brood food theory is generally accepted in Europe, and has been the basis of many methods of swarm control; an extreme example is Blumenhagen's (1950) modification of the building frame, arranged so that the bees are encouraged to expend their surplus bee milk in feeding large quantities of drone brood. It is a feature of Gerstung's theory that the rearing of drone brood should provide an outlet for energies which would otherwise be directed towards swarming, but Huber (1792) considered that the rearing of drones was essential to swarming, and Langstroth (1890) held that their presence was very conducive to the process. No experimental evidence relevant to this conflict of opinion is known to the writer.

The congestion theory

While Huber (1792) believed that the presence of drone brood in the hive was a condition of swarming, he added that 'this single condition is not sufficient; it is also necessary that the bees be very numerous in the hive; they should be superabundant, and it may be said that they are aware of it, for if the hive is scantily peopled they will not build any royal cells at the time of the laying of the male eggs'. The importance of congestion was appreciated by others, including Langstroth (1890); Newell (1913) found that swarming was either delayed for several weeks, or prevented entirely, by adding an empty super which increased the broodnest capacity by 50%, providing that this operation was undertaken before queen-cell construction had been commenced.

Demuth (1921) claimed that one condition only was invariably present when bees prepared to swarm—congestion of their broodnest. Demuth (1922, 1931) subsequently emphasized this attitude and considered that congestion was the actual cause of swarming. The congestion occurred in the area occupied by the bees, but not necessarily throughout the hive itself.

Demuth (1921) considered that the incidence of swarming could be reduced by (*a*) a large broodchamber, (*b*) providing extra space between the frames in the broodnest, (*c*) more ventilation, (*d*) shading the hives, (*e*) providing additional supers; that it was increased by (*a*) very rapid increase in the size of colonies in spring, (*b*) large numbers of young bees, (*c*) idle foragers. Demuth (1921) stated that there was less swarming during meagre seasons; this does not hold for Britain, where meagre seasons are usually seasons of bad weather and confinement to hives—Demuth (1922) rightly emphasized that confinement always results in excessive swarming. Demuth (1921) concluded that a break in the continuity of emergence of young bees was an essential part of any successful treatment for swarming and that all effective methods of treatment came under three headings: (a) taking away the brood, (b) taking away the queen, or (c) separating the brood and the queen within the hive.

Demuth (1931) took away all the brood combs from 'hundreds' of colonies with sealed queen cells, and substituted combs which contained only sealed brood; 'in practically every case they immediately gave up swarming' (this procedure should have aggravated any surplus of bee milk).

Important support for the congestion theory of swarming is contained in the work of Perepelova (1947), who reduced the spacing between the combs of the broodnest in spring from 0·47 inch to 0·33 inch, in an effort to increase breeding; she found that the narrowed spacing increased the rate of egg laying for only about 10–12 days, and that preparations for swarming were then made (although the recently increased laying would tend to decrease any surplus of bee milk). It is curious that a quarter of a century elapsed before the publication of this experimental evidence relative to Demuth's postulate that increased spacing in the broodnest was an aid to swarm prevention, and that even this evidence was not deliberately acquired.

In addition, Nestervodsky (1939) observed 288 colonies, some facing north and shaded, others facing south and standing in the sun. Swarming preparations occurred in only 27% of the former group, and none swarmed, but 80% of the latter group prepared to swarm and 66% did so. Kotogyan (1941) found that swarm cells were induced by warming hives with gadgets containing hot water. These results are consistent with Demuth's views of the effects of shading and ventilation on swarm reduction.

The congestion theory of swarming was criticized by Hamilton (1932), who suggested that swarming might be a consequence of dissatisfaction resulting from some reduction in the odour of the laying queen. Hamilton said that congestion would not explain starvation swarms, nor why colonies could be made to swarm by dequeening them at the height of the season. Neither difficulty appears important: starvation swarms only occur when foraging conditions are very bad and they must therefore be associated with idle foragers; queenless colonies would rear more than one new queen, and conditions at the time of emergence would determine whether swarming would occur.

It is not easy to design decisive experiments on swarming, and many existing observations could be interpreted with equal readiness in terms of either the brood food theory or the congestion theory. The swarming recorded by Perepelova (1928*b*) when she repeatedly robbed colonies of unsealed brood and returned to them their own sealed brood together with the sealed brood of another colony, and that recorded by Morland (1935) when he repeatedly added newly emerged bees to two colonies, comes into this category; so do the observations of Taranov & Ivanova (1946).

The amount of available bee milk might reasonably be expected to depend not only upon the number of bees of nursing age, but also upon the food supply available to them, yet in uncramped colonies swarming preparations are not made in optimum foraging conditions. They are usually commenced, however, in bad conditions, when diminishing food supply might be expected to lower the bee-milk output of the nurse bees, and to congest the broodnest with idle foragers.

The cause of swarming: opinion

I conclude that there are many well-established facts which are quite inconsistent with Gerstung's brood food theory, or any modification of it. On the other hand, most of the available evidence is consistent with Demuth's congestion theory.

Yet one may still wonder whether congestion is the only important factor. Simpson (1953), who is using various methods in attempts to produce swarming, has so far only been able to achieve this end in overcrowded conditions—but he finds that even with artificially produced overcrowding of a very high order (much greater than in any non-experimental colony) swarming does not occur inevitably.

One might guess that the onset of swarming is related to a decline in the quantity of the substance, derived from the queen, the circulation of which apparently inhibits the development of laying workers (p. 282) and makes the bees behave as if queenright (pp. 285–7). Perhaps the production of this substance usually bears some relation to the quantity of eggs laid, and egg laying is suddenly diminished by congestion?

The function and evolution of the swarming habit are discussed in Chapter 38 (p. 316).

Chapter 34
SEASONAL DIFFERENCES IN PHYSIOLOGICAL CONDITION AND LENGTH OF LIFE

Life is not measured by the time we live.
CRABBE, *The Village*

Seasonal differences in physiological condition

RECENT INVESTIGATIONS INTO variations in the physiological state of honeybees are relevant to their behaviour. Seasonal differences were described by Adrienne Köhler (1921), who found that in winter the fat bodies of honeybees contained very many globules of protein, as well as fat, although these protein globules were never found in the fat bodies of honeybees in summer. They appeared in September and started to disappear in March, so Köhler considered that bees kept their winter protein reserves in their fat bodies. Himmer (1927c) found that protein reserves were required for spring brood rearing, and that fat bodies were reduced and altered in the spring; he too concluded that the main reserves for brood rearing were stored in the fat bodies.

The results of Kratky (1931) were more surprising: during his investigation of the glands of the honeybee he discovered that in the winter the pharyngeal glands of a majority of the bees were fully developed, just as if they were nurse bees. Himmer (1930), when investigating the division of labour among bees, had found that nuclei could be manipulated so that their nurse bees had to continue brood rearing for a very long time, and that the nurse bees' expectation of life was then considerably lengthened. He had concluded that one might almost speak of a life-prolonging function of the pharyngeal glands. Kratky's work could be fitted with Himmer's conclusion, since bees live very much longer in winter than in summer.

Kratky's results were confirmed by Evenius (1937), who marked groups of newly emerged bees on 31 August and 23 October, and dissected samples at intervals during the winter. He found that, despite the $7\frac{1}{2}$ weeks difference in age, there was no significant difference between the pharyngeal gland development of the two groups.

These results led Ruth Lotmar (1939) to carry out a thorough analysis of the state of development of both the pharyngeal glands and the fat bodies of a colony of bees, during two successive winters. She also determined the nitrogen content of the abdomens (including gut) and the heads of the bees. Her analyses showed that the protein reserves were increased in autumn by the consumption of pollen; this was mainly eaten in September, but some was consumed in October, and a little in November and December. Pollen consumption increased again in January and February, rose to a maximum in March, and then fell off again; the spring consumption could be related to spring brood rearing.

Lotmar found that the enlargement of the pharyngeal glands slightly preceded the development of the winter fat bodies. During the winter there was a close relation, in individual bees, between the development of both structures, but when brood rearing commenced this relation ceased to hold. In September 20% of the bees had winter fat bodies and 58% had well-developed pharyngeal glands; in October these percentages rose to 56 and 84%; from November to January about 75% of the bees had winter fat bodies and about 2–3% more had well-developed pharyngeal glands; in February only 60% had winter fat bodies but over 80% had well-developed pharyngeal glands; in March the values were 15 and 50% and in April 0 and 50%. Analyses showed that the total nitrogen content of the winter bees was actually greater than that of 10-day-old nurse bees in summer, but much of this nitrogen was in the gut in pollen grains being digested, and the average amount in the head (0·39 mg.) was rather less than that of nursing bees (0·43 mg.).

An interesting feature of Lotmar's results was the difference (statistically highly significant) between the bees on the boundary of the winter cluster and those at its centre; in October and February the bees on the boundary had glands and fat bodies much less well developed than those within, but in March, when brood rearing had begun, this was reversed.

In experiments with caged bees, Bertholf (1942) found that the average expectation of life of caged winter bees was slightly longer than that of summer ones; the caged bees in Anna Maurizio's (1946) experiments survived for about twice as long as those in Bertholf's, and she obtained substantial differences between winter bees (average life 36 days) and summer bees (average life 24 days). She also found that the life of summer bees could be prolonged by adding pollen to their diet, but that of winter bees was not affected by this; these results might therefore be due solely to the differences in the protein reserves of the two kinds of bees in conditions in which protein is essential.

The work of Corkins & Gilbert (1932) on the metabolism of honeybees suggests that there are other differences between summer and winter bees which are not structurally obvious. An apparatus was designed for measuring the respiration of a cluster of honeybees exposed to constant low temperature for prolonged periods. They used two clusters, both containing some 12,000 bees; experiments were made in March (with a broodless winter cluster) and repeated in July (with a broodless cluster formed by shaken bees). Their results showed that at the same temperatures the carbon dioxide output of the summer cluster was always more than twice that of the winter cluster.

Effect of brood rearing upon length of life

Investigations on the division of labour have shown that the expectation of working life of summer bees can be substantially increased if all the sealed brood is regularly removed from a colony, so that the same nurse bees continue with brood rearing indefinitely. In such circumstances summer lives of 72 days have been reported (Moskovljevic, 1939). Conversely, Ribbands (1950b) showed that the expectation of adult life of bees which commenced foraging at an early age was significantly less (30.1 ± 1.2 days) than that of those starting to forage later (37.1 ± 0.6 days), although the expectation of *foraging* life of the early foragers was greater (15.0 ± 1.2 days compared with 10.8 ± 0.8 days).

Maurizio (1950) has investigated the length of life of honeybees living in a normally breeding queenright colony and of those living in a colony with a caged queen but without brood (and without laying workers). One thousand to two thousand newly emerged bees were marked and liberated in each colony and samples of them were captured at intervals; some were caged and others dissected. One experiment commenced in mid-June. After 10 days the pharyngeal glands of both groups were equally well developed but the fat bodies of the broodless bees were larger. After 27 days the broodless bees had pharyngeal glands fully developed, fat bodies building up, and an increased expectation of caged life, but the other bees had exhausted glands and fat bodies and shortened lives. After 38 days there were no more marked bees in the normal colony, but plenty in the broodless one. In this broodless colony the expectation of caged life only reached its maximum in a sample taken on the 58th day; the last 5 bees lived until the 166th day and dissection then showed that they still had relatively well-developed pharyngeal glands and fat bodies. This experiment was repeated, and gave similar results.

These experiments have an additional value in the understanding of queenless colonies. They enable a distinction to be drawn between the primary effect of queenlessness—upon ovary development—and the secondary effects—upon pharyngeal glands, fat bodies, length of life and foraging. The latter are the consequence of broodlessness.

Maurizio also found that the pharyngeal glands and fat bodies developed more, and the expectation of caged life increased, among bees introduced into a colony which swarmed; here again it seems that the pause in breeding caused the retention of protein reserves and so contributed towards longevity.

Effect of anaesthesia on length of life

Ribbands (1950b) found that treatment of young bees with either nitrogen or carbon dioxide decreased their expectation of life. This was not a consequence of artificial ageing, but of a change in their behaviour patterns. Brood rearing and other hive duties were curtailed or eliminated, and the treated bees commenced to forage at an early age; in foraging their pollen-gathering activities were markedly reduced. Fyg (1950) found that queenless carbon-dioxide treated bees did not draw foundation, build queen cups, or store food; their pharyngeal glands and fat bodies were less developed than those of controls.

Length of life of bees in summer

Maurizio's comparison of the length of life of bees in normal and in broodless colonies, in summer, provided results which confirm the impression which many beekeepers have that bees in a queenless colony live longer than those in normal queen-right colonies. Her results of maximum life in two normal colonies (less than 38 days and less than 31 days) are slightly lower than others. Rösch (1925) recorded the ages at which 13 individually marked bees were last seen alive; these ages ranged

PHYSIOLOGICAL CONDITION AND LENGTH OF LIFE 273

from 20 to 55 days and the mean age was 32·1 days. Morland (1938), who introduced known numbers of marked bees into two colonies, reported that their expectation of life was about 3 weeks and that bees seldom survived more than 45 days. Rockstein (1950a) introduced 2,700 marked bees into a colony, removed a total of 176 for sampling on various dates, and found

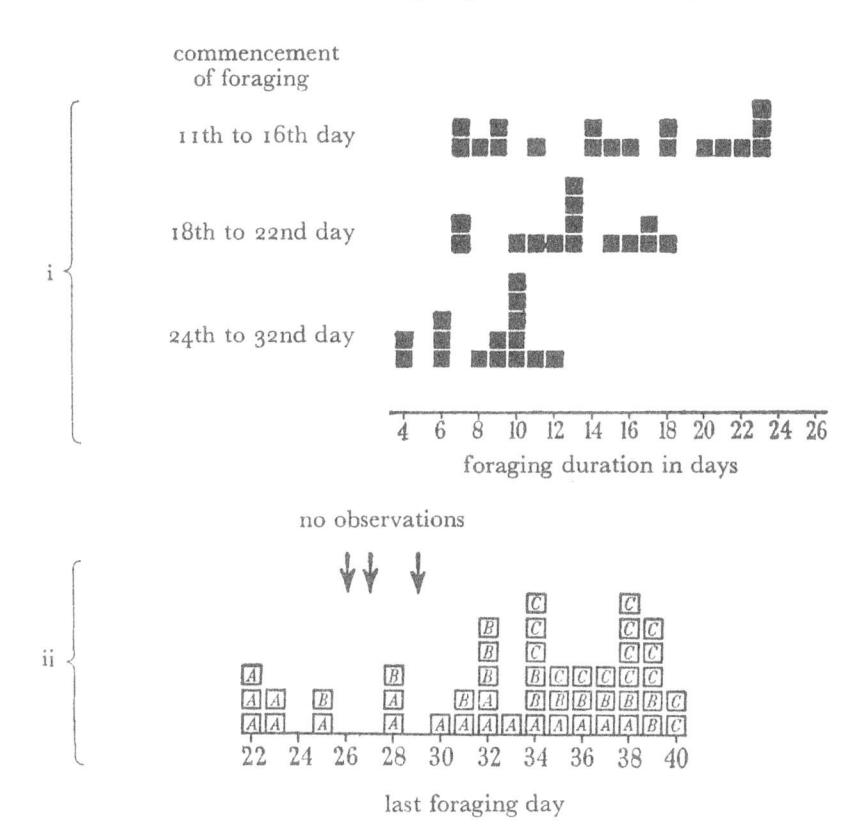

Fig 59 (i) FORAGING DURATION. (ii) LONGEVITY. Records of 47 bees, all with the same birthday and living in the same colony. A=bees which commenced foraging when 11 to 16 days old. B=bees which commenced foraging when 18 to 22 days old. C=bees which commenced foraging when 24 to 32 days old. From Ribbands (1952b).

only 11 of the remainder alive after 51 days. In one experiment involving 47 individually marked bees, Ribbands (1952b) recorded that their mean length of life was 33·6±0·2 days (range 22 to 40 days); in another, involving 52 bees, their mean length of life was 34·8±0·5 days (range 17 to 40 days).

Length of life of bees in winter

Bees in winter are known to live much longer. Root (1910) instanced a colony of black bees, requeened with Italians in September, in which some black bees were seen until the following May, and Anderson (1931) cited a similar observation on a completely isolated colony of bees, on the Isle of Lewis, which was queenless from 20 August until 24 September, 1912, and was then given an Italian queen. This colony still contained a few black bees on 12 July, 1913, 304 days after the loss of the original queen. Farrar (1949) marked large numbers of newly emerged bees at various dates in autumn and overwintered them in a colony which was short of pollen reserves and therefore broodless, and which was situated at 7,200 feet in a mountain valley, where there was a long winter and no fresh pollen was available until mid-June. In these circumstances a very small number of bees survived for at least 320 days. Evenius (1937), who marked 440 newly emerged bees on 31 August and 694 more on 19–23 October, concluded that the length of life of both groups was the same, the younger bees dying earlier in the spring.

Length of life and senility

Hodge (1894) claimed that sections of the brains of young and old bees showed that the former contained nearly three times as many cells as the latter, and that in the latter the cytoplasm of the cells was reduced and vacuolated. His claim was disputed by Smallwood & Phillips (1916), but partly confirmed by Helen Pixell-Goodrich (1920). Rockstein (1950b), who studied this problem in more detail, found that the average number of cells counted in sections across the brains of newly emerged bees was 522, and that this average decreased to 445, 434 and 369 respectively in bees 2, 4 and 6 weeks old.

Ribbands (1952b) noted that the mean length of foraging life of 32 bees which commenced foraging when 11 to 22 days old was 14.4 ± 0.9 days, but among 15 bees (of the same age and living in the same colony) which commenced foraging when 24–32 days old it was only 8.3 ± 0.6 days. In a second experiment 44 bees which commenced foraging when 9–22 days old had a mean foraging life of 17.2 ± 0.8 days, but 8 similar bees which commenced to forage when 28–35 days old had a mean foraging life of 7.2 ± 0.5 days. This result indicates that deaths were not entirely attributable to accidents (as has been postulated for some insect populations), but that senility played a part in the mortalities of at least the later foragers.

Conclusions

Winter bees differ from normal summer bees in the greater development of their pharyngeal glands and their fat bodies. This development results from autumn consumption of pollen, in excess of the requirements for immediate brood rearing. In queenright colonies in summer, prevention of brood rearing can produce similar consequences, and in pre-swarming colonies temporary interruption of brood rearing produces conditions different only in degree.

In all these instances the enhanced development of the pharyngeal glands and fat bodies is associated with a considerable increase in the expectation of life. There is also a marked decrease in the rate of working; in wintering colonies foraging and brood rearing are both greatly reduced; in pre-swarming colonies brood rearing ceases and the bees cluster on the combs instead of foraging normally (Taranov, 1947); Maurizio makes no mention of the relative foraging abilities of normal and broodless queenright colonies, but the foraging of queenless colonies is known to be reduced and one might infer that broodless queenright colonies are similar in this respect.

In winter bees, the reduced rate of work could be attributed to external factors, but in broodless and pre-swarming bees the reduced foraging must be derived from the change in the physiological condition of the bees. It is therefore probable that the reduced metabolism which Corkins & Gilbert (1932) observed in winter bees (p. 271) is characteristic of broodless bees and a consequence of the hypertrophy of the pharyngeal glands and fat bodies.

In winter bees and swarming bees, at least, stores of food within the body have some clear advantages over external stores; the stores of the former are more protected and those of the latter are more mobile. It seems that in both they also contribute towards quiescence and longevity, at a time when these two qualities are of most value.

Are the increased protein stores eventually associated with an increased total output of work as well as longevity? The reduced activity associated with the increased stores is sufficient to prevent a clear answer to this question, but Maurizio (1950) found that the mean caged life of broodless queenright bees continued to increase until they were 58 days old at sampling, and at this time the expectation was substantially greater than that of normal summer bees of any age. This evidence certainly suggests that this question could be answered in the affirmative. The increased expectation of life of bees which are forced to continue indefinitely with brood rearing does not solve the problem; such bees are active and not equivalent to winter bees (cf. fat bodies); their comparative longevity can be attributed to their less hazardous and less strenuous lives.

If increased pollen consumption does in fact increase the total output of work, why is it that normal summer bees do not lengthen their lives and their usefulness to some extent in this manner? One would think that the increased forage collected by a longer-lived bee would more than repay a slight increase in her pollen consumption. Perhaps a larger but less durable population is usually of more value to a summer colony than a smaller but more durable one?

Chapter 35
LAYING WORKERS

Something there was in her life,
Incomplete, imperfect, unfinished.
LONGFELLOW, *Evangeline*

THE TERM 'LAYING WORKER' is sometimes extended to include not only those workers which actually lay eggs, but also those which possess ovaries either partially or fully developed. Perepelova (1926) classified egg-laying workers as 'physiological laying workers', and those with developing ovaries as 'anatomical laying workers'; the latter group will here be referred to as 'ovary-developed workers'.

Production of laying workers

An important characteristic of laying workers is that different species, races and strains of honeybees vary greatly in their propensity for producing them. Thus Millen (1942) found that in the United States *Apis mellifera* did not produce laying workers when either eggs, young brood or 'anything which the bees regarded as a queen' was present in the hive, and that in established colonies no workers laid eggs until 10–26 days after removal of the queen and 5–13 days after removal of the last queen cell. When he went to India he found that *Apis indica* often had laying workers when brood and queen cells were in the hive, and that such bees might continue laying for some time after the virgin had emerged. Gertrud Hess (1942), who studied ovary development in the workers of eleven stocks which she dequeened, found that it usually commenced in 10% or more of the bees within a week, but there were two exceptions (which she was unable to account for) in which it had not commenced within a fortnight. These variations make it possible that some reported results may not be generally applicable.

Verlaine (1929) suggested that workers often laid eggs in normal honeybee colonies, and that they were responsible for the production of a proportion of the drones; the complete broodlessness of dequeened colonies, and many dissections of workers in normal summer colonies, suffice to contradict this hypothesis.

Perepelova (1926) made up a nucleus on 29 June with bees 1–3 days old and supplied it with a virgin queen which commenced to lay 9 days later; dissections revealed ovary-developed workers when the nucleus was 5 days old; their numbers increased to about 60% a few days after the virgin commenced to lay, and they remained at this level for about a fortnight. Perepelova (1928*b*) also reported worker ovary development in colonies from which young unsealed brood was regularly removed, and in swarming colonies; she stated that it increased from the moment of decreased brood rearing.

Perepelova (1926) found that workers 8 days old might have developed ovaries; Leuenberger (1927) found 70% of laying workers and workers with developed ovaries in a colony which had been 6 weeks queenless, and noted that the proportion of bees with developed ovaries was similar among pollen-gathering foragers, store bees and nurse bees. Perepelova (1928*a*) and Gertrud Hess (1942) dequeened a colony which contained groups of marked bees of known ages; they both found ovary development among workers of any age; its incidence was very irregular and bore no relation to age.

The cause of worker ovary development

The cause of worker ovary development has been the subject of much speculation. Huber (1814), who did not realize that laying workers are usually produced only in queenless colonies, suggested that they were bees which had developed in cells adjoining queen cells and had accidentally received some of the nutriment intended for the queen larvae. Other hypotheses (e.g. Dadant, 1893; v. Buttel-Reepen, 1915; Leuenberger, 1928) have related their development to the surplus of bee milk which is usually found in queenless colonies; the researches of Gertrud Hess (1942) show that this conjecture is insufficient.

Hess compared the development of the pharyngeal glands of dequeened bees with the development of their ovaries. She found that the glands usually became well developed very quickly after dequeening, and that ovary development followed a few days later. The pharyngeal gland development may not have been related to the queenlessness, but to the ensuing brood food surplus. Her data showed that the pharyngeal glands of the ovary-developed workers were, in general, fuller than those of the rest of the colony, but that every possible combination of gland and ovary development was likely to be found in each colony.

In one colony, which Hess dequeened on 7 October, 95% of a sample of 53 bees had well-developed pharyngeal glands on that date. Subsequent samplings at intervals during November and December (which showed worker ovary development) showed that in this colony the proportion of bees with well-developed pharyngeal glands had been reduced to about 10%. This gland reduction was contrary to the seasonal trend (pp. 269–70) and appears to indicate that the ovaries of some of the bees were developing at the expense of the glands of all of them. A second experiment, with a nucleus dequeened in mid-September and examined in October and January, showed a similar trend.

Perepelova (1926), Peterka (1929) and Müssbichler (1952) reported the rapid development of ovaries in queenless honey-bees which had been kept in cages and fed on sugar and water only. These experiments provide additional grounds for the conclusion, which Hess derived from her own results, that ovaries sometimes develop in conditions in which an accumulation of surplus bee milk is unthinkable, so that this cannot be the cause of their development. Developed ovaries, however, were completely resorbed when bees were starved for 40 hours (Drösi, 1929).

Gertrud Hess also confined 60 bees individually in separate cages and kept them in an incubator for a fortnight. During this time she fed them with rich food, but their ovaries did not develop.

Dönhoff (1859) supplied two colonies with virgin queens confined so that they could not mate. No eggs were laid in either colony for six weeks, and the length of this period led him to suggest that some non-nutritional factor played a part in laying-worker development. Haydak (1940b) pointed out that superabundance of nourishment could not sufficiently explain the cause of laying workers, since these do not occur in colonies with failing or unproductive queens, despite the reduction or absence of brood in such circumstances. He made up two 3-lb. packages of bees, and left one of them queenless but supplied the other with a queen which was caged and unable to lay. After 31 days there was an abundance of drone brood in the queenless nucleus, but the other one, with a caged queen, contained neither eggs nor brood for 65 days. The queen laid normally when she was then released. From this experiment Haydak concluded that some factor other than nourishment was operative, and he suggested that it might be psychological.

Hess (1942) designed an experiment in which 250 bees on combs without brood were placed in the hive of a queenright colony, but separated from it by a double partition of two sheets of gauze, 5 mm. apart. Separate entrances were provided for the 2 groups. Eighteen days later young drone brood, richly supplied with brood food, was present in the queenless portion, and 8% of the queenless bees had developed ovaries. She therefore concluded that neither temperature nor scent was the operative factor, but that some material substance which inhibited ovary development was probably produced by the queen and circulated in food exchange.

Hess assumed that laying workers would not have developed if the queenless group had only been separated by a single partition, but this assumption was not justified. Müssbichler (1952) experimented with bees which were separated from their queen by a single partition of 1½ or 4 mm. mesh wire gauze; the separated bees reared cells as if queenless; after 21 days 68% of them had enlarged ovaries and after 5 weeks they had laid drone eggs. In other experiments the queen was caged alone in the midst of her bees; their ovaries enlarged although usually queen cells were not built, but the experiments were not continued to egg laying (p. 286).

Müssbichler carried out a series of experiments to determine the relative importance of the queen and the unsealed brood. In three experiments with cages of queenless and broodless bees the percentage of ovary-developed workers ranged from 41 to 69%; in six experiments with queenright broodless bees the mean was 40%; in three experiments with queenless bees repeatedly supplied with an abundance of unsealed brood the range was 10–30%.

Müssbichler also fed caged queenless bees on various diets. Enlarged ovaries were found in bees which were taken from their colony when 3 days old and fed for 21 days on sugar solution only, but newly emerged bees showed no ovary development after similar treatment. Addition of casein to the diet of the newly emerged bees produced ovary enlargement, but an extract of the vitamin-B complex did not alter the effect of either the sugar or the sugar-casein diet.

The corpora allata are a pair of small glandular organs which produce an internal secretion concerned with growth and reproduction (Wigglesworth, 1950). Müssbichler noted that in her experiments enlargement of corpora allata paralleled the development of ovaries, but she found that additional corpora allata from either workers or queens produced no noticeable effect upon the ovaries of workers into whose thoraces they had been grafted.

Evidence of the existence of a chemical substance with a positive influence on ovary development in worker bees was provided by Altmann (1950). Groups of 20 worker bees were injected with various extracts, caged for 3 days, and their ovaries examined. Control groups were injected with Ringer's solution only. No ovary development was induced by extracts of worker bees, worker larvae 3-4 days old, food of worker larvae 2 or 4 days old or royal jelly from older queen cells; on the other hand significant ovary enlargement was induced by extracts of queens or queen larvae, worker larvae 2 days old, royal jelly from cells of queen larvae 1-2 days old or the heads and thoraces of nurse bees. The effective material was of a protein nature, soluble in water and destroyed by heat.

We must now try to fit together the varied evidence concerning the origin of laying workers. Nutrition is important. Protein is essential, but protein reserves within the bees themselves may suffice; starvation causes retrogression of the ovaries. In queenless colonies which have to feed an abundance of unsealed brood, ovary enlargement is restricted and is actually less than in queenright broodless colonies. Altmann's results indicate a special role for a factor associated with, or derived from, bee milk.

However, nutrition does not provide a complete explanation. In queenless colonies the incidence of ovary development seems to be fortuitous and not related to either the age or the duties of the bees concerned, and there may be every possible combination of ovary and gland development. Moreover, the ovaries of richly fed isolated individuals do not enlarge.

The results of Dönhoff, Haydak and Hess would all be explicable with equal readiness in terms of inhibition either by a chemical or a purely psychological factor, and Müssbichler's demonstration of ovary enlargement in bees separated from their queen by a single screen of gauze would invalidate neither explanation; chemical transmission is much the more likely alternative. The effect could not be produced by an odour. Some substance might be obtained from the queen by the workers who caress and attend her (p. 284), or, as Hess suggested, an ovary-inhibiting factor might be contained in the excrement of the queen. Any such substance would be a kind of social hormone.

The development of laying workers would seem to depend upon at least two factors—a positive nutritional influence and an inhibitory influence derived from the queen. Worker ovary development may be produced either in the presence of the one or the absence of the other, but in *Apis mellifera* there is as yet no evidence that actual laying workers are produced by either factor acting alone.

Behaviour of laying workers

Dönhoff (1857) recorded that there was no queenly antagonism between laying workers, and that colonies with laying workers tried to produce queen cells, so they were aware of their queenlessness. Laying workers defended their colony against robbers, and there were large eggs in the ovaries of some pollen gatherers.

Perepelova (1928a) watched 6 laying workers, each for 1 day, and found that in that time each produced from 19 to 32 eggs; eggs were laid one at a time in selected cells, and the average time of oviposition was 78 seconds (about 8 times as long as a queen takes.) Intervals between laying might be from several minutes up to 3 hours, and during intervals the laying worker carried on 'with any of the usual duties of ordinary workers, sometimes even flying out of the hive and bringing in nectar or pollen'.

Behaviour towards laying workers

Perepelova (1928a) noted that several days before any workers laid eggs, some bees began to behave strangely—rushing round the combs and examining the cells intently. Several other bees gathered round them, cleaning them assiduously; sometimes they submitted, but at other times they ran away and were chased by the cleaners. When a worker was actually laying an egg there were always 3 or 4 bees caressing her with their antennae; they cleaned her and then left her when the egg had been laid. Perepelova did not see laying workers being fed.

Behaviour of colonies with laying workers

The difficulties of requeening colonies with laying workers are well known (p. 289). Yet such colonies try to rear queen cells, and they pay more attention to brood (either normal brood supplied to them by the beekeeper, or drone brood from their laying workers) than normal colonies do (Perepelova, 1928a).

Chapter 36
QUEEN RECOGNITION; QUEEN INTRODUCTION

And for their monarch Queen—an egg-casting machine,
Helpless without attendance as a farmer's drill.
ROBERT BRIDGES, *The Testament of Beauty*

Behaviour of the workers towards laying queens

THE LAYING QUEEN receives special attention from the workers near her. About a dozen of them form a loose circle facing her (Pl. 1); according to Rösch (1925), bees 1–28 days old were seen in her retinue, which changed continuously as the queen moved around and was formed immediately on any part of the comb to which she was transferred; Taranov & Ivanova (1946) agreed that the entourage was always changing, but said that it was composed of nurse bees only and that if the queen went to a part of the hive where there were no nurse bees she gradually lost all her suite.

The workers in front of the queen frequently feed her; the rest of the suite touch and caress the queen with their antennae, and sometimes lick her. Perepelova (1928*d*) observed that when in full lay the queen was fed every 20–30 minutes; 32 feeds were watched, and the donors were all of nursing age (6–13 days). The process lasted $1\frac{2}{3}$–$3\frac{1}{4}$ minutes. When laying was reduced, the queen received less attention and less food. At the end of the summer the workers stopped feeding the queen, who fed herself on honey; at this time any eggs which she laid were eaten by the workers.

Some writers have stated that the workers pay no special attention to a virgin queen, but that they form a suite around her from the time when she returns from a successful mating flight. I have not been able to trace precise observations on this point.

Behaviour of the colony after the removal or caging of the queen

When a queen is removed from her colony the bees do not at first seem to miss her, and it may even be several hours before disquiet commences. Agitation commences in one part; the disturbed bees run over the combs and meet others, and the antennae are reciprocally crossed and lightly struck; the recipients become agitated in their turn, and the disorder rapidly spreads through the entire colony (Huber, 1792). The loss is appreciated more quickly in a small colony (v. Buttel-Reepen, 1900); bees run wildly over the combs and the agitation quickly spreads from bee to bee; they become very irritable and vibrate their wings to produce what has been described as 'a low, mournful lament', an activity which may have some other significance for the bees.

We do not know how the agitation commences, but odour plays an important part. V. Buttel-Reepen observed that a queen cage from which a queen had just been removed would cause the bees to fan, and that for at least 24 hours the bees of a swarm, the queen of which had been crushed, remained quiet and licked her body. Yet he emphasized that no agitation commenced when the queen of a colony was kept isolated in a honey-filled super well away from the broodnest, and that the excitement of a queenless colony died away 'almost at once' when a queen was introduced in such a position. In a control experiment the queen was caged and placed on a stick, at the height of the entrance and 14 inches in front of the alighting board, without effect upon the queenless colony and without being noticed by any of the numerous bees which were flying in and out.

V. Buttel-Reepen, contending that odour could not penetrate or be lost from the colony quickly enough to account for his observations, considered that news of the presence or absence of the queen was conveyed by sound—lament or otherwise—but odours quickly directed by fanning would provide an alternative explanation of the results.

Colonies deprived of their queens always construct emergency queen cells if eggs or suitable larvae are available, but such cells may also be constructed in less extreme circumstances.

Huber (1814) divided a colony into two parts by a grated partition, so that the bees of one half were separated from the queen 'by a space not exceeding a third of an inch', through which smell and sound could pass; the queenless half soon became agitated and constructed cells. There was no distress when the queen was isolated in a cage through which the bees thrust their antennae and mouthparts, but agitation commenced when the queen was confined in a double cage and could not be touched.

Müssbichler (1952) experimented with an observation hive which contained combs side by side; the combs could be separated by a single wire screen of either 1½-mm. or 4-mm. mesh, and the hive could be rotated so that either side could fly. When the screen was put in position bees could contact one another through it, but the queenless half started to construct queen cells within 24 hours (Fig. 60a). Alternatively, the queen was confined *alone* on one of the combs in a cage made from wire mesh like that of the screen (Fig. 60b); although she could not lay she was regularly fed by the bees, which showed no signs of queenlessness and fed and sealed their brood without constructing queen cells. When the cage and screen were both used (Fig. 60c), the bees on the queenless side of the screen produced queen cells but those on the side with the caged queen did not.

Müssbichler, who was primarily concerned with the production of laying workers (see p. 280), does not say whether the bees screened off from the queen became agitated. These results appear to have been produced when there was only restriction (not complete prevention) of contact. They may be taken to imply that the queenright

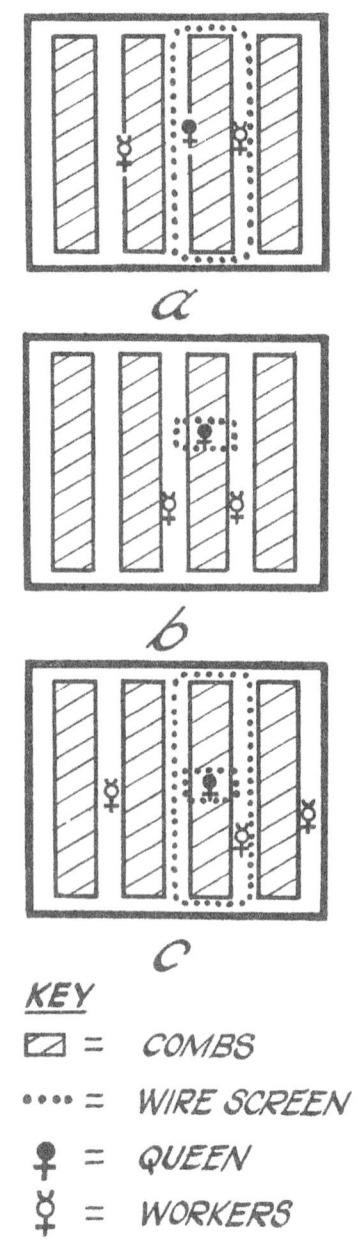

Fig 60

DIAGRAMS TO ILLUSTRATE EXPERIMENTS CONCERNING THE EFFECT OF QUEEN CAGING ON QUEEN-CELL PRODUCTION (after Müssbichler, 1952).

KEY
- ▨ = COMBS
- •••• = WIRE SCREEN
- ♀ = QUEEN
- ☿ = WORKERS

state of a colony depends upon a material which is derived from the queen, and which is transmitted by food sharing or by antennal contact. It is probably the same material as that which inhibits the development of laying workers (pp. 280-2).

Antagonism between queens

Queens which meet usually fight, and one or the other is quickly stung and killed. Piping (p. 54) is considered to be a challenge between rival queens; Huber (1792) observed that the piping of queens stilled surrounding workers,[1] and Landois (1867) found that queens could be encouraged to pipe by confining them separately in small cages placed close together.

Colonies with more than one queen

There are conditions in which a colony will tolerate more than one queen. The most usual circumstance is after supersedure, when the old queen often continues to lay until her daughter has mated and begun to lay also; there are observations (e.g. Doolittle, 1908; Kelsall, 1940) on colonies in which both queens have survived, laying well, until the following spring. It is likely that the old queen often survives until her daughter happens to meet and kill her, as witnessed by Davis (1908), but the hostility does not seem to be inevitable, as there are records of mother and daughter meeting and touching each other without enmity (e.g. Taranov, 1951).

In exceptional circumstances several queens may be found living together; Smith (1923) produced such a colony: when he put 12 laying queens together in a jar, 5 of them were soon killed, but the remaining 7 settled down together, showed no animosity, and were introduced to a comb of emerging brood; they laid many eggs and the nucleus built up rapidly. However, one morning Smith noticed a dead queen outside the hive, and later another, and another. He examined the colony and found that the bees were balling another queen, and the process continued until only one queen remained, so he concluded that the workers would frustrate attempts to build up permanent multiple-queen colonies.

More complicated methods for producing multiple-queen colonies have recently been described by Kovtun (1949-50),

---

[1] Hansson (1951) found that bees in the hive were quietened by sounds of 25-1,500 c.p.s.; this could be observed if one produced a creaking noise by rubbing a wet finger on the glass of an observation hive.

Melnik (1951) and Barykin (1951). These Russian beekeepers have attempted to find a use for old queens (more than 1½ years old) in this manner, and their methods indicate that the opposition of the workers to the innovation was of more account than the rivalry of the queens. Kovtun stated that multiple-queen colonies provided 'about 5 times as much' surplus honey as normal ones, did not raise drones and never swarmed; Melnik's experience was less satisfactory—he produced a colony which immediately started to rear cells, and all 4 of the old queens were found dead outside the hive on the day the virgin emerged, so he considered that the method might be useful for preserving additional queens but not for increasing honey production. Doolittle (1908) reported swarming from more than half of a number of 2-queen colonies (mostly separated by excluders), but he thought that swarming 'seemed to be delayed to quite an extent'.

Queen introduction: favourable and unfavourable circumstances

Queen introduction is a problem of considerable practical importance, but there have been few experiments concerning the relative value of the alternative methods which are advocated. In these circumstances the following account is necessarily a selection of the opinions of leading experts, which vary to some extent.

i. *The colony.* All beekeepers agree that the risks of queen introduction vary greatly with circumstances. Queens are most readily accepted by queenless colonies during a nectar flow, and in dearth conditions it is advisable to feed the colony (e.g. Langstroth, 1890; Root, 1902; Smith, 1923). Small nuclei are said to be more easily requeened than large colonies (e.g. Langstroth; Hutchinson, 1905), and very young bees are much less hostile than older ones (e.g. Langstroth; Smith).

Huber (1792) found that queens were well received in a colony 24–30 hr. queenless but were attacked after a shorter period, and most subsequent writers have recommended that the queen should not be liberated for at least 24 hours after removal of her predecessor. For instance, Langstroth recommended caging for 1–2 days; Root would supply a caged queen as soon as possible and not more than 4–5 days after queen removal, but said that she should not be liberated for at least 24 hr.; Doolittle (1915) said that queens could often be run into colonies which had been queenless for 4–5 days; Hutchinson recommended that big colonies should be hopelessly queenless, with cells destroyed and no open brood; Smith (1949) recommended caging for at least 3–4 days. If the colony is left queenless for too long it will develop laying workers (p. 278) and become very difficult to requeen (e.g. Janscha, 1771; Root; Snelgrove, 1940).

Hutchinson's (1905) opinion that broodless bees are more easily requeened was shared by Laidlaw & Eckert (1950), but has not been generally stressed. Snelgrove (1940) said that the presence of eggs and young brood is a condition which favours queen introduction, but he contrarily added on the next page that the removal of all brood and eggs a short time before the introduction of a new queen creates a condition specially favourable to her reception. It is agreed that shaken bees, which are confused, are more easily requeened (e.g. Root; Hutchinson; Doolittle), and that stupefied bees are also more amenable—e.g. Thorley's (1744) allegedly dangerous use of puffball smoke; Alley (1883) recommended tobacco smoke, which Hutchinson endorsed.

Snelgrove endorsed Langstroth's view that bees made irritable by clumsy manipulation are less likely to accept a new queen, and emphasized that colonies irritated by robbing were in 'the worst possible condition for the reception of a new queen'.

ii. *The queen.* The state of the queen should match the state of the bees. Laying queens in similar colonies can usually be easily exchanged (e.g. Sechrist, 1944; Brother Adam, 1950); a colony recently deprived of a laying queen is not likely to accept a virgin, but the latter could be introduced into a colony which has built queen cells and so is prepared to receive a queen in that condition (e.g. Doolittle; Snelgrove; Sechrist). Newly emerged virgins, which are the only sort it would be advantageous for a natural colony to receive, are accepted more readily than older ones (e.g. Langstroth; Root), and various methods for introducing newly emerged virgins into queenright stocks have been suggested, and are sometimes successful (e.g. Alley; Doolittle; Snelgrove). Queenless stocks shaken into an empty box, and thereby confused and also put in a state similar to that of a swarm, will readily accept virgins (Snelgrove).

Laying queens which have been mailed are much less acceptable than those in full lay (e.g. Doolittle; Hutchinson); they are more excited and in a state dissimilar from that of the receiving colony. This brings us to a second consideration—the behaviour of the queen. She will be most readily accepted if she walks about in a quiet and queenly manner and goes on with her egg laying (e.g. Hutchinson; Smith, 1949; Brother Adam, 1950), and half an hour of starvation will lead her to solicit food and encourage her acceptance (Simmins, 1914).

iii. *The strain of bee.* Wedmore (1952a, c) has reported survey results which show that queens of a particular strain were more readily introduced (1 failure in 59) than queens of various other strains (11 failures in 65). The difference was highly significant, but the cause of the difference is in doubt because in the less successful group the apparent differences between the queens and the colonies to which they were to be introduced were usually greater.

Methods of queen introduction

A thorough review of methods of queen introduction has been provided by Snelgrove (1940).

The methods can be subdivided thus:

(a) Direct liberation among undisturbed bees,
(b) Direct liberation among confused bees,
(c) Cage introduction, acquaintance and subsequent liberation.

(a) Direct liberation among undisturbed bees is possible in favourable conditions, and a laying queen can then often be put immediately in the place of another on the comb on which the latter was found. The introduced queen may be dipped in syrup (Cheshire, 1888) or water (Snelgrove). Direct introduction of a queen to a nucleus of emerging brood, without adult bees, is considered quite safe (e.g. Root; Smith, 1923).

(b) The queen may be liberated among confused or stupefied bees. Farrar (1938) reported only 1·7% losses from direct introduction among 355 packages of bees sprayed with syrup, but 2·9% loss from a cage-release method used for 555 packages.

(c) Huish (1815) recorded that German beekeepers, convinced that bees recognized strangers by their odour, used to confine the new queen for 3 days in a small cage in the colony to be requeened. The invention of movable frames for hives led to the perpetuation of colonies, with need for periodic requeening, and many types of cage were invented both for mailing and for introducing queens.

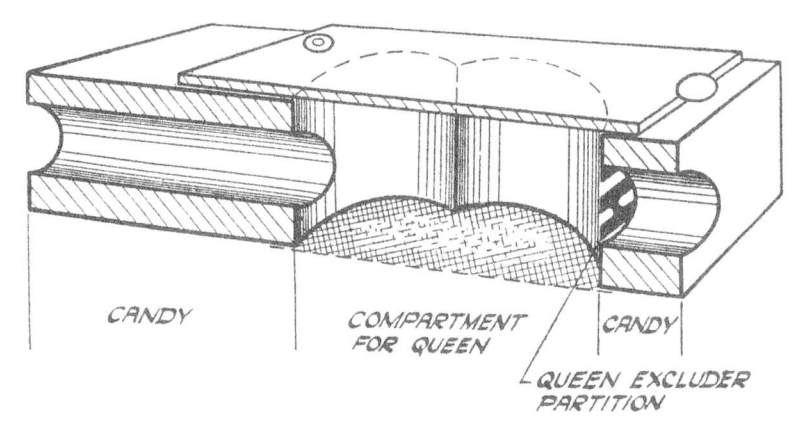

Fig 61 SECTION THROUGH THE MIDDLE OF A QUEEN CAGE DESIGNED BY WEDMORE (1952b). The cage is of wood, the queen compartment is covered with gauze on one side and transparent plastic on the other. Two compartments are filled with candy, and queen excluder separates the shorter of these from the queen compartment; thus the queen is reached before she is released.

In simple kinds of cage the queen cannot be liberated by the bees until they have eaten their way through a compartment filled with candy. In a modified version, ascribed to Chantry (Root, 1929), the queen is separated from the colony by a long and a short compartment, both containing candy, as in Wedmore's cage (Fig. 61); the candy is removed from the short compartment in about 24 hours, but this compartment is covered with queen excluder so that the bees can get inside with the queen but the queen is still confined in the cage. The bees which get in, and which are more confused than the queen, are not hostile to her; the queen is not liberated into the colony until the candy is eaten out of the long compartment, several days later—Smith (1949) says that bees eat good candy at the rate of $\frac{3}{4}$ inch in 24 hours.

A third kind of cage consists of a folded piece of gauze which is lightly pushed into a comb of emerging brood; the confined queen mixes only with newly emerged bees, commences to lay in the vacated cells, and if the cage has been pushed in to the correct extent the bees of the colony may tear the comb away and liberate her within a suitable time. A cage combining the principles of the push-in and Chantry cages was devised and at one time used by Brother Adam (Herrod-Hempsall, 1930).

The various instructions which have been given for the use of the cages have sometimes included the recommendation to remove all attendants and to insert the queen alone (e.g. Doolittle; Sladen, 1913; Wankler, 1924; Laidlaw & Eckert, 1950). It has been suggested that the attendants should be removed because the bees in the colony may show animosity toward them, but Müssbichler's results (p. 286) suggest that the presence of attendants tends to defeat the object of the caging. Lack of attendants will not adversely affect the queen—Woodrow (1941) found that queens without attendants survived 14–32 days when placed in isolated cages supplied with suitable food.

Protheroe (1923) suggested that the cage should be smeared with the corpse of the removed queen.

Introduction of queen cells

Queenless nuclei often tear down cells presented to them, but Smith (1923) reported that unsealed cells were nearly always accepted by queenless bees, and that sealed cells were more likely to be accepted if the nucleus had been liberally fed; he obtained 11 acceptances of sealed cells by 100 unfed nuclei, but 96 by 100 fed ones. Smith wrote that nuclei would accept sealed cells within a few hours of being made both queenless and broodless. Manley (1936) stated that nuclei might accept sealed cells within 2–3 hours of the loss of their queen, while they were still highly confused, but that otherwise such cells should not be presented for 2–3 days.

Snelgrove (1940) said that the chance of acceptance of sealed cells was not good until the third day of queenlessness. Yet nuclei commenced their own queen cells within 12 hours of queen removal, and within 24 hours unsealed queen cells could be safely introduced (Snelgrove, 1946).

These statements indicate the close parallel between queen introduction and queen-cell introduction—the cell is more readily accepted if it is in a stage of development which the nucleus might expect; acceptance is facilitated either by confusion or by an abundance of food.

Theories of queen introduction

Three major hypotheses seek to account for the difficulties of queen introduction. They are ascribed to (i) differences between the odour of the queen and the colony, or (ii) differences between the state or balance of the new colony and that of the colony from which the queen came (Sechrist, 1944), or (iii) the unsettled behaviour of the queen at the time of introduction (Brother Adam, 1950–51).

Brother Adam agreed that queens possess a particular odour, by which they are recognized as queens, but he held that they probably do not possess a distinctive individual odour and that, even if they do, this odour 'has nothing whatever to do with the acceptance of a queen'. In his view, laying queens which were quickly liberated and which immediately proceeded with their normal activities were invariably accepted, by virtue of their condition and behaviour. Newly mated or virgin queens were more readily alarmed, and so more frequently attacked, while mailed queens had to be caged in a colony for 3 days (and fed during that time through the wires of their cage) in order to return them to laying condition. Brother Adam considered that the condition of the colony and the disposition of the bees were of minor importance, requiring consideration only during the introduction of immature queens (those which had been laying for less than a month and were still of nervous disposition).

Sechrist defined a theory of colony balance in relation to queen introduction, holding that for ready acceptance of the queen she had to be in the condition which the colony expected, e.g. a virgin into a colony with late queen cells, or a laying queen into a colony from which a queen in about the same egg-laying condition had just been removed.[1] This view was accepted by Cale (1946), another beekeeper of wide experience.

The third and oldest theory, conventional among beekeepers but recently despised by scientists, suggests that the difficulties of queen introduction occur because attacked queens possess an odour which the bees recognize as foreign. This theory is based

[1] Janscha (1771), who lived in the swarm-production era of beekeeping, noted that if two swarms had mingled in the air they would tolerate one another—unless they had dissimilar queens (one a virgin, the other fertilized), when the bees gathered into nut-sized clusters and fought one another.

on observations like that of Huber (1792), who removed the queen from a hive, waited until the colony became agitated, and noted, 'As soon as she is returned to them, calm is restored instantly among them, and what is very singular is that they recognize her; this expression, sir, must be accepted strictly. The substitution of another queen does not produce the same effect, if she is introduced in the hive within the first 12 hours after the removal of the reigning queen. In such case, the agitation continues, and the bees treat the stranger just as they do when the presence of their own queen leaves nothing for them to desire. They seize her, envelop her all around . . . usually she dies.'

Let us now compare these three theories and see how far they are mutually compatible.

That part of the theory of colony balance which concerns the acceptable type of queen (virgin or mated) or queen cell (unsealed or sealed) is generally agreed; it would be explicable in terms of odour, because the laying queen possesses an odour (p. 285), and one would expect the odour of a virgin to be different because her diet and metabolism are different (cf. distinctive worker odours, p. 177). The more controversial part of the theory is the suggestion that differences between laying queens, dependent on their egg laying, are distinguishable by the workers—which prefer the kind to which they have been recently accustomed. This suggestion is not far fetched; one might postulate two extremes with dissimilar odours—the virgin and the queen in full lay—and suppose that the bees are able to recognize intermediates by the proportions of these two odours (p. 42). To me, the difficulty is to understand *why* the workers should be hostile to the queen on account of any such difference.

The theory of distinctive queen odour implies that the bees must be accustomed to the individual odour of the queen, or her odour be changed to match the colony odour. The latter would now seem the more likely, as the worker odours are not inherited (p. 175). Protagonists of this theory would accept the idea that, if individual queens have different odours, the differences would be much greater between two queens in different states; their views thus coincide to some extent with those of the advocates of the theory of colony balance. Lack of any understanding of how or why distinctive queen odour could exist has hitherto provided the most weighty objection to the theory.

The recent demonstration of the existence and origin of distinctive worker odour (pp. 174–8) does not imply that distinctive queen odours must exist. The worker odours are derived from the distinctive flower scents contained in their diet; the queens are fed on bee milk, which is derived from various pollens which have been processed by nurse bees, and it is possible that all odorous materials were extracted during the processing. We do not know. However, the experiments with worker odours do show us how distinctive queen odours, if they exist, could originate.

We may also consider whether the presence of individual queen odours and their recognition by the bees would be of any advantage to the colony. In apiary conditions the hazards of queen introduction are disadvantageous to both bees and beekeepers, but the fact that these hazards occur in all races and strains of honeybees, without exception, seems to me to point to some deep-seated function which has possessed survival value. Comparison with other bees suggests an answer. There are many genera of parasitic solitary bees, and all subfamilies of social bees 'except the honeybees have such satellites, and in each case it is evident that the parasite is descended from the host' (Wheeler, 1928). Therefore I think that parasitism, which may have existed in earlier stages of their evolution, has been frustrated by the ability of the workers to recognize and reject all queens other than their own, and thus to defend themselves against usurpers. In the absence of such an ability, parasitic habits would receive strong encouragement; the production of virgin queens able to enter and survive in colonies other than their own would have a high selective value, and this habit could start as a consequence of misorientation (drifting).

Moreover, in the primitive state of the honeybee community nectars probably formed a substantial proportion of the diet of the queen, so she might be expected to have possessed a distinctive odour,[1] as the workers now do; such an odour, with survival value, would be unlikely to be lost during evolution. Thus one can understand how distinctive queen odours could have originated and also why they might survive.

In important respects the behaviour of queen and workers during queen introduction is parallel to the behaviour of workers towards the entry of strange workers (pp. 179–83).

[1] A carpenter bee recognizes her nest by the distinctive odour which she has imparted to it (p. 86).

On both occasions antagonism is (i) minimal during a nectar flow, (ii) increased by disturbance or alerting of the colony, (iii) reduced or eliminated if the bees are thoroughly confused. Moreover, even in unfavourable conditions a proportion of queen introductions are successful, as a proportion of strange workers succeed in entering a hive; bees seem to prefer their own queen to any other, but they may welcome another if they are thoroughly queenless, as they are attracted to a lesser degree by the odour of strange workers. There is another similarity: opponents of the concept of individual queen odour have stressed the importance of the behaviour of the introduced queen (p. 293), in the same way that those who have ignored or minimized the function of distinctive odour at the hive entrance (p. 180) have stressed the importance of the behaviour of the strange worker.

Let us now consider the importance of queen behaviour. Several experienced beekeepers agree that this behaviour plays an important role, but why should the queen sometimes act in a manner which invites attack? In so far as she does so, is not the most likely reason that the foreign odour of the colony has frightened her? If this is so, her behaviour compares with that of the robber bee (pp. 181–2). The parallel difficulties of queen-cell introduction cannot be attributed to misbehaviour.

Moreover bees are less likely to become suspicious of a queen which behaves quietly, but there is evidence which suggests that they do not attack their own queen even if she behaves abnormally. Huber (1792) cut off both antennae of a queen, and noted that she appeared tormented and ran all over the combs, dropping her eggs haphazardly. 'Notwithstanding these symptoms of delirium, the bees did not cease to render her the same attention as they ever pay to their queens, but she received it with indifference.'

On the other hand, Dönhoff (1858) reported an occasion (certainly exceptional) on which workers were hostile to their own queens which he had held for a minute in his sweaty hand, although they were not attacked if he had lifted them up with forceps for the same length of time.

The balling of accepted queens

Bees occasionally envelop and kill their own queen. This phenomenon, which 'has mystified beekeepers from the earliest days' (Manley, 1948) may occur in colonies which are disturbed

(a) in early spring (e.g. Cheshire, 1888; Cowan, 1924; Manley), (b) when a young queen is just beginning to lay after a long break in breeding (e.g. Cowan; Manley), or (c) when a queen has recently been introduced (e.g. Langstroth, 1890; Latham, 1923). In addition, occasionally queens are balled on their return from a successful mating flight (Alley, 1883), but not after an unsuccessful one (Manley, 1946). Humphreys (1883) recorded two instances of swarms which issued, balled their queen, and then returned to their old hive. When two queenright colonies are united both queens may be balled for some time; one may be killed and the other eventually released.

The cause of balling

The varied circumstances in which bees ball their own queen have this in common—it is possible to interpret them all in terms of response to an unaccustomed odour. The balling could have been initiated in the united colonies by bees from the opposite colony, in the swarms by bees from other colonies attracted into the swarm, in early spring by bees which had been on the outside of the cluster and thus apart from their queen and comrades within; after successful mating or recent introduction, or when just beginning to lay, the queen's odour might also have been unfamiliar.

For this reason I am more inclined to agree with Cheshire (1888), who considered that the process usually starts in an attack on the queen by a few aliens, than with those beekeepers who think that balling is an attempt to protect the queen.

Although it may be an unaccustomed odour from the queen which causes the attack on her, either by her own bees or by bees to which she is being introduced, there are some differences between the treatment then accorded to her, and that given to strange workers. Strange queens are not stung—but they do not attempt to retaliate, and stinging of workers is reserved for those which defend themselves. Strange workers are seldom attacked by more than two or three bees at a time, while queens are balled by many; this difference might possibly be attributable to a difference in the quantity of the offensive odour.

When a queen is balled, small subsidiary balls containing workers only may also be formed. This was observed by Perepelova (1928a), who tried to introduce a young mated queen into a hostile colony; it has also been seen by Mr. N. E. Ellement, apiarist at Rothamsted (unpublished). Workers which have participated in the balling may perhaps carry an offensive scent, which precipitates attack by their own comrades. Janscha's observation (p. 293) also indicates that some property of the queen may be conveyed to the workers.

Chapter 37
DIVISION OF LABOUR

> *I'm house and pantry-maid, and cook, and nurse,*
> *And more.*
> *From humans, birds and insects all, I guard*
> *The door.*
> *Sip nectar from the flowers, and honey make,*
> *And store.*   M.

The relation between age and occupation

AN EARLY OBSERVATION showed that the age of workers helped to determine their occupation. Dönhoff (1855b) requeened a colony of black bees with a yellow queen on 18 April and noted that her first workers emerged on 10 May, and many more on the following day. These bees flew from the hive on their seventh day, but none were observed on feeders in the apiary garden until they were 15 days old. Meanwhile they were in the brood area, and when 10 days old they had repaired broken comb and built new comb.

Gerstung (1891–1926), noting that the younger bees of the colony remained in the broodnest and that the older ones went foraging, postulated a detailed scheme for allocation of duties according to age. He based his brood food theory of swarming (p. 264) upon this division into age groups, and he supposed that the division was rigid and that the workers could only carry out whatever duties their gland development fitted them for at that time.

A pioneer study of division of labour was undertaken by Rösch (1925), who used a colony housed in a 6-comb observation hive, in which the bees could carry out all their usual tasks except comb building. A number of newly emerged bees were individually marked, and on occasion their subsequent activities were recorded. These intermittent observations are summarized in Table 17; Rösch postulated that each worker bee carried out all the tasks in the colony, in a sequence of duties which was the same for every bee; he considered that adaptability was achieved because the periods for the various tasks were not fixed, and that the task of a bee at any particular time was strongly influenced by her age.

TABLE 17. *The ages at which various worker duties are carried out*

Reference	Duty	Number of bees observed	Number of records	Age range, days	Mean age, days
Rösch (1925)	Feeding older larvae (>4 days)	10	24	3–11	4·6
Perepelova (1928d)	Feeding older larvae (>4 days)		40	3–12	5·2
Rösch (1925)	Feeding younger larvae (<4 days)	9	25	6–13	8·6
Perepelova (1928d)	Feeding younger larvae (<4 days)		47	6–16	9·2
Rösch (1925)	Receiving nectar from foragers	18	19	8–14	11·2
Rösch (1925)	Cleaning debris from the hive	7	7	10–23	14·7
Rösch (1927)	Comb building cluster	736		2–52*	15·8
Rösch (1925)	First flight from the hive	41		5–15	7·9
Rösch (1925)	First foraging trip	34		10–34	19·5
Ribbands (1925b)	First foraging trip	47		10–32	20·1
Ribbands (1925b)	First foraging trip	52		9–35	19·2

* Wax glands were not developed in bees more than 24 days old.

Rösch watched individually marked bees which had been introduced into his observation hive in small groups on different occasions; subsequent results (p. 307) have shown that in very unbalanced colonies there are wide variations in the usual sequence of duties, and so it was possible that the latitude which Rösch observed in the age-duty scale was a consequence of changes in the balance of the nucleus.

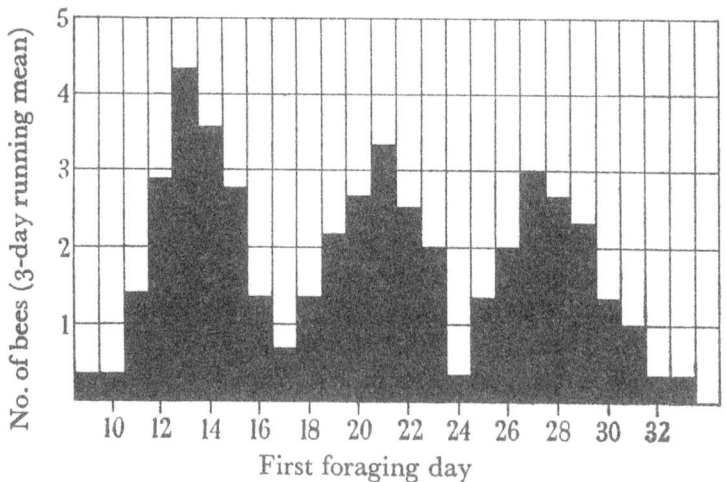

Fig 62 AGE AT WHICH FORAGING COMMENCED. Three-day running means from records of 47 bees, with the same birthday and living in the same colony (Ribbands, 1952b).

To investigate this point, on two occasions Ribbands (1952b) individually marked large numbers of one-day-old bees, introduced them into a normal colony, and recorded their subsequent foraging activities. These results are also incorporated in Table 17; the age at which bees commenced foraging ranged from 9 to 35 days, although all had the same birthday and were living together in the same colony. The wide latitude in the time at which the bees changed to this occupation confirmed that, although division of labour may be related to age, it cannot be determined by it.

The activities of nurse bees at different ages

Rösch (1925) concluded from his results that the younger nurse bees only fed old larvae. However, the older nurses did not attend only to young larvae; Rösch observed a bee 11 days old which fed larvae 5, 4 and 1 days old in succession. Perepelova's (1928d) observations, also summarized in Table 17, support this conclusion. In addition, in 26 days she distinctively

TABLE 18. Ages of bees found on different kinds of inserted comb (Perepelova, 1928d)

Age of bee	Kind of comb						
	Eggs	1–3-day larvae	4–5-day larvae	Sealed brood (2 pieces)	Pollen	Honey	Foundation
1–2 days	19	0	18	32	0	0	0
3–5 days	12	7	52	43	9	14	0
6–12 days	27	112	66	27	26	15	20
13–18 days	6	31	0	20	14	0	40
19–24 days	13	18	0	21	17	41	17

marked 21 groups of about 250 newly emerged bees and put them in an observation hive; the combs were then removed from this hive and replaced by variously filled pieces of comb; 12 and 24 hours later the pieces were taken out, and the marked bees on them were counted; bees under 6 days old congregated on eggs, old larvae and capped brood, while bees 6–12 days old were most numerous among the young larvae, and bees more than 12 days old were predominant on honey and foundation (Table 18). This result implies that a division of labour related to age occurred among these bees, and seems to confirm the view that the younger nurses attended to the old larvae.

However, Lindauer (1952) obtained contrary results when he put a succession of distinctively marked groups of newly emerged bees into an observation hive and then observed which bees fed larvae of which age. Some of these results are presented in Fig. 63: they seem to have a random distribution and to indicate that any nurse can feed larvae of any age. The distribution of the feeding activities was also random in space (p. 192).

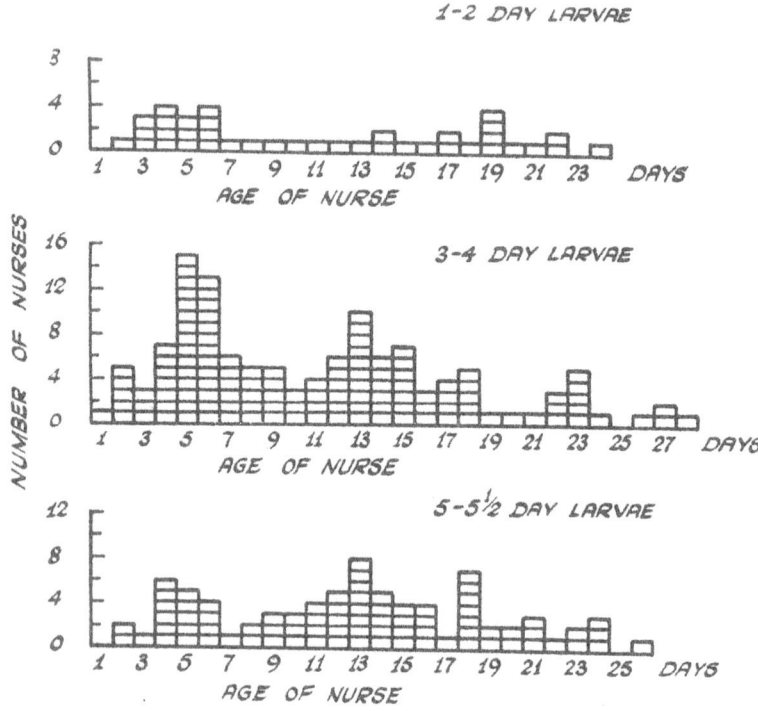

Fig 63 THE AGES OF NURSE BEES FEEDING WORKER LARVAE OF DIFFERENT AGES (after Lindauer, 1952).

Perepelova (1928*d*) observed that the marked bees which fed newly emerged bees, queens and drones were all of nursing age (120 records, range 6–15 days), and one may suppose that they were supplying bee milk.

Cell preparation

Rösch's (1925) observations indicated that cell cleaning was carried out by the youngest bees in the colony, and the jagged edges of cells smoothed by bees 15–20 days old. Perepelova (1928*c*) divided cell preparation into four phases; she found that marked bees between 1 and 21 days old participated in it, in about equal numbers for bees of any age but with the older bees spending more time on the work. There appeared to be a division of labour according to the type of cell preparation undertaken (Table 19). Perepelova pointed out that the same marked bees often removed the remains of the capping and cleared the bottom of the cell, and that these were bees which also cleaned other parts of the hive.

TABLE 19. The number of bees of various ages engaged in preparing 30 cells (Perepelova, 1928b)

Kind of work	Age of bee in days						
	1–3	4–6	7–9	10–12	13–15	16–18	19–21
Removal of remains of the capping	0	1	25	36	39	45	23
Smoothing the edges of the cells	37	18	12	32	32	39	17
Cleaning the cell walls	59	36	16	12	9	2	0
Cleaning the cell bottom	2	2	21	36	36	37	27

Lindauer (1952) did not subdivide cell preparations in this way. His bees spent a higher proportion of their time on the process during their first week, but the activity was carried out by marked bees up to 25 days old.

Wax-working activities

Rösch (1925) had provided no opportunity for comb building, but he set out to remedy this by a separate study of this one activity (1927). He found that the wax glands were small in very young bees and that their mean thickness increased to a maximum of 53µ in bees 16–18 days old, and then rapidly deteriorated with advancing age (19µ at 22 days, 3µ in bees more than 24 days old). However, there were considerable variations; not all the bees of wax-secreting age had developed glands, and the wax glands of one bee 24 days old were 60µ thick. He concluded that wax production came at the end of a sequence of hive duties. Taranov (1936) marked wax-working bees which were building new comb and returned them to the colony; on the following day only 5–7% of them were still building this comb.

Lindauer (1952) emphasized that comb production accounted for only a small portion of the time spent in wax manipulation.

Most time was occupied in capping cells of brood and stores (p. 200), and this activity was carried out by nurse bees as it was required. In addition, he observed nurse bees manipulating their wax scales during intervals between brood tending. As this latter conclusion ran counter to that of Rösch, Lindauer took special steps to verify it. He examined both the pharyngeal glands and the wax glands from a large number of individuals, and found that both glands were often developed in the same individual at the same time. He also found that very young bees could secrete wax; there were wax scales in the pockets of some bees 3 days old, the wax glands of which were only 20µ thick.

Lindauer did not provide his bees with large spaces in which to build comb, as Rösch (1927) had done. Cell capping is an activity ancillary to brood rearing, so one might expect it to be carried out by the nurse bees as required, using either old wax (p. 203) or some relatively small amount of new wax which they had themselves produced. The production of the much larger quantities of wax sometimes required for comb construction may become a full-time job to which the older nurse bees are diverted, as Rösch's results indicate.

Guarding and foraging activities

Lindauer (1952) pointed out that it would not be economic for more than a small proportion of the bees to take part in guard duties, and he found that only 23 out of 159 marked and watched individuals undertook this duty at all, although one or two of them carried it out for a long period—one for 9 days.

The youngest guard bee observed by Butler & Free (1952) was 11 days old; they concluded that honeybees would undertake guard duties before they became foragers, but that many of the guards were foragers as well and alternated between these two duties.

We have already seen that there is no usual sequence in the foraging activities of an individual (pp. 121–3).

The activities of two individual bees

Lindauer (1952) observed for long periods the activities of one individually marked bee, which was watched for several hours each day during each of the first 24 days of her life. These observations, which are summarized in Fig. 64, indicated that for this bee there was no sequence of hive duties. She carried out several different tasks each day, and there seemed to be only

DIVISION OF LABOUR

three differentiated stages in her adult life—(a) a short period after emergence during which she cleaned cells but did not take part in other activities, (b) a period when she participated in any

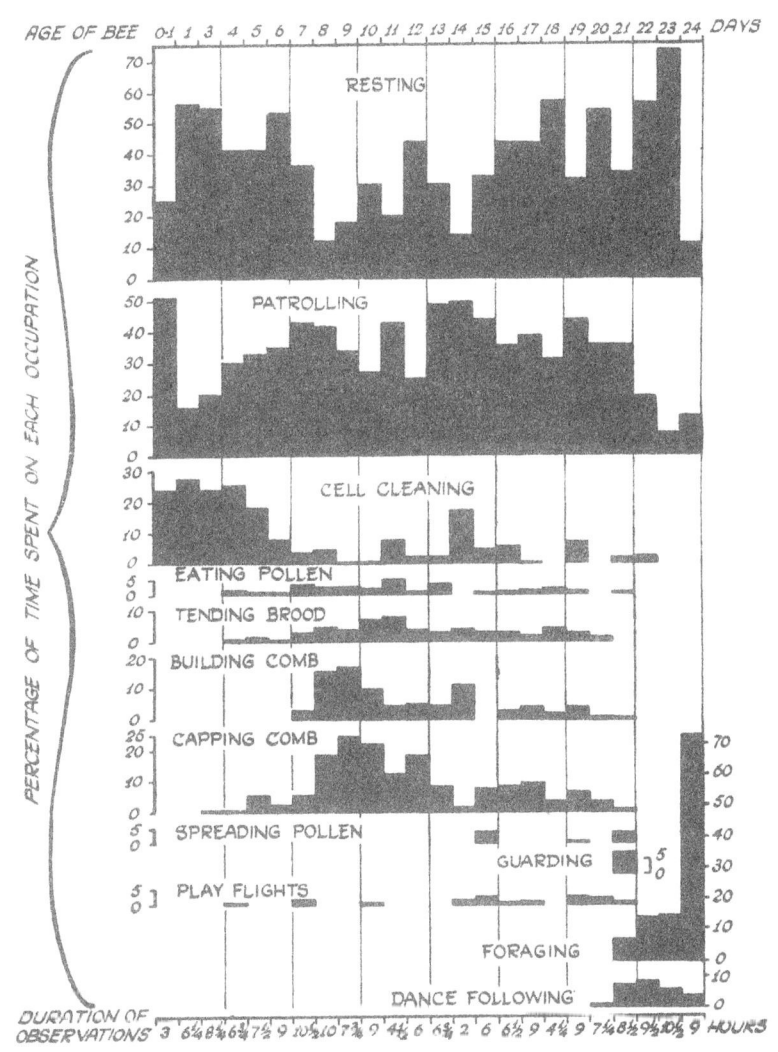

Fig 64 ACTIVITIES OF ONE MARKED BEE DURING THE FIRST 24 DAYS OF HER ADULT LIFE (data from Lindauer, 1952).

required hive duties, (c) one when she forsook inside duties and became a guard and forager. The variety and irregularity of the hive work is illustrated by the continuous chart of the activities of this bee during 10 hours of her eighth day of adult life (Fig. 65).

306 LIFE WITHIN THE COMMUNITY

Lindauer also watched another individual, and obtained a record of her activities through the whole 24 hours of her first 8 days of adult life. This record, which is summarized in Fig. 66, is similar to the record in Figs. 64 and 65, and shows in addition that the rhythm of activity was similar throughout both day and night.

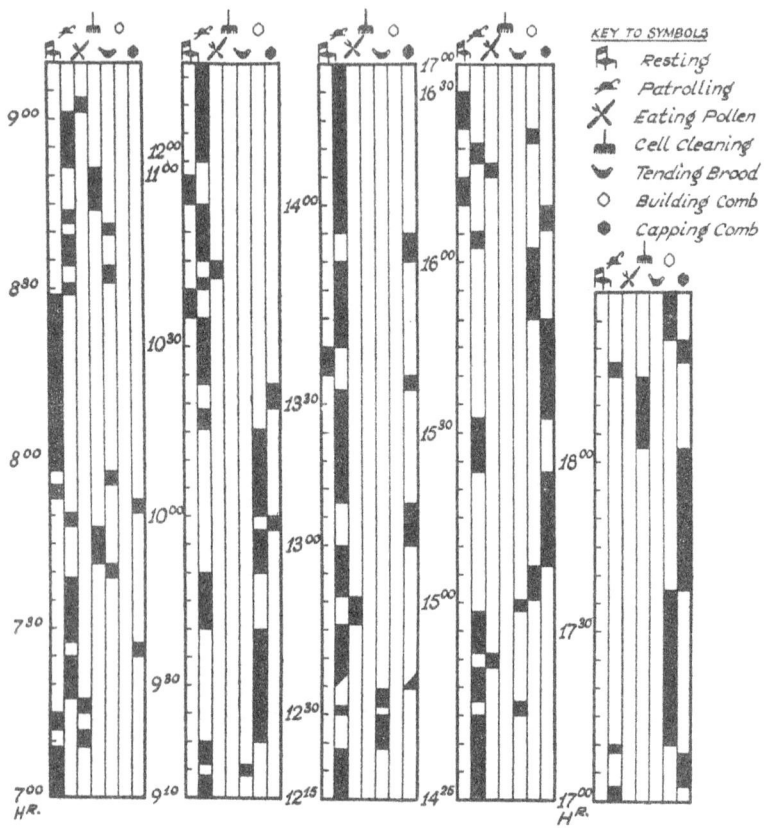

Fig 65 ACTIVITIES OF ONE BEE DURING HER EIGHTH DAY OF ADULT LIFE (data from Lindauer, 1952).

Figs. 64–66 show that both of these bees spent a considerable proportion of their time either at rest (various cleaning movements, not separately recorded, occupied about one quarter of this resting time) or patrolling over the combs; these two occupations shared more than two-thirds of the total time. Moreover, preparation of the cell for a larva, or capping that cell prior to pupation, took much longer than nursing (p. 231). Lindauer pointed out that the bee milk would be prepared during resting and patrolling.

Division of labour in abnormal colonies

A greater diversity of behaviour can be expected among bees living in unusual circumstances. The most popular experiment of this type has been the formation of nuclei in which all the workers were newly emerged. The nucleus has

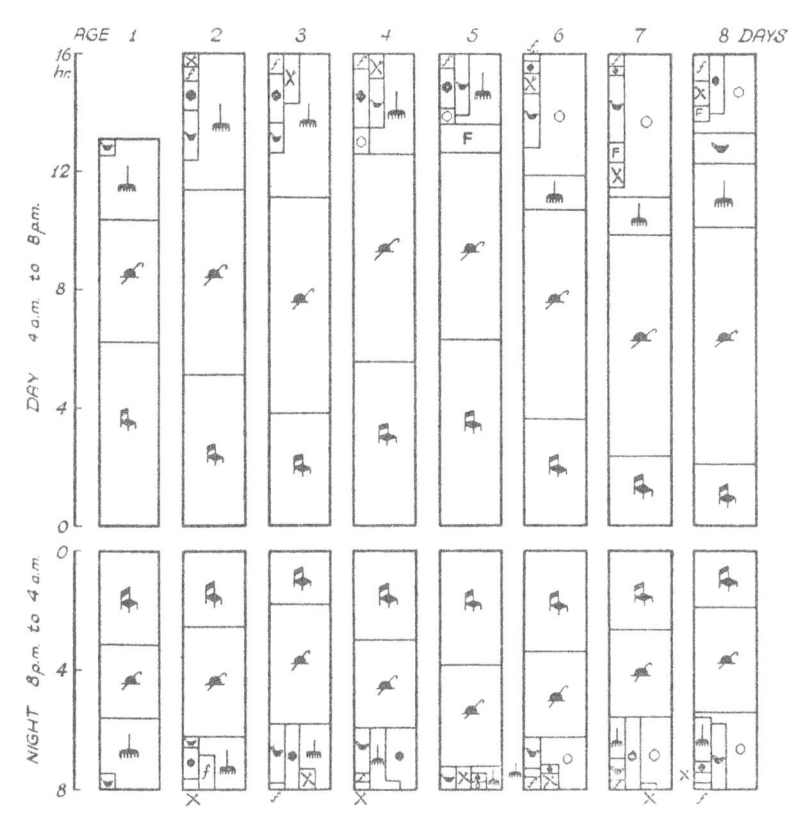

Fig 66 ACTIVITIES OF ANOTHER MARKED BEE THROUGHOUT THE WHOLE OF HER FIRST EIGHT DAYS OF ADULT LIFE (data from Lindauer, 1952). Symbols as in Fig. 65.

either been supplied with brood in all stages of development (Haydak, 1930–32), or with empty combs and a queen (Wiltze, 1882; Nelson, 1927; Himmer, 1930; Gontarski, 1949a). In the former instance the adaptation was more extreme; bees 1 day old completed the sealing of cells, and when 2 days old they fed the older larvae and produced wax for repairing combs (Lindauer has now recorded these activities by bees of these ages in normal colonies); the bees guarded the entrance and commenced orientation flights on their third day, when they also built queen cells; they collected pollen loads and danced on the combs when they were 4 days old, and one day afterwards they brought in nectar. In the other instances flighting commenced on the fourth (Nelson, Gontarski) or fifth (Himmer) day, nectar was collected on the sixth (Wiltze, Nelson) or eleventh (Himmer) day, and pollen on the fourth (Gontarski), eighth (Nelson, Himmer) or eleventh (Wiltze) day. Brood rearing did not commence until the eighth (Nelson) or the fourteenth (Gontarski) day, a later time than in normal colonies.

A less extreme method of experiment was adopted by Rösch (1930). He divided an observation hive into two separate portions, established a small nucleus in one half (19 June), and individually marked about 20 newly emerged bees each day. Then (26 August) the halves of the observation hive were reversed, so that the flying bees were diverted into the empty portion. These older bees were supplied with the original queen, and with a comb containing both sealed and unsealed brood; the portion with the younger bees possessed a comb which contained all stages of brood but very little food, and these bees were supplied with a new queen. Rösch wished to see whether the older bees would be able to nurse their brood and whether the younger ones would be able to feed themselves.

In the young portion, bees from 7 to 15 days old went foraging, while other bees *of the same ages* remained as nurses. The pharyngeal glands of the young foragers were well developed when they commenced this task, but they degenerated within 4 days; the young foragers never returned to nursing duties.

The bees in the old portion were all at least 8 days old. During the first two days the younger larvae (up to 4 days old) were fed, but newly hatched ones were neglected, and some of the brood died. On the fifth day the comb was replaced by another, and nurse bees (now 16–28 days old) only fed larvae 4–5 days old. Four or five days later the older larvae were all sealed, and the nurses were feeding young larvae. Rösch observed 7 nurse bees, whose ages ranged from 17 to 33 days, which were feeding 4–5-day larvae on 1 September, and feeding 1–3-day larvae on 4–6 September. He therefore concluded that the state of pharyngeal gland development, and not the age of the nurse bee, was the determining factor in brood rearing.

Rösch's experiment did not provide conclusive evidence that the pharyngeal glands could be regenerated for a second brood rearing, because the glands in a proportion of the foragers which were switched into the older nucleus had probably not then degenerated. Moskovljevic (1940), however, formed a nucleus from 503 marked foragers, 28 days old, and forced them to rear brood for 3 weeks; a sample showed that at the beginning of her experiment 86% of the bees had reduced pharyngeal glands, but at the end 86% of the 200 bees which remained had them well developed.

Rösch (1930) cited experiments to prove that older bees could, if necessary, develop their wax glands a second time, and he believed that they used materials from their ventral fat bodies for this purpose. Moskovljevic (1940) found that bees even 50 days old could regenerate their wax glands and build comb, and Jordan *et al.* (1940) showed that the wax from regenerated glands did not differ materially from normal wax.

Yet another kind of abnormal situation was investigated by Himmer (1930), who forced some bees to tend brood for 42 days, by repeatedly removing all the sealed brood from their nucleus so that they could not be relieved by younger recruits. Moskovljevic (1939) repeated this experiment for a longer period; her nucleus finally contained marked bees which were all 72-75 days old, and 70% of them were still active nurses with fully developed pharyngeal glands.

The extent of the division of labour

The preceding observations have shown that her age plays some part in determining the task carried out by each worker, but there is a wide latitude in the age at which different workers commence the same duty and most of the workers never undertake all the duties. Any particular task may be either unduly prolonged or skipped completely, and in abnormal colonies the worker may perform any required duties, quite irrespective of age, with the limitation that after sudden unbalancing of a colony some days may elapse before the glands of the diverted workers become adequate to equip them for their new tasks.

The various results are generally in good agreement concerning the usual age range for carrying out particular duties in normal colonies (Table 17), and some of their inconsistencies can be resolved; the most serious conflict, which concerns the age at which nurse bees begin to feed the younger larvae, might be explicable in terms of different proportions of young and old brood present during the opposing observations.

Rösch's (1925) results implied that the various hive duties were carried out in a sequence, but Perepelova (1928d) saw nurse bees cleaning other bees, cleaning cells, and performing other duties, and she emphasized that bees were not confined to one kind of work at one period. Lindauer's (1952) observations seem to indicate that throughout her pre-foraging life, after the first day or two, each bee changes from one hive duty to another at short intervals of time. This should not be surprising;[1] solitary bees would perforce lead such a varied existence, there are obvious advantages in retaining a large measure of adaptability, and the production of bee milk and wax are neither mutually incompatible nor incompatible with the carrying out of other non-strenuous activities.

Nevertheless, the observations of Rösch and Perepelova which suggested some kind of sequence of hive duties are too solid to be ignored; they become compatible with Lindauer's results if they are considered to indicate a marked preference for one or other of the various tasks, and that this *preference* changes with age although all the tasks can still be carried out. Alternatively, nectar handling and comb building could be associated with increasing desire to leave the vicinity of the broodnest, a process which culminates in foraging.

Foraging is a strenuous activity, and one which should be carried out continuously whenever circumstances are suitable. Both of these considerations afford good reason for a division of labour between foragers and non-foragers, and there is no doubt that there is a well-marked division into at least these two main groups.

Foraging soon causes the pharyngeal glands to retrogress, and it is incompatible with extensive wax production; conversely there is evidence that in queenless or swarming bees gland development reduces foraging activity (p. 275). Incidentally, foraging is the most hazardous activity, so it is advantageous for it to be carried out by the older and more expendable members of the community, and for them not to be filled with valuable food materials.

How the division of labour is arranged

Lindauer (1952) and Ribbands (1952b) both concluded that the allocation of duties is arranged to meet the needs of the

[1] It is more surprising that the contrary view, of a rigid sequence of duties, has been so readily accepted.

community, whatever they may be. Now that the division of labour appears to be less rigid than was supposed, it can be more readily accounted for.

Let us first discuss what determines the needs of the community. Ribbands (1952b) considered that the incoming food supply determined the proportion of workers which would be required for each of the main tasks—brood rearing, wax secretion and comb building, handling and storing food, foraging for pollen or nectar, defence of the community. Evidence for this proposition can be found elsewhere (brood rearing, pp. 238–43; comb building, p. 203; foraging, p. 124; defence, p. 183).

Can food supplies *directly* influence the allocation of duties, thus catering for the needs which they have created? The following evidence is suggestive:

(1) Among female honeybees the primary division of labour, into reproductive and worker castes, is determined by the food received during the larval stage (Ch. 31).

(2) In some social insects a secondary division of labour, between various worker duties, seems to be partly determined in a similar way, and emerging worker adults then differ in structure. (Rigidity of adult function is a corollary of such early differentiation, but the determination of non-reproductive duties in honeybees is not subject to this disadvantage; all emerging workers are capable of all worker duties, and the secondary division of labour is determined by the exigencies of adult life and associated with the waxing and waning of glands and fat bodies.)

(3) Food supply controls the development of the pharyngeal glands of the nurse bees, which do not enlarge unless the bees are fed abundantly on suitable protein foods (p. 56).

(4) Changes in the behaviour of broodless bees, queenless bees, swarming bees, and winter bees are associated with differences in food supplies received or given, and in gland development (Ch. 34).

(5) Worker bees unable to deposit nectar loads might be diverted to wax production by the food supply (p. 203).

(6) The proportion of pollen gatherers is readily adjusted to colony needs (p. 124); this must follow an appreciation of the quantity of food available.

(7) Recruitment to crops, which helps to determine the food supply, is associated with transfer of a sample of food from

forager to recruit (p. 148). The extent of the dancing and the success of the recruiting depend upon the availability of food from other sources (p. 161).

(8) Food carried into a colony is evenly shared among all its members (p. 193), so each bee can appreciate any changes in the supply.

Ribbands supposed that the extensive food transmission might enable individual bees to adjust their duties to meet the needs of the community. The quantity of bee milk in circulation, for instance, might determine whether a nurse bee would eat sufficient pollen to produce more, or whether she would refrain from eating pollen and her pharyngeal glands would regress. In a group of bees engaged on one particular task (or group of tasks) the older members might leave and go to a more senior duty if a surplus of labour were reflected in the food transmitted among the group.

Another view was put forward by Lindauer (1952), who pointed out a possible significance for the high proportion of time which nurse bees spend on patrolling; this activity could enable the individual to become aware of the variety and relative abundance of the tasks required, and Lindauer suggested that simple stimuli would prompt each individual to carry out any work which required attention. He gave several examples of circumstances in which individual bees seemed to act without much co-operation with other members (e.g. p. 200).

I do not think that there need be an incompatibility between Lindauer's views and my own. The worker is first an individual, secondly a part of the community. By their responses to simple stimuli the solitary ancestors and living relatives of the modern honeybee did and do react satisfactorily to many similar circumstances, and such arrangements can be expected to persist into social life.

They can be made more efficient for the latter purpose if responses to other simple stimuli, based on the quality and quantity of the food in circulation among them, are incorporated into the system. We know that individual bees can appreciate such differences in their food supply, and that responses of this kind have revolutionized foraging behaviour; it would be surprising if they had not affected work within the hive.

Chapter 38

THE EVOLUTION OF THE HONEYBEE COMMUNITY

*The cheese-mites asked how the cheese got there,
And warmly debated the matter:
The orthodox said it came from the air,
And the heretics said from the platter.*

ANON

MAJOR PROGRESS IN evolution does not come from many unrelated changes, but from those which are stages in some definite trend. Such progress is usually reflected in structural modifications, e.g. the story of horse evolution can be told in a series of adaptations of legs and teeth, and the evolution of man in terms of increasing brain development.

Developments of the food-processing glands of the honeybee (Ch. 7) distinguish them from their non-social relatives, and from this we may infer that food has played an important role in their evolution.

THE FEEDING OF LARVAE

Early stages in social life

As in other social insects, the family is the social unit and there has been no integration above this level. Most kinds of bees are solitary, but some are social, and among the latter the bumble bees and honeybees have a well-organized social life; comparison between bees at different social levels helps us to guess how this social life evolved.

The first requisite, characteristic of solitary bees, was the collection by the mother of food for her offspring—in contrast to the habits of most insects, which merely lay their eggs in the vicinity of the larval food supply.

Some kinds of solitary bees lay their eggs singly, each in a small chamber which is provided with sufficient food to nourish the offspring until it becomes adult.[1] Other kinds lay a

[1] 'mass provisioning.'

number of eggs at once; so much food may then be required that some of the eggs hatch before the mother has collected it all, and she then takes food to the young larvae[1]—an important stage on the road toward social life.

A provided food supply rich enough to enable the young to develop rapidly helped towards the second requisite, that the mother should survive until her offspring were adult (among other insects evolution has often progressed in the opposite direction, with a long immature existence and a short adult life, making family life impossible). If the first young adults which emerge stay in the nest and help to nurse the rest of the family, a primitive form of social life develops. The communities so formed are small, and soon break up.

Production of a worker caste

A further stage in the development of social life is the production of sexually undeveloped adults, smaller than their mother. These 'workers' are produced by underfeeding during the larval stages, the mother of the colony having laid more eggs than she can adequately provision; in a separate existence they would be valueless, but they remain attached to the nest, collect food and nurse larvae; when there are sufficient of them their younger sister larvae are adequately fed and are thus able to attain sexual maturity.

Bumble-bee communities are still in this stage of evolution; the colony breaks up in the autumn, the fertile females hibernate, and they found colonies of their own in the spring. Although the queen undertakes every kind of task, there is a division of her labour in time; soon after her first offspring emerge she ceases to forage and devotes herself to egg laying.

In honeybee communities the process is carried further; the production of workers is still determined by the food supply (Ch. 31), but this is controlled by the worker bees independently of the environment. Worker bees are reared in small cells of equal size (built from a converted food); when a new queen is required she can be reared at any time in a specially constructed and much bigger cell, and she is fed more lavishly.

The use of predigested food

All bees supply their larvae with regurgitated or masticated food, which contains some digestive secretion; worker honeybees

---

[1] 'progressive provisioning.'

have extended the process, and produce in their salivary glands a predigested food, rich in protein, which enables their larvae to grow very rapidly. This food can appropriately be called 'bee milk'.

THE FEEDING OF ADULTS

Feeding of the reproductive castes

The rich salivary secretion of the nurse bees is also fed to queens and drones, and it stimulates the development of their sexual organs (p. 281), with the result that they can produce the great quantities of eggs or of sperm which enable the colony to grow to so large a size.

Food transmission

This brings us to another development—the food-supplying instinct of honeybee workers is extended so that all adult members of the colony are fed. This food sharing is of a different order from that found in bumble-bee colonies, whose members help themselves from a common store but do not feed one another. The feeding of workers by workers has important social consequences; the food serves as a method of communication, and welds the colony into a unit for both productive and defensive purposes.

Through this food transmission, potential foragers are informed of the quality and scent of the crops gathered by their comrades, and in response to abundance of food the more successful foragers dance excitedly and convey information about the whereabouts of the supply (Ch. 19); this considerably increases the foraging efficiency of the colony. The food supplies acquired by neighbouring colonies are usually different, because the colony—not the individual—has become the foraging unit (p. 167).

The food sharing has become so extensive that the diet of each individual in any one colony contains almost the same proportions of the various nectars brought in (Ch. 24); the scents in these nectars become converted into odorous waste products which can be given out from the scent gland, so each member produces the same scent. However, because the colony is the foraging unit, neighbouring colonies acquire different proportions of the various nectars, and thus the workers of neighbouring colonies produce different scents. The acute scent perception of honeybees (pp. 40–3) enables them to detect such differences, and in this manner the bees are able to recognize their companions and to distinguish them from other honeybees (Ch. 21). Thus they can defend their community against robbers (Ch. 22) and usurpers (p. 295).

The above-mentioned advantages of food sharing have been demonstrated beyond dispute, but there is another likely advantage which is not yet so firmly established. The bees carry out many different tasks, and the proportion of bees required for each task depends upon the food supply received in the colony; however these proportions vary, the workers carry them out efficiently (Ch. 37); food sharing could enable the members to appreciate changes in the needs of the community, and could ensure that the most suitable bees undertake each of the various tasks, in the right proportions (pp. 310–12).

In addition, food transmission helps the colony to respond to major environmental changes; through the effects of food upon the pharyngeal glands and fat bodies the metabolism and the expectation of life of the worker bees are controlled, so that they adjust to wintering, swarming and broodless conditions (Ch. 34).

In queenless stock the ovaries of the workers enlarge and function, and this development is probably inhibited in queen-right stocks by some substance which is produced by the queen and circulated among the workers in food transmission (Ch. 35).

Food storage

The storage of ample food supplies, sufficient to satisfy the needs of the community during long periods of adversity, may well be another consequence of successful food sharing—feeding of the queen expands her egg laying and this increases the size of the colony, communication between foragers increases their efficiency, and the two things together would improve the primitive habits of food storing (at first for larvae, then for adults) to its present level.

Swarming

The differentiation between the queen and the workers could not become complete until the queen ceased to found a new colony alone. This was made possible by development of the swarming habit, new colonies being founded by a queen accompanied by a large group of attendant workers.

Swarming probably originated as an act of migration, in

response to unsuitable conditions. This would explain why, contrary to what one might expect in an act of reproduction, the first swarm is led by the old queen, as Janscha (1771) observed. Honeybees evolved in the Indo-Malay region, the only part of the world where *Apis* species other than *A. mellifera* exist; in this environment *A. dorsata*, *A. indica* and *A. florea* often migrate from the plains to the hills in the hot season, and return towards the plains when the weather cools. They follow suitable climatic conditions, and suitable food supplies correlated with those conditions, the effective stimulus not having been clearly determined; opposing climatic factors would have to operate to cause this to-and-fro migration, but food supplies could be effective in either direction. Exhaustion of their crop would cause foragers to be idle in the hive, and so produce overcrowding which might trigger off the swarming impulse (p. 266).

Apis indica is farmed in India, and Indian beekeepers distinguish between swarming (which occurs during spring nectar flows) and absconding (which occurs during summer dearth). As listed by Rahman (1945), the causes of the two phenomena are the same—heat, lack of ventilation in the hive, a poorly laying queen or lack of space in which to lay—except that absconding is precipitated by lack of stores and swarming by lack of storage room. These similarities indicate that swarming is a modification of absconding.

Swarming is made possible by a method of communication (p. 168) which is one of the consequences of food transmission, and the habit may perhaps be regarded as an extension of the process of communicating the whereabouts of good crops.

Swarming enables the honeybee to perfect the separation between reproductive and worker castes—the pollen-collecting apparatus, mouthparts and glands of the reproductive castes can be reduced, because these bees never lead an independent existence. Swarming also reduces the hazards of reproduction, thereby contributing towards the independence and success of the honeybee.

CONCLUSION

This analysis indicates that the evolution of the social life of honeybees can be understood in terms of a series of adaptive responses to food supplies. The source of these adaptions is food sharing. The earlier stages of social evolution, achieved by many kinds of bees, involve the feeding of larvae, and a refinement in which some larvae are fed with a limited quantity of food, and develop into workers. The later stages of evolution, achieved among bees only by the honeybees, involve the feeding of adults—feeding of the queen to promote egg laying and of workers as a method of communication between them. The latter process, 'food transmission', enables the bees to forage as a community, to recognize companions and to defend the colony, and probably to organize a suitable division of labour among themselves. The swarming habit, which may be derived indirectly from these advances, and the habit of storing ample supplies of food which is another consequence of them, have together enabled *Apis mellifera* to perfect its social life and to live far beyond its original home.

The social life of ants and termites, the only other insects which live in highly organized perennial communities, seems to rest on the same foundation. A less restricted diet has enabled these insects to reap other advantages from their social life, and both groups have produced many different species, each adapted to a particular way of life. The achievement of social life by a series of adaptations to food supplies is in line with an insect characteristic; almost every kind of substance of plant or animal origin may nourish some insect or other, and the variety of adaptations to this wide range of food is largely responsible for the success of the whole insect class.

References

References

Abbreviations of names of journals are from the *World List of Scientific Periodicals*. Volume numbers are in heavy type, page numbers in ordinary type. Part numbers, in brackets, are only included where the pages of each part are separately numbered.
Translated titles are in square brackets. Both titles are included wherever an English translation of a foreign publication is known to be in print.
Publications which the author has not been able to obtain are marked with an asterisk. Library references follow after the square bracket at the end of the reference.

M = publication in the Moir Library of the Scottish Beekeepers' Association;
B = publication in the Bee Research Association Library;
T = translation in the Bee Research Association Library;
the numbers refer to an abstract of the publication in *Apicultural Abstracts*.
I am indebted to the Librarians for this information.

ADAM, BROTHER (1950). Das Zusetzen von Königinnen. *Schweiz. Bienenztg.* **73**: 267–73, 314–16. [M, B, T, 18/51
—— (1951). Introduction of queens. *14th Int. Beekeep. Congr.* Paper 10: 1–5. [M, B, 187/51
ALFONSUS, A. (1933). Zum Pollenverbrauch des Bienenvolkes. *Arch. Bienenk.* **14**: 220–3. [B
ALFONSUS, E. C. (1932a). Swarming and supersedure. *Wis. Beekeeping* **8**: 34–6. [B
—— (1932b). The rocking movements of bees. *J. econ. Ent.* **25**: 815–20.
—— (1933). Some sources of propolis. Methods of gathering and conditions under which this work is done. *Glean. Bee Cult.* **61**: 92–3. [M, B
ALLEY, H. (1883). *The beekeepers' handy book; or, twenty-two years' experience in queen-rearing.* Wenham: Alley. [M
ALPATOV, V., & SAF'YANOVA, V. (1950). [A new method of studying the mutual feeding relations of bees.] *Priroda* (4): 60. [B, 159/51
ALTENBERG, E. (1926). A working model for demonstrating the mosaic theory of the compound eye. *Brit. J. exp. Biol.* **4**: 38–45.
ALTMANN, G. (1950). Ein Sexualwirkstoff bei Honigbienen. *Z. Bienenf.* **1**: 24–32. [B, 154/51
ANDERSON, E. J. (1948). Hive humidity and its effect upon wintering of bees. *J. econ. Ent.* **41**: 608–15. [B
—— (1952). Capping pattern may be beautiful and strong. *Amer. Bee. J.* **92**: 388–9. [M, B, 83/53
ANDERSON, J. (1931). How long does a bee live? *Bee World* **12**: 25–6. [M, B
ANDREAE, E. (1903). Inwiefern werden Insekten durch Farbe und Duft der Blumen angezogen. *Beih. Bot. Zbl.* **15**: 427–70.
ARMBRUSTER, L. (1919). Messbare phaenotypische und genotypische Instinktveränderungen, Bienen und Wespengehirne. *Arch. Bienenk.* **1**: 145–84. [B
—— (1923a). *Die Warmehaushalt im Bienenvolk.* Berlin: Pfenningstorff. [M, B
—— (1923b). The heat economy of bees in winter. *Bee World* **4**: 207. [M, B
AUTRUM, H. (1949). Neue Versuche zum optischen Auflösungsvermögen fliegender Insekten. *Experientia* **5**: 271–7.
AUTRUM, H. & SCHNEIDER, W. (1948). Vergleichende Untersuchungen über den Erschütterungssinn der Insekten. *Z. vergl. Physiol.* **31**: 77–88.
AUTRUM, H. & STUMPF, HILDEGARD (1950). Das Bienenauge als Analysator für polarisiertes Licht. *Z. Naturf.* **5b**: 116–22. [B, T, 78/52
BACHMETJEV, P. (1899). Uber die Temperatur der Insekten nach Beobachtungen in Bulgarien. *Z. wiss. Zool.* **66**: 521–604. [T
—— (1907).* *Experimentelle entomologische Studien vom physikalisch-chemischen Standpunkte aus.* Sophia.
BAILEY, L. (1953). Not yet published.
BARLOW, H. B. (1952). The size of ommatidia in apposition eyes. *J. exp. Biol.* **29**: 667–74. [B
BARYKIN, D. J. (1951). [Multiple-queen colonies.] *Pchelovodstvo* (3): 37–8. [B, T, 150/53

BATEMAN, A. J. (1947a). Contamination in seed crops. I. Insect pollination. *J. Genet.* **48**: 257–75.
—— (1947b). Contamination in seed crops. III. Relation with isolation distance. *Heredity* **1**: 303–36.
—— (1951). The taxonomic discrimination of bees. *Heredity* **5**: 271–8.   [B, 64/53
BAUMGARTNER, F. (1948). Zweimalige Begattung einer jungen Königin. *Schweiz. Bienenztg.* **71**: 26–7. Cited *Bee World* **29**: 69.   [M, B
BAUMGARTNER, H. (1928). Der Formensinn und die Sehschärfe der Bienen. *Z. vergl. Physiol.* **7**: 56–143.
BECKER, F. (1925). Die Ausbildung des Geschlechtes bei der Honigbiene, II. *Erlanger Jb. Bienenk.* **3**: 163–223.   [B
BEECKEN, W. (1934). Ueber die Putz- und Säuberungshandlungen der Honigbiene. *Arch. Bienenk.* **15**: 213–75.   [M, B
BELING, INGEBORG (1929). Ueber das Zeitgedächtnis der Biene. *Z. vergl. Physiol.* **9**: 259–338.   [B
—— (1931). Beobachtungen über das Pollensammeln der Honigbiene. *Arch. Bienenk.* **12**: 76–83.   [M, B
BERLEPSCH, A. v. (1860).* *Die Biene und die Bienenzucht.* Mülhausen: Heinrich-shofensche Buchhandlung.
BERTHOLF, L. M. (1927). The relative sensitivity of honeybees to light of different wavelengths. *J. econ. Ent.* **20**: 521.
—— (1931a). Reactions of the honeybee to light. *J. agric. Res.* **42**: 379–419.   [M, B
—— (1931b). The distribution of stimulative efficiency in the ultra-violet spectrum for the honeybee. *J. agric. Res.* **43**: 703–13.   [B
—— (1942). Effect of certain biological factors on the longevity of caged bees. *J. econ. Ent.* **35**: 887–91.   [B
BETHE, A. (1898). Dürfen wir Ameisen und Bienen psychische Qualitäten zuschrieben? *Pflüg. Arch. ges Physiol.* **70**: 15–100.
BETTS, ANNIE D. (1920). The constancy of the pollen-collecting bee. *Bee World* **2**: 10–11.   [M, B
—— (1935). The constancy of the pollen-collecting bee. *Bee World* **16**: 111–13.   [M, B
—— (1939a). Drone comb in the supers. *Bee World* **20**: 1–2.   [M, B
—— (1939b). The mating flight: a summary of present knowledge. *Bee World* **20**: 20–4, 33–6.   [M, B
—— (1943). Temperature and food consumption of wintering bees. *Bee World* **24**: 60–2.   [M, B
BEUTLER, RUTH (1930). Biologisch-chemische Untersuchungen am Nektar von Immenblumen. *Z. vergl. Physiol.* **12**: 72–176.   [B
—— (1936). Ueber den Blutzucker den Bienen. *Z. vergl. Physiol.* **24**: 71–115.   [B
—— (1949). Ergiebigkeit der Trachtquellen. *Imkerfreund* **4**: 207–8.   [B, 67/50
—— (1950). Zeit und Raum im Leben der Sammelbiene. [Time and distance in the life of the foraging bee.] *Naturwissenschaften* **37**: 102–5. Transl. *Bee World* **32**: 25–7.   [M, B, 136/53
BEUTLER, RUTH & SCHÖNTAG, ADELE (1940). Uber die Nektarabscheidung einiger Nutzpflanzen. *Z. vergl. Physiol.* **28**: 254–85.   [B
BIERENS de HAAN, J. A. (1928). Experiments on the determination of the choice of bees by absolute or relative characteristics. *Tijdschr. n.-dierk.*, Ver. **3**, **1**: 45–7.
BISHOP, G. H. (1920). Fertilisation in the honeybee. *J. exp. Zool.* **31**: 225–86.
BLUMENHAGEN, R. (1950). Verhindert der Baurahmen das Schwärman? *Leipzig. Bienenztg.* **64**: 80–1.   [M, B, 85/50
BODENHEIMER, F. S. (1937). Population problems of social insects. *Biol. Rev.* **12**: 393–430.   [B
BODENHEIMER, F. S. & NERYA, A. BEN (1937). One-year studies on the biology of the honeybee in Palestine. *Ann. appl. Biol.* **24**: 385–403.   [B
BOËTIUS, J. (1948). Ueber den Verlauf der Nektarabsonderung einiger Blütenpflanzen. *Bieh. Schweiz. Bienenztg.* **2**: 257–317.   [B, T
BONNIER, G. (1879a). Étude anatomique et physiologique des nectaires. *C.R. Acad. Sci., Paris* **88**: 662–5.
—— (1879b). Les nectaires. Étude critique, anatomique et physiologique. *Ann. Sci. nat., Bot.* Sér. 6, **8**: 1–212.   [T
—— (1905). L'accoutumance des abeilles et la couleur des fleurs. *C.R. Acad. Sci., Paris* **141**: 988–94.
—— (1906). Sur la division du travail chez les abeilles. *C.R. Acad. Sci., Paris* **143**: 941–6.
BORDAS, L. (1895a). Appareil glandulaire des Hyménoptères. *Ann. Sci. nat., Zool.* **19**: 1–362.
—— (1895b). Glandes salivaires des Apinae. *Bull. Soc. philom., Paris.* Sér. 8, **7**: 9–26.

BÖTTCHER, F. K. (1950). Soll der Imker seine Stöcke beheizen? *Imkerfreund* **5**: 23. [B
BOZLER, E. (1926). Experimentelle Untersuchungen über die Funktion der Stirnaugen der Insekten. *Z. vergl. Physiol.* **3**: 145–82.
BRAUN, E. (1942). One year and two year old queens. *Amer. Bee J.* **82**: 356–7. [M, B
BRAUN, E. & GEIGER, J.-E. (1953). Pertes dans les colonies durant l'hiver et facteurs entrant en cause. *Abeille et l'érable*, Sér. 2, **22**: 12–16, 26–8. [B
BRIAN, ANN D. (1952). Divison of labour and foraging in *Bombus agrorum* Fabricius. *J. Anim. Ecol.* **21**: 223–40. [B, 131/53
BRIAN, M. V. & BRIAN, ANN D. (1952). The wasp, *Vespula sylvestris* Scopoli: feeding, foraging and colony development. *Trans. roy. ent. Soc. Lond.* **103**: 1–26. [M, B, 130/53
BRIANT, T. J. (1884). On the anatomy and functions of the tongue of the honeybee (worker). *J. linn. Soc. (Zool.)* **17**: 408–17. [B
BRITTAIN, W. H. (1933). Field studies in the role of insects in apple pollination. *Bull. Dep. Agric. Can. N.S.*, No. 162: 91–157.
BRITTAIN, W. H. & NEWTON, DOROTHY E. (1933). A study in the relative constancy of hivebees and wild bees in pollen gathering. *Canad. J. Res.* **9**: 334–49. [B
—— (1934). Further observations on the pollen constancy of bees. *Canad. J. Res.* **10**: 255–63. [B
BRUMAN, F. (1928). Die Luftzirkulation im Bienenstock. *Z. vergl. Physiol.* **8**: 366–70.
BRUN, R. (1914). *Die Raumorientierung der Ameisen und das Orientierungs-Problem im allgemeinen*. Jena: Gustav Fischer.
BRÜNNICH, K. (1922). The influence of the age of the queen on the honey crop. *Bee World* **4**: 6–7. [M, B
—— (1922–23). Graphische Darstellung der Legetätigkeit einer Bienenkönigin. [A graphic representation of the oviposition of a queen bee.] *Arch. Bienenk.* **4**: 137–47 and *Bee World* **4**: 208–10, 223–4. [M, B
BRUNSON, M. H. (1938). Influence of Japanese beetle instar on the sex and population of the parasite *Tiphia popilliavora*. *J. agric. Res.* **57**: 379–86.
BUDDENBROCK, W. v. (1937).* *Vergleichende Physiologie*. Berlin.
BÜDEL, A. (1948). Der Wasserdampfhaushalt im Bienenstock. *Z. vergl. Physiol.* **31**: 249–73. [B
—— (1949). Bee physics: its aims and methods. *Bee World* **30**: 74–6. [M, B
BUGNION, E. (1928). *Les glandes salivaires de l'Abeille et des Apiaires en général*. Vaucluse: Montfavet. [B
BULMAN, G. W. (1892). The constancy of the bee. *Sci. Gossip* **329**: 98–9.
—— (1902). The constancy of the bee. *Zoologist* **6**: 220–2.
BURTOV, V. (1950). [Influence of temperature on the development of bees' eggs.] *Pchelovodstvo*: 399–400. [B, 187/52
BUTLER, C. (1609). *The feminine monarchie*. Oxford: Joseph Barnes. [M
BUTLER, C. G. (1939). The drifting of drones. *Bee World* **20**: 140–2. [M, B
—— (1940*a*). The ages of the bees in a swarm. *Bee World* **21**: 9–10. [M, B
—— (1940*b*). The choice of drinking water by the honeybee. *J. exp. Biol.* **17**: 253–61. [M, B
—— (1945). The behaviour of bees when foraging. *J. roy. Soc. Arts* **93**: 501–11. [M, B
—— (1946). The provision of supplementary food to hive bees. *Ann. appl. Biol.* **33**: 307–9. [B
—— (1951). The importance of perfume in the discovery of food by the worker honeybee. *Proc. roy. Soc. B* **138**: 403–13. [M, B, 124/52
BUTLER, C. G. & FINNEY, D. J. (1942). An examination of the relationship between honeybee activity and solar radiation. *J. exp. Biol.* **18**: 206–13. [M, B
BUTLER, C. G. & FREE, J. B. (1952). The behaviour of worker honeybees at the hive entrance. *Behaviour* **4**: 263–92. [M, B, 135/53
BUTLER, C. G., JEFFREE, E. P. & KALMUS, H. (1943). The behaviour of a population of honeybees on an artificial and on a natural crop. *J. exp. Biol.* **20**: 65–73. [M, B
BUTTEL-REEPEN, H. v. (1900). Sind die Bienen Reflexmaschinen? [Are bees reflex machines?] *Biol. Zbl.* **20**: 1–82. Transl. (1908) *Glean. Bee Cult.* **36**: 223 et. sequ. [M, B
—— (1915). *Leben und Wesen der Bienen*. Braunschweig: Vieweg. [M, B
—— (1923). Memory of location in queens. *Amer. Bee J.* **63**: 25–7. [M, B

BUXTON, P. A. (1932). Terrestrial insects and the humidity of the environment. *Biol. Rev.* **7**: 275–320.
BUYSSON, R. DU (1903). Monographie des Guêpes ou Vespa. *Ann. Soc. ent. Fr.* **72**: 260–88.
BUZZARD, C. N. (1936). De l'organisation du travail chez les abeilles. *Bull. Soc. Apic. Alpes-Marit.* **15**: 65–70. [B, T
CALE, G. H. (1946). In *The hive and the honeybee*, Ch. XIV. Hamilton, Ill.: Dadant. [M, B
CAMERON, A. T. (1947). The taste sense and the relative sweetness of sugars and other sweet substances. *Sci. Rep. Ser. Sug. Res. Fdn.* No. 9: 1–72.
CASTEEL, D. B. (1912a). The behaviour of the honeybee in pollen collecting. *Bull. U.S. Bur. Ent.* No. 121: 1–36. [M, B
—— (1912b). The manipulation of the wax scales of the honeybee. *Circ. U.S. Bur. Ent.* No. 161: 1–13. [M, B
CHADWICK, P. C. (1922). Ventilation. *Amer. Bee J.* **62**: 158–9. [M, B
—— (1931). Ventilation of the hive. *Glean. Bee Cult.* **59**: 356–8. [M, B
CHELDELIN, V. H. & WILLIAMS, R. J. (1942). The B vitamin content of foods. *Univ. Tex. Publ.* No. 4237: 105–24.
CHERIAN, M. C., RAMACHANDRAN, S., & MAHADEVAN, V. (1947). Studies in bee behaviour. *Indian Bee J.* **9**: 98–100, 116–24. [B
CHESHIRE, F. (1886). *Bees and beekeeping*, Vol. 1. London: Upcott Gill. [M, B
—— (1888). *Bees and beekeeping*, Vol. II. London: Upcott Gill. [M, B
CHEWYREUV, I. (1913). Le rôle des femelles dans la determination du sexe de leur descendance dans le groupe des Ichneumonides. *C.R. Soc. Biol., Paris* **64**: 695–7.
CIESIELSKI, T. (1895).* Cited Himmer (1932).
CLEMENTS, F. E. & LONG, F. L. (1923). *Experimental pollination, an outline of the ecology of flowers and insects.* Washington: Carnegie.
COOPER, B. A. (1952). Colony odour and choice of forage. *Bee World* **33**: 189–90. [M, B
CORKINS, C. L. (1930). The metabolism of the honeybee colony during winter. *Bull. Wyo. agric. Exp. Sta.* No. 175: 1–54. [M, B
—— (1932a). The temperature relationship of the honeybee cluster under controlled temperature conditions. *J. econ. Ent.* **25**: 820–5.
—— (1932b). Drifting of honeybees. *Bull. Wyo. agric. Exp. Sta.* No. 190: 1–24. [M, B
CORKINS, C. L. & GILBERT, C. S. (1932). The metabolism of honeybees in winter cluster. *Bull. Wyo. agric. Exp. Sta.* No. 187: 1–30. [B
COWAN, T. W. (1924). *The British bee-keeper's guide book.* 25th edn. London: Larby. [M, B
CRANE, E. EVA (1950). The effect of spring feeding on the development of honeybee colonies. *Bee World* **31**: 65–72. [M, B
CRANE, M. B. & MATHER, K. (1943). The natural cross-pollination of crop plants with particular reference to the radish. *Ann. appl. Biol.* **30**: 301–8.
CRAWSHAW, L. S. (1914). Cappings of comb. *Brit. Bee J.* **42**: 255–6. [M, B
CURRIE, G. A. (1932). Research on bees: a progress report. *J. Coun. Sci. industr. Res. Aust.* **5**: 81–7. [M, B
—— (1935). Research on bees: progress report II. *J. Coun. Sci. industr. Res. Aust.* **8**: 14–20. [M, B
DADANT, C. (1893). *Langstroth on the hive and the honeybee.* 2nd edn. Hamilton, Ill.: Dadant.
DADANT, H. C. (1926). The behaviour of bees in comb building. *Amer. Bee J.* **66**: 224–6.
DAHLBERG, A. C. & PENCZEK, E. S. (1941). The relative sweetness of sugars as affected by concentration. *Tech. Bull. N.Y. St. agric. Exp. Sta.* No. 258: 1–12.
DARWIN, C. (1859). *The origin of species by natural selection.* London: Murray. [M
—— (1876). *The effects of cross and self fertilisation in the vegetable kingdom.* London: Murray.
DAVIS, J. L. (1908). Queen killed by a rival queen. *Glean. Bee Cult.* **36**: 1259–60. [M, B
DEMUTH, G. S. (1921). Swarm control. *Fmrs' Bull. U.S. Dep. Agric.* No. 1198: 1–28. [M, B
—— (1922). The cause of swarming. *Glean. Bee Cult.* **50**: 371–3. [M, B
—— (1931). Cause of swarming is known. *Amer. Bee J.* **71**: 419. [M, B

DIETERICH, K. (1911). Weitere Beiträge für Kenntnis des Bienenharzes. *Pharm. Zentralh.* **52**: 1019–27. [T

DOBKIEWICZ, L. V. (1912). Beitrag zur Biologie der Honigbiene. *Biol. Zbl.* **32**: 664–94.

DOLGOVA, L. P. (1928). [The effect of some factors upon the visiting of honey plants by bees.] *Opuit. Pas.*: 176–9, 248–53. [B

DÖNHOFF, E. (1855a). Ueber das Geruchsorgan der Biene.

—— **(1855b).** Ueber das Herrschen verschiedener Triebe in verschieden Lebensattern bei den Bienen.

—— **(1857).** Erhalten die fruchtbaren Arbeiter ausser ihrer Eigenschaft, zu legen, noch etwas von der Natur, dem Geruch und Instinkt der Königen?

—— **(1858).** Ueber das leichte Annehmen eines fremdartigen Geruchs von Seiten der Bienenkönigen.

—— **(1859).** Hypothese über den Einfluss psychischer Aufregung auf Erzeugung von Eiern bei den Arbeitsbienen.

All reprinted in *Beiträge zur Bienenkunde*, Berlin: Pfenningstorff. [B

DOOLITTLE, G. M. (1901). Cited Miller, C. C. *Glean. Bee Cult.* **29**: 930. [M, B

—— **(1908).** Plurality of queens; when profitable and when not. *Glean. Bee Cult.* **36**: 1307–8. [M, B

—— **(1915).** *Scientific queen-rearing.* Hamilton, Ill.: Dadant. [M

DREYLING, L. (1903). Ueber die wachsbereitenden Organe der Honigbiene. *Zool. Anz.* **26**: 710–15.

—— **(1905).** Die wachsbereitenden Organe bei den gesellig lebenden Bienen. *Zool. Jb.*, Abt. 2, **22**: 289–330.

DRÖSI (1929).* Studien an eierlegenden Arbeitsbienen. *Allatani Közlemennek* **16**. Cited *Schweiz. Bienenztg.* **52**: 510. [B

DUFOUR, L. (1901). Recherches sur la ponte de la reine. *Ann. Féd. Soc. franç. Apic.* **10**: 18–34. Reprinted, *Apiculteur* **83**: 271 et sequ. [B, T

DUJARDIN, F. (1852). Quelques observations sur les abeilles, et particulièrement sur les actes qui, chez les insectes peuvent être rapportés a l'intelligence. *Ann. Sci. nat.* (Zool.) 3 Sér. **18**: 231–40.

DUNHAM, W. E. (1929a). The influence of external temperature on the hive temperatures during the summer. *J. econ. Ent.* **22**: 798–801.

—— **(1929b).** Relation of heat to brood rearing. *Glean. Bee Cult.* **57**: 359–62. [M, B

—— **(1930).** Temperature gradient in the egg-laying activities of the queen bee. *Ohio J. Sci.* **30**: 403–10. [B

—— **(1931a).** The effect of low external temperatures on the brood nest temperatures of a normal colony of bees during summer. *J. econ. Ent.* **24**: 638–43. [B

—— **(1931b).** A colony of bees exposed to high external temperatures. *J. econ. Ent.* **24**: 606–11. [B

—— **(1933).** Hive temperatures during the summer. *Glean. Bee Cult.* **61**: 527–9. [M, B

DZIERZON, J. (1845).* Eichstädt. *Bienenztg.* **1**: 113.

ECKERT, A. (1927). Das Werden des Wachses nach der biologischen Seite. *Bienenpflege, Ludwigsburg* **49**: 193–8. [B

ECKERT, J. E. (1933). The flight range of the honeybee. *J. agric. Res.* **47**: 257–85. [B

—— **(1942).** The pollen required by a colony of honeybees. *J. econ. Ent.* **35**: 309–11. [B

EGGERS, F. (1923). Ergebnisse von Untersuchungen am Johnstonchen Organ der Insekten und ihre Bedeutung für die allgemeine Beurteilung der stiftführenden Sinnesorgane. *Zool. Anz.* **57**: 224–40.

ELSER, E. (1929). Die chemische Zusammensetzung der Nährungsstoffe der Biene. *Märk. Bienenztg.* **19**: 210–15, 232–5, 248–52. [B

EMERY, J. (1875). Ants and bees. *Nature, Lond.* **12**: 25–6.

EVANS, H. M., EMERSON, G. A. & ECKERT, J. E. (1937). Alleged vitamin E content in royal jelly. *J. econ. Ent.* **30**: 642–6. [B

EVENIUS, C. (1929). Zur Frage der Herkunft des Kittharzes im Bienenstock. *Biol. Zbl.* **49**: 257–61.

—— **(1937).** Beobachtungen an der Schlunddrüse der Honigbienen während der Winterruhe. *Dtsch. Imkerführer* **11**: 128–33. [B, T

FABRE, J. H. (1879). *Souvenirs entomologiques*, I. Paris: Delagrave.

—— **(1882).** *Souvenirs entomologiques*, II. Paris: Delagrave.

FAHN, A. (1949). Studies in the ecology of nectar secretion. *Palest. J. Bot. Jerusalem* **4**: 207–24. [B, 164/53

FARRAR, C. L. (1934). Bees must have pollen. *Glean. Bee Cult.* **62**: 276–8. [M, B

—— **(1936).** Influence of pollen reserves on the surviving population of over-wintered colonies. *Amer. Bee J.* **76**: 452–4. [M, B

FARRAR, C. L. (1937). The influence of colony populations on honey production. *J. agric. Res.* **54:** 945–54. [B
—— **(1938).** New recommendations for the installation of package bees, using a spray and direct-release method. *Circ. U.S. Bur. Ent.* No. E-427: 1–7. [M, B
—— **(1949).** (No title). *Bee World* **30:** 51. [M, B
—— **(1952).** Ecological studies on overwintered honeybee colonies. *J. econ. Ent.* **45:** 445–9. [M, B, 148/53
FIELDE, ADELE M. (1904). Power of recognition among ants. *Biol. Bull., Wood's Hole* **7:** 227–50.
—— **(1905).** The progressive odour of ants. *Biol. Bull., Wood's Hole* **10:** 1–16.
FILMER, R. S. (1932). Brood area and colony size as factors in activity of pollination units. *J. econ. Ent.* **25:** 336–43.
FIRSSOV, J. G. (1951). Die Dressur der Bienen auf Rotklee. *Dtsch. Imkerztg.* **1:** 238–41. [B, 156/52
FISCHER, G. (1871).* Ueber die Funktion der Speicheldrüsen der Biene nach Schiemenz. *Eichstadt. Bienenztg.* (130). Cited Schiemenz (1883).
FLANDERS, S. E. (1950). Control of sex in the honeybee. *Sci. Monthly* **71:** 237–40. [B, 114/51
FOREL, A. (1874). *Fourmis de la Suisse.* Zurich; Zürcher und Furrer.
—— **(1904).** *Ants and some other insects.* Chicago: Open Court.
—— **(1906).*** Mémoire du temps et association des souvenirs chez les abeilles. *Bull. Inst. gén. psychol.*: 257–9. Cited Forel (1908).
—— **(1908).** *The senses of insects.* London: Methuen. [M
—— **(1910).** *Das Sinnesleben der Insekten.* München: Reinhardt.
FRANÇON, J. (1939). *The mind of the bees.* London: Methuen. [M, B
FREUDENSTEIN, K. (1932). Das 'Hobeln' der Bienen. *Biol. Zbl.* **52:** 343–9. [B
FRINGS, H. (1944). The loci of olfactory end-organs in the honeybee. *J. exp. Zool.* **97:** 123–34.
—— **(1951).** Sweet taste in the cat and the taste spectrum. *Experientia* **7:** 424–6. [B
FRISCH, K. v. (1914). Der Farbensinn und Formensinn der Biene. *Zool. Jb.*, Abt. 3, **35:** 1–188.
—— **(1919a).** Zur alten Frage nach dem Sitz des Geruchsinnes bei Insekten. Versuche an Bienen. *Verh. zool-bot. Ges. Wien.* **69:** 17–26.
—— **(1919b).** Ueber den Geruchsinn der Bienen und seine blütenbiologische Bedeutung. *Zool. Jb.*, Abt. 3, **37:** 1–238.
—— **(1920–2).** *München med. Wochenschr.* **1920:** 566–9. **1921:** 509–11. **1922:** 781–2.
—— **(1921).** Ueber den Sitz des Geruchsinnes bei Insekten. *Zool. Jb.*, Abt. 3, **38:** 1–68.
—— **(1923).** Ueber die 'Sprache' der Bienen. *Zool. Jb.*, Abt. 3, **40:** 1–186.
—— **(1924).** Sinnesphysiologie und 'Sprache' der Bienen. *Naturwissenschaften* **12:** 981–7.
—— **(1927).** Versuche über den Geschmackssinn der Bienen. *Naturwissenschaften* **15:** 321–7. [B
—— **(1928).** Versuche über den Geschmackssinn der Bienen. II. *Naturwissenschaften* **16:** 307–15.
—— **(1930).** Versuche über den Geschmackssinn der Bienen. III. *Naturwissenschaften* **18:** 169–74.
—— **(1934).** Ueber den Geschmackssin der Biene. Eine Beitrag zur vergleichenden Physiologie des Geschmacks. *Z. vergl. Physiol.* **21:** 1–156. [B
—— **(1942).** Die Werbetänze der Bienen und ihre Auslösung. *Naturwissenschaften* **30:** 269–77.
—— **(1943).** Versuche über die Lenkung des Bienenfluges durch Duftstoffe. *Naturwissenschaften* **31:** 445–460. [T
—— **(1944).** Weitere Versuche über die Lenkung des Bienenfluges durch Duftstoffe. *Biol. Zbl.* **64:** 237–66.
—— **(1946a).** Die Tänze der Bienen. [The dances of the Bee.] *Osterr. Zool.* **1:** 1–48. Transl. (1947) *Bull. Anim. Behav.* No. 5: 1–32. [M, B
—— **(1946b).** Die 'Sprache' der Bienen und ihre Nutzanwendung in der Landwirtschaft. *Experientia* **2:** 397–404. [B, T
—— **(1947).** *Duftgelenkte Bienen im Dienste der Landwirtschaft und Imkerei.* Wien: Springer. [B
—— **(1948).** Gelöste und ungelöste Rätsel der Bienensprache. [Known and unknown questions regarding the language of the bees.] *Naturwissenschaften* **35:** 12–23 and 38–43. Transl. (1951) *Bull. Anim. Behav.* No. 9: 2–25. [B
—— **(1949).** Die Polarisation des Himmelslichtes als orientierender Faktor bei den Tänzen der Bienen. [Polarisation of light from the sky as an orientation factor in the dances of bees.] *Experientia* **5:** 142–8. Trans. (1951) *Bull. Anim. Behav.* No. 9: 26–32. [B
——**(1950a).** Die Sonne als Kompass im Leben der Bienen. *Experientia* **6:** 210–21. [B, T, 77/51.

FRISCH, K. v. (1950b). *Bees, their vision, chemical senses and language.* Ithaca, N.Y.: Cornell. [M, B, 197/51
—— (1951). Orientierungsvermögen und Sprache der Bienen. [Orienting ability and communication among bees.] *Naturwissenschaften* 38: 105–12. Transl. (1952) *Bee World* 33: 19–25. [M, B, 144/53
FRISCH, K. v. & RÖSCH, G. A. (1926). Neue Versuche über die Bedeutung von Duftorgan und Pollenduft für die Verständigung im Bienenvolk. *Z. vergl. Physiol.* 4: 1–21.
FYG, W. (1944). Zweimaliger Hochzeitsflug? *Schweiz. Bienenztg.* 67: 85–91. [M, B
—— (1950). Beobachtungen über die Wirkungen der Kolhensäure-Narkose auf Arbeitsbienen. *Schweiz. Bienenztg.* 73: 174–84. [B, 66/50
—— (1952). The process of natural mating in the honeybee. *Bee World* 33: 129–39. [M, B, 133/53
GATAULIN, F. (1945). [How to make a backward colony work a crop.] *Pchelovodstvo* (3): 37–8. [145/50
GATES, B. N. (1914). The temperature of the bee colony. *Bull. U.S. Dep. Agric.* No. 96: 1–29. [M, B
GERSTUNG, F. (1891–1926). *Der Bien und Seine Zucht.* 7 edns. Berlin: Pfenningstorff. [B
GILLETTE, C. P. (1897). Weights of bees and the loads they carry. *Proc. Soc. Prom. agric. Sci.* 14: 60–3.
GILTAY, E. (1904). Ueber die Bedeutung der Krone bei den Blüten und über das Farbenunterscheidungsvermögen der Insekten. *Jb. wiss. Bot.* 40: 368–402.
GOETZE, G. (1926). Zur Züchtungsbiologie, Variabilitätsstudien an der Honigbiene. *Preuss. Bienenztg.* 71: (cited *Amer. Bee J.* 68: 115–17).
GONTARSKI, H. (1935a). Leistungsphysiologishe Untersuchungen an Sammelbienen. *Arch. Bienenk.* 16: 107–26. [M, B
—— (1935b). Wabenzellmaasze bei *Apis mellifica. Z. vergl. Physiol.* 21: 682–98. [B
—— (1949a). Wandlungsfähige Instinkte der Honigbiene. *Umschau* 49: 310–12. [B, 35/1
—— (1949b). Mikrochemische Futtersaftuntersuchungen und die Frage der Königinentstehung. *Hess. Biene* 85: 89–92. [B
—— (1950). Betäubungsmittel und ihre Wirkung auf Königinnen und Arbeiterinnen. *Bienenzucht* 3 (12).
GRABENSBERGER, W. (1933). Untersuchungen über das Zeitgedächtnis der Ameisen und Termiten. *Z. vergl. Physiol.* 20: 1–54.
—— (1934a). Experimentelle Untersuchungen über das Zeitgedächtnis von Bienen und Wespen nach Verfütterung von Euchinin und Jodthyreoglobulin. *Z. vergl. Physiol.* 20: 338–42.
—— (1934b). Der Einfluss von Salicylsäure, gelbem Phosphor und weissem Arsenik auf das Zeitgedächtnis der Ameisen. *Z. vergl. Physiol.* 20: 501–10.
GRAHAM, C. H. & HUNTER, W. S. (1931). Thresholds of illumination for the visual discrimination of direction of movement and for the discrimination of discreteness. *J. gen. Psychol.* 5: 178–90.
GROOT, A. P. de (1950). The influence of temperature and kind of food on the increase in the nitrogen content of the young worker honeybee. *Proc. Acad. Sci. Amst.* 53 (4): 1–8. [B, 100/52
GROSDANIC, S. (1931). Zwei Beispiele des rhythmischen Verhaltens der Arbeiterinen von *Apis mellifica* L., die ineinander übergehen Können. *Bienenvater* 63: 233–6. [M, B
—— (1951). 'Hipnoza' medonosne pcele. *Pcelarstvo* 6: 77–84. [B, T, 233/52
GUBIN, A. F. (1936).* *Pchelovodstvo* (6), cited v. Frisch (1943).
—— (1938).* *Pchelovodstvo* (5). 40–44, cited v. Frisch (1943).
—— (1939).* *Pchelovodstvo* (6): 31, cited v. Frisch (1943).
—— (1945a). Cross pollination of fibre flax. *Bee World* 26: 30–1. [M, B
—— (1945b). Bee training for pollination of cucumbers. *Bee World* 26: 34–5. [M, B
GUBIN, A. F. & ROMASHOV, G. I. (1933). [*Utilization of the responses of the bee for pollination of plants,* Ch. 8.] Moscow. [T
GWIN, C. M. (1936). Further developments concerning wax production by the honeybee colony. I. A study of the production of wax scales and comb building. *J. econ. Ent.* 29: 318–21. [B
HAAN, J. A., BIERENS de (1928). Experiments on the determination of the choice of bees by absolute and relative characteristics. *Tijdschr. n. dierk.*, Ver. 3, 1: 45–7.
HAMBLETON, J. I. (1925). The effect of weather upon the change in weight of a colony of bees during the honey flow. *Bull. U.S. Dep. Agric.* No. 1339: 1–52. [B

HAMILTON, W. (1932). The probable cause of swarming in the honeybee. *Bee World* **13**: 2–5. [M, B
HANSSON, A. (1945). Lauterzeugung und Lautauffassungsvermögen der Bienen. *Opusc. ent.* **6**: 1–122. [B
—— (1951). Om ljudproduktion och ljuduppfatning hos bina. *Nord. Bitidskr.* **3**: 68–76.
HARTRIDGE, H. (1950). Recent advances in the physiology of vision. Part III. *Brit. Med. J.* (I): 1331–40.
HAVILAND, G. D. (1882).* Cited Cheshire (1886).
HAYDAK, M. H. (1930).* *Čes. Včelař* **64**: 166–8 (cited Haydak, 1932).
—— (1932). Division of labour in the colony. *Wis. Beekeeping* **8**: 36–9. [B
—— (1933). Der Nährwert von Pollenersatzstoffen bei Bienen. *Arch. Bienenk.* **14**: 185–219. [M, B
—— (1935). Brood rearing by honeybees confined to a pure carbohydrate diet. *J. econ. Ent.* **28**: 657–60. [B
—— (1937). Further contribution to the study of pollen substitutes. *J. econ. Ent.* **30**: 637–42. [B
—— (1940a). Comparative value of pollen and pollen substitutes. II. Bee bread and soybean flour. *J. econ. Ent.* **33**: 397–9. [B
—— (1940b). Laying workers. *Glean. Bee. Cult.* **68**: 615. [M, B
—— (1943). Larval food and development of castes in the honeybee. *J. econ. Ent.* **36**: 778–92. [B
—— (1945). Value of pollen substitutes for brood rearing of honeybees. *J. econ. Ent.* **38**: 484–7.
—— (1949). Causes of deficiency of soybean flour as a pollen substitute for honeybees. *J. econ. Ent.* **42**: 573–9. [B
HAYDAK, M. H. & PALMER, L. S. (1938). Vitamin E content of royal jelly and bee bread. *J. econ. Ent.* **31**: 576–7. [B
—— (1940). Vitamin content of bee foods. II. Vitamin B_1 content of royal jelly and bee bread. *J. econ. Ent.* **33**: 396–7. [B
—— (1942). Royal jelly and bee bread as sources of vitamins B_1, B_2, B_6, C, and nicotinic and pantothenic acids. *J. econ. Ent.* **35**: 319–20. [B
HAYDAK, M. H. & TANQUARY, M. C. (1943). Pollen and pollen substitutes in the nutrition of the honeybee. *Tech. Bull. Minnesota agric. Esp. Sta.* No. 160: 1–23. [B
HAYDAK, M. H. & VIVINO, A. E. (1943). Changes in vitamin content during the life of the worker honeybees. *Arch. Biochem.* **2**: 201–7. [B
—— (1950). The changes in the thiamine, riboflavin, niacin and pantothenic acid content in the food of female honeybees during growth, with a note on the vitamin K activity of royal jelly and bee bread. *Ann. ent. Soc. Amer.* **43**: 361–7. [B, 196/51
HAZELHOFF, E. H. (1941). De Luchtverversching van een Bijenkast gedurende den zomer. *Maandschr. Bijent.* **44**: 1–16. [B, T
HECHT, S. & WOLF, E. (1929). The visual acuity of the honeybee. *J. gen. Physiol.* **12**: 707–60. [B
HEIN, G. (1950). Ueber richtungsweisende Bienentänze bei Futterplätzen in Stocknähe. *Experientia* **6**: 142–4. [B, T, 62/52
HENKEL, C. (1938).* Cited von Frisch (1942).
HERAN, H. (1952). Untersuchungen über den Temperatursinn der Honigbiene (*Apis mellifica*) unter besonderer Berücksichtigung der Wahrnehmung strahlender Wärme. *Z. vergl. Physiol.* **34**: 179–206. [B, 159/53
HERAN, H. & WANKE, L. (1952). Beobachtungen über die Entfernungsmedung der Sammelbienen. *Z. vergl. Physiol.* **34**: 383–93.
HERROD-HEMPSALL, W. (1930). *Bee-keeping, new and old.* London: British Bee Journal. [M, B
HERTZ, MATHILDE (1929a). Die Organisation des optischen Feldes bei der Biene, I. *Z. vergl. Physiol.* **8**: 693–748.
—— (1929b). Die Organisation des optischen Feldes bei der Biene, II. *Z. vergl. Physiol.* **11**: 107–45.
—— (1931). Die Organisation des optischen Feldes bei der Biene, III. *Z. vergl. Physiol.* **14**: 629–74.
—— (1933). Ueber figurale Intensitäten und Qualitäten in der optischen Wahrenehmung der Biene. *Biol. Zbl.* **53**: 10–40.
—— (1934a). Eine Bienendressur auf Wasser. *Z. vergl. Physiol.* **21**: 463–7. [T
—— (1934b). Zur Physiologie des Formen- und Bewegungssehens, II. Auflösungsvermögen des Bienenauges und optimotorische Reaktion. *Z. vergl. Physiol.* **21**: 579–603.

HERTZ MATHILDE (1934c). Zur Physiologie des Formen- und Bewegungssehens, III. Figurale Unterscheidung und reziproke Dressuren bei der Biene. *Z. vergl. Physiol.* **21**: 604–15.
—— (1935). Die Untersuchungen über den Formensinn der Honigbiene. *Naturwissenschaften* **23**: 618–24.
—— (1937). Beitrag zum Farbensinn und Formensinn der Biene. *Z. vergl. Physiol.* **24**: 413–21.
—— (1939). New experiments on colour vision in bees. *J. exp. Biol.* **16**: 1–8. [M, B
HESELHAUS, F. (1922). Die Hautdrüsen der Apiden und verwandter Formen. *Zool. Jb.*, Abt. 2, **43**: 369–464.
HESS, C. v. (1910). Neue Untersuchungen über den Lichtsinn bei wirbellosen Tieren. *Pflüg. Arch. ges. Physiol.* **136**: 282–367.
—— (1916). Messende Untersuchung des Lichtsinnes der Biene. *Pflüg. Arch. ges. Physiol.* **163**: 289–320.
—— (1918). Beiträge zur Frage nach einem Farbensinn bei Bienen. *Pflüg. Arch. ges. Physiol.* **170**: 337–66.
—— (1920). Neues zur Frage nach einem Farbensinn bei Bienen. *Naturwissenschaften* **8**: 927–9.
HESS, GERTRUD (1942). Ueber den Einfluss der Weisellösigkeit und des Fruchtbarkeitsvitamins E auf die Ovarien der Bienenarbeiterin. *Beih. Schweiz. Bienenztg.* **1**: 33–109. [B
HESS, W. R. (1926). Die Temperaturregulierung im Bienenvolk. *Z. vergl. Physiol.* **4**: 465–87. [B, T
HESSE, A. (1900–1).* Ueber ätherisches Jasminblütenöl. *Ber. Dtsch. chem. Ges.* **33**: 1585–91; **34**: 291–6 and 2916–33.
HIMMER, A. (1925). Körpertemperaturmessungen an Bienen und anderen Insekten. *Erlanger Jb. Bienenk.* **3**: 44–115. [B
—— (1926). Der soziale Wärmehaushalt der Honigbiene, I. Die Wärme im nichtbrütenden Wintervolk. *Erlanger Jb. Bienenk.* **4**: 1–51. [B, T
—— (1927a). Der soziale Wärmehaushalt der Honigbiene, II, Die Wärme der Bienenbrut. *Erlanger Jb. Bienenk.* **5**: 1–32. [B
—— (1927b). Ein Beitrag zur Kenntnis des Wärmehaushalts im Nestbau sozialer Hautflüger. *Z. vergl. Physiol.* **5**: 375–89.
—— (1927c). Anatomie und Biologie der Bienen. *Erlanger Jb. Bienenk.* **5**: 69–104. [B
—— (1930). Von der Arbeitsteilung im Bienenstaat. *Leipzig. Bienenztg.* **45**: 39–43, 64–7. [B
—— (1932). Die Temperaturverhältnisse bei den sozialen Hymenopteren. *Biol. Rev.* **7**: 224–53. [B
HIRSCHFELDER, H. (1951). Quantitative Untersuchungen zum Polleneintragen der Bienenvölker. *Z. Bienenforsch.* **1**: 67–77. [B
HODGE, C. F. (1894). Changes in the ganglion cells from birth to senile death; observations on man and the honeybee. *J. Physiol.* **17**: 129–34.
HODGES, DOROTHY (1952). *The pollen loads of the honeybee.* London: Bee Research Association. [B
HOLLICK, F. S. J. (1940). The flight of the dipterous fly *Muscina stabulans* Fallen. *Philos. Trans. B.* **230**: 357–90.
HOMANN, H. (1924). Zum Problem der Ocellenfunktion bei den Insekten. *Z. vergl. Physiol.* **1**: 13–29.
HOWELL, D. E. & USINGER, R. L. (1933). Observations on the flight and length of life of drone bees. *Ann. ent. Soc. Amer.* **26**: 239–46.
HUBER, F. (1792). *Nouvelles observations sur les abeilles, I.*
—— (1814). *Nouvelles observations sur les abeilles*, II. [New observations on bees, I and II]. Transl. (1926) Hamilton, Ill.: Dadant. [M, B
HUISH, R. (1815). *A treatise on the nature, economy and practical management of bees.* London; Henry Bohn. [M
HUMPHREYS, W. (1883). Queen-encasement. *Brit. Bee J.* **11**: 139 [M
HUNTER, J. (1792). Observations on bees. *Philos. Trans.* **82**: 128–96.
HUTCHINSON, W. Z. (1905). *Advanced bee culture.* Medina, Ohio: Root. [M, B
INGLESENT, H. (1940). Zymotic function of the pharyngeal, thoracic and postcerebral glands of *Apis mellifica*. *Biochem. J.* **34**: 1415–8. [B
JACOB, N. (1568). Gründlicher und nützlicher Unterricht von Wartung der Bienen, aus wahrer Erfahrung zusammengetragen. Halberstadt. [T
JACOBS, W. (1924). Das Duftorgan von *Apis mellifica* und ähnliche Hautdrüsenorgane sozialer und solitärer Apiden. *Z. Morph. Ökol. Tiere* **3**: 1–80.
JANET, C. (1903). *Observations sur les guêpes.* Paris: Naud.

JANSCHA, A. (1771). *Abhandlung vom Schwärmen der Bienen.* Wien: Kurzbock. Reprinted (1925), Berlin; Pfenningsstorff. Summarised in *Anton Janscha on the swarming of bees*, by Fraser, H. M. (1951), Royston, Herts: Apis Club. [B

JAUBERT, G. F. (1927). Sur l'origine de la coloration de la cire d'abeilles et la composition de la propolis. *C. R. Acad. Sci., Paris* **184**: 1134–6.

JESSUP, J. (1924a). Ventilation by the bee colony. *Rep. Ia. St. Apiar.* 1924: 35–7. [M, B
—— (1924b).* Unpublished thesis, cited Park (1949).

JONES, G. D. GLYNNE (1952). The responses of the honeybee to repellent chemicals. *J. exp. Biol.* **29**: 372–86. [B, 140/53

JONESCU, C. N. (1909). Vergleichende Untersuchungen über das Gehirn der Honigbiene. *Jena Z. Naturw.* **45**: 111–80.

JONGBLOED, J. & WIERSMA, C. A. G. (1934). Der Stoffwechsel der Honigbiene wahrend des Fliegens. *Z. vergl. Physiol.* **21**: 519–33.

JORDAN, R. (1928). Experiments refuting the pressure theory and the explanation of the insemination of the egg based upon it. *Bee World* **9**: 50–54. [M, B

JORDAN, R., TISCHER, J. & ILLNER, W. (1940). Vergleich des von Altbienen erzeugten Wachses mit 'Jungfernwachs' und gewöhnlichen Bienenwachs. *Z. vergl. Physiol.* **28**: 353–7. [T

KALMUS, H. (1934). Ueber die Natur des Zeitgedächtnis der Bienen. *Z. vergl. Physiol.* **20**: 405–19. [B
—— (1937). Vorversuche über die Orientierung der Biene im Stock. *Z. vergl. Physiol.* **24**: 166–87.
—— (1938). Der Füllungszustand der Honigblase entscheidet die Flugrichtung der Honigbiene. *Z. vergl. Physiol.* **26**: 79–84. [B
—— (1941). The defence of a source of food by honeybees. *Nature, Lond.* **148**: 228.
—— (1945). Correlations between flight and vision, and particularly between wings and ocelli, in insects. *Proc. roy. ent. Soc. Lond.* A, **20**: 84–96.
—— (1953). *Not yet published.*

KALMUS, H. & RIBBANDS, C. R. (1952). The origin of the odours by which honeybees distinguish their companions. *Proc. roy. Soc.* B, **140**: 50–9.
[M, B 145/53

KALNITZKI, A. (1949). [Queens mated successfully with overwintered drones.] *Pchelovodstvo*: 370. [B, 113/51

KALTOFEN, R. S. (1951). Das Problem des Volkesduftes bei der Honigbiene. *Z. vergl. Physiol.* **33**: 462–75. [B, 65/53

KAPUSTIN (1938).* *Pchelovodstvo* (8, 9): 37–8. Cited v. Frisch (1943).

KATHARINER, L. (1903). Versuche über die Art der Orientierung bei der Honigbiene. *Biol. Zbl.* **23**: 646–60.

KELSALL, A. (1940). A multi-queened colony of bees. *Amer. Bee J.* **80**: 170. [M, B

KERNER, A. v. MARILAUN (1902). *The natural history of plants.* Vol. II. Transl. 1902. London: Blackie.

KHALIFMAN, T. A. (1950). [New facts about foraging behaviour of bees.] *Pchelovodstvo*: 415–18. [B, 94/51

KING, G. E. (1928).* Unpubl. thesis, cited by Park (1949).
—— (1932). Drifting bees may make production records of little value. *Amer. Bee J.* **72**: 141–2. [M, B

KITZES, G., SCHUETTE, H. A. & ELVEHJEM, C. A. (1943). The B vitamins in honey. *J. Nutr.* **26**: 241–50.

KLEBER, ELISABETH (1935). Hat das Zeitgedächtnis der Bienen biologische Bedeutung? *Z. vergl. Physiol.* **22**: 221–62. [B

KLEIN (1904).* Cited Becker (1925).

KLEIST, F. V. (1919). Nahrungsaufnahme und Kälte beim Bienenvolk. *Arch. Bienenk.* **1**: 103–40. [M, B

KNIGHT, T. A. (1807). On the economy of bees. *Philos. Trans.* **97**: 234–44.

KNUTH, F. (1891). Die Einwirklung der Blutenfarben auf die photographische Platte. *Bot. Zbl.* **48**: 161–5.
—— (1898). *Handbuch der Blütenbiologie.* [*Handbook of flower pollination.*] Transl. (1906). Oxford: Clarendon Press.

KOCH, P. (1934). Farbe der Wohnung und Honigertrag. *Kurmärk. Imker* **24**: 333–5.

KOCH, R. K. (1934). Die Entstehungsurfache des Schwärmens ist doch erkannt. *Rhein. Bienenztg.* **85**: 234–5. [B

KOCHER, V. (1942). Untersuchungen über den Aneuringehalt (Vitamin B_1) von Honig, Pollen und Futtersaft mit Hilfe des Phycomyces-Testes. *Beih. Schweiz. Bienenztg.* **1**: 155–207. [B

KÖHLER, ADRIENNE (1921). Weist die Bienen in ihrem Körper Reservestoffe für die Winterruhe auf? *Schweiz. Bienenztg.* **44**: 224-8. [B
—— (1922). Neue Untersuchungen über den Futtersaft der Bienen. *Verh. dtsch. zool. Ges.* **27**: 105-7. [B
KÖRNER, ILSE (1939). Zeitgedächtnis und Alarmierung bei den Bienen. *Z. vergl. Physiol.* **27**: 445-59.
KOSMIN, NATALIE P., ALPATOV, W. W. & RESNITCHENSKO, M. G. (1932). Zur Kenntnis des Gaswechsals und des Energieverbrauchs der Bienen in Beziehung zur deren Aktivität. *Z. vergl. Physiol.* **17**: 408-22. [B
KOSMIN, NATALIE P. & KOMAROV, P. M. (1932). Ueber das Invertierungsvermogen der Speicheldrüsen und des Mitteldarmes von Bienen verschiedenen Alters. *Z. vergl. Physiol.* **17**: 267-79. [B
KOTOGYAN, A. M. (1941).* Cited Taranov (1947).
KOVTUN, F. N. (1949). [How to make and use multiple-queen colonies.] *Pchelovodstvo*: 29-30. [B, 48/51
—— (1950). [Multiple-queen colonies.] *Pchelovodstvo*: 112. [B, 80/51
KRATKY, E. (1931). Morphologie und Physiologie der Drüsen in Kopf und Thorax der Honigbiene. *Z. wiss. Zool.* **139**: 120-200.
KRIJGSMAN, B. J. (1930). Reizphysiologische Untersuchungen an blutsaugenden Arthropoden im Zusammenhang mit ihrer Nahrungswahl. I. *Stomoxys calcitrans. Z. vergl. Physiol.* **11**: 702-29.
KRIJGSMAN, B. J. & WINDRED, G. L. (1930). Reizphysiologische Untersuchungen an blutsaugenden Arthropoden im Zusammenhang ihrer mit Nahrungswahl. II. *Lyperosia epigna. Z. vergl. Physiol.* **13**: 61-73.
KROGH, A. (1948). Determination of temperature and heat production in insects. *Z. vergl. Physiol.* **31**: 274-80.
KRÖNING, F. (1925). Ueber die Dressur der Biene auf Töne. *Biol. Zbl.* **45**: 496-507.
KÜHN, A. (1924). Versuche über das Unterscheidungsvermögen der Bienen und Fische für Spektrallichter. *Nachr. Ges. Wiss. Göttingen*: 66-71.
—— (1927). Ueber den Farbensinn der Bienen. *Z. vergl. Physiol.* **5**: 762-800.
KÜHN, A. & FRAENKEL, G. (1927). Ueber das Unterscheidungsvermögen der Bienen für Wellenlängen im Spektrum. *Nachr. Ges. Wiss. Göttingen*: 330-5.
KÜHN, A., & POHL, R. (1921). Dressurfähigkeit der Bienen auf Spektrallinien. *Naturwissenschaften* **9**: 738-40.
KUNZE, G. (1927). Einige Versuche über den Geschmackssinn der Honigbiene. *Zool. Jb.*, Abt. 3, **44**: 287-314.
—— (1933). Einige Versuche über den Antennengeschmacksinn der Honigbiene. *Zool. Jb.*, Abt. 3, **52**: 465-512.
KUPETZ, F. (1931). Ein zweiter Hochzeitsflug? [A second mating flight?] *Bienenvater* **10**: 236. Transl. (1935) *Bee World* **16**: 123. [M, B
KUWABARA, M. (1947). Ueber die Regulation im weisellosen Volke der Honigbiene, besonders die Bestimmung des neuen Weisels. *J. Fac. Sci. Hokkaido Univ.*, Ser. 6, **9**: 359-81. [B, 212/52
LAIDLAW, H. H. & ECKERT, J. E. (1950). *Queen rearing.* Hamilton, Ill.: Dadant. [M, B, 50/51
LANDOIS, H. (1867). Die Ton- und Stimmapparate der Insekten in anatomischphysiologischer und akustischer Beziehung. *Z. wiss. Zool.* **17**: 105-86.
LANGER, J. (1912).* *Bienenwirt. Zbl.* **48**.
—— (1929). Der Futtersaft die Kost der Bienenkindes. *Bienenvater* **61**: 25-30, 45-8. [B, T
LANGSTROTH, L. L. (1890). *The hive and the honeybee.* Rev. edn. Hamilton, Ill.: Dadant. [M, B
LATHAM, A. (1923). Introduction of queens. *Amer. Bee J.* **63**: 398-9. [M, B
LECOMTE, M. J. (1950). Sur le déterminisme de la formation de la grappe chez les abeilles. *Z. vergl. Physiol.* **32**: 499-506. [B, 92/51
—— (1951). Recherches sur le comportement agressif des ouvrières d'*Apis mellifica. Behaviour* **4**: 60-6. [B, 208/52
LEFEBVRE, A. (1838).* Note sur le sentiment olfactif des antennes. *Ann. Soc. ent. Fr.* **7**: 395-9.
LEHNART, A. (1935). Die Schwarmursache. *Rhein. Bienenztg.* **86**: 143-5. [B
LEUENBERGER, F. (1927). Afterköniginnen. *Schweiz. Bienenztg.* **50**: 323-6. [M, B
—— (1928). *Die Biene.* Aarau: Sauerlander. [M, B
LEVIN, M. D. & HAYDAK, M. H. (1951). Seasonal variation in weight and ovarian development in the worker honeybee. *J. econ. Ent.* **44**: 54-7. [6/53
LIGHT, S. F. (1942). The determination of the castes of social insects. *Quart. Rev. Biol.* **17**: 312-26.

LINDAUER, M. (1949). Ueber die Einwirkung von Duft- und Geschmacksstoffen sowie anderer Faktoren auf die Tänze der Bienen. *Z. vergl. Physiol.* **31**: 348–412. [B, 200/51
—— (1951a). Die Temperaturregulierung der Bienen die Stocküberhitzung. *Naturwissenschaften* **38**: 308–9. [B, 151/52
—— (1951b). Bienentänze in der Schwaumtraube. *Naturwissenschaften* **38**: 509–13. [B, T, 103/52
—— (1952). Ein Beitrag zur Frage der Arbeitsteilung im Bienenstaat. [A contribution on the question of division of labour in the bee colony]. *Z. vergl. Physiol.* **34**: 299–345. Transl. *Bee World* **34**: 63–73, 85–90. [B, T
LINEBURG, B. (1923a). What do bees do with brood cappings? *Amer. Bee J.* **63**: 235–6 [M, B
—— (1923b). Conservation of wax by the bees. *Amer. Bee J.* **63**: 615–16. [M, B
—— (1924a). Comb building. *Amer. Bee J.* **64**: 271–2. [M, B
—— (1924b). The feeding of honeybee larvae. *Bull. U.S. Dep. Agric.* No. 1222: 25–37. [B
LISSMANN, H. W. (1950). Proprioceptors. *Symp. Soc. exp. Biol.* **4**: 34–59.
LIVENETZ, T. P. (1951). [The drifting of drones.] *Pchelovodstvo*: 25–30. [B, T, 150/52
LOEW, E. (1886). Beobachtungen über den Blumenbesuch von Insekten an Freilandpflanzen des botanischen Gartens zu Berlin. *Jb. Bot. Gart., Berlin* **4**: 93–178.
LOTMAR, RUTH (1933). Neue Untersuchungen über den Farbensinn der Bienen mit besonderer Berücksichtigung des Ultraviolets. *Z. vergl. Physiol.* **19**: 673–723. [B
—— (1939). Der Eiweiss-Stoffwechsel im Bienenvolk während der Ueberwinterung. *Landw. Jb. Schweiz.* **53**: 34–70. [B
LOVELL, J. H. (1910). The colour sense of the honeybee; can bees distinguish colours? *Amer. Nat.* **44**: 673–92. [M
—— (1912). The colour sense of the honeybee; the pollination of green flowers. *Amer. Nat.* **46**: 83–107. [M
LUBBOCK, J. (1874). Observations on bees and wasps, I. *J. linn. Soc., (Zool.)* **12**: 110–39.
—— (1875). Observations on bees and wasps, II. *J. linn. Soc., (Zool.)* **12**: 227–51.
—— (1876). Observations on ants, bees and wasps, III. *J. linn. Soc., (Zool.)* **12**: 445–514.
—— (1882). Observations on ants, bees and wasps, IX. *J. linn. Soc. (Zool.)* **16**: 110–21.
LUNDEN, W. (1914).* Cited by Herberle, J. A. (*Glean. Bee Cult.* **42**: 904). [B
LUNDIE, A. E. (1925). The flight activities of the honeybee. *Bull. U.S. Dep. Agric.* No. 1328: 1–38. [B
LUTZ, F. E. (1924). Apparently nonselective characters and combinations of characters, including a study of ultra-violet in relation to the flower-visiting habits of insects. *Ann. N.Y. Acad. Sci.* **29**: 181–283.
MANLEY, R. O. B. (1936). *Honey production in the British Isles*. London: Faber & Faber. [M, B
—— (1946). *Honey farming*. London: Faber & Faber. [M, B
—— (1948). *Bee-keeping in Britain*. London: Faber & Faber. [M, B
MANUILOVA, A. I. (1938). [The presence of vitamin C in bee bread.] *Vop. Pitan.* **7**: 151 (*Chem. Abst.* **34**: 3392).
MARCHAL, P. (1896). La reproduction et l'évolution des guêpes sociales. *Arch. Zool. exp. gén.*, Sér. 3, **4**: 1–100.
MARSHALL, J. (1935a). On the sensitivity of the chemoreceptors of the antenna and fore-tarsus of the honeybee. *J. exp. Biol.* **12**: 17–26. [M
—— (1935b). The location of olfactory receptors in insects: a review of experimental evidence. *Trans. roy. ent. Soc., Lond.* **83**: 49–72.
MATHER, K. (1947). Species crosses in *Antirrhinum*. I. Genetic isolation of the species *majus, glutinosum*, and *orantium*. *Heredity* **1**: 175–86.
MATHIS, M. (1947). Fécondité, oviposition et rhythme de la ponte chez la reine. *Gaz. apic.* **48**: 149–50. [B
MAURIZIO, ANNA (1946). Beobachtungen über die Lebensdauer und den Futterverbrauch gefangen gehaltener Bienen. *Beih. Schweiz. Beinenztg.* **2**: 1–48. [B, T
—— (1950). Untersuchungen über den Einfluss der Pollennahrung und Brutpflege auf die Lebensdauer und den physiologischen Zustand von Bienen. [The influence of pollen feeding and brood rearing on the length of life and physiological condition of the honeybee.] *Schweiz. Bienenztg.* **73**: 58–64. Transl. *Bee World* **31**: 9–12. [M, B, 137/51

MAURIZIO, ANNA (1953). Weitere Untersuchungen an Pollenhöschen. *Beih. Schweiz. Bienenztg.* 2: 485–556. [B

MAYER, A. M. (1874). Researches in acoustics, V. *Amer. J. Sci. Arts*, Ser. 3, 8: 81–109.

McCOOK, H. C. (1877). Mound-making ants of the Alleghenies. *Trans. Amer. ent. Soc.* 6: 253–96.

McGREGOR, S. E. (1952). Collection and utilization of propolis and pollen by caged honeybee colonies. *Amer. Bee J.* 92: 20–1. [M, B, 161/53

McGREGOR, S. E. & TODD, F. E. (1952). Cantaloup production with honeybees. *J. econ. Ent.* 45: 43–7. [B, 177/53

McINDOO, N. E. (1914a). The olfactory sense of the honeybee. *J. exp. Zool.* 16: 265–346. [M, B

—— (1914b). The scent-producing organ of the honeybee. *Proc. Acad. nat. Sci. Philad.* 66: 542–55. [M, B

—— (1916). The sense organs of the mouth-parts of the honeybee. *Smithson. misc. Coll.* 65: No. 14, 1–55. [B

MECKEL, H. (1846).* Cited Cheshire (1886) and Bugnion (1928).

MEINERT, F. (1860).* Bidtag til de danske Myrers Naturhistorie. *K. danske vidensk. Selsk.* 5.

MELAMPY, R. M. & JONES, D. B. (1939). Chemical composition and vitamin content of royal jelly. *Proc. Soc. exp. Biol. N.Y.* 41: 382–8. [B

MELAMPY, R. M. & MASON, K. (1936). Absence of vitamin E in the royal jelly of bees. *Proc. Soc. exp. Biol. N.Y.* 35: 459–63.

MELAMPY, R. M. & WILLIS, E. R. (1939). Respiratory metabolism during larval and pupal development of the female honeybee. *Physiol. Zool.* 12: 302–11. [B

MELAMPY, R. M., WILLIS, E. R. & McGREGOR, S. E. (1940). Biological aspects of the differentiation of the female honeybee. *Physiol. Zool.* 13: 283–93.

MELNIK, M. I. (1951). [Managing multiple-queen colonies.] *Pchelovodstvo* (9): 36–7. [B, 108/52

MENDLESON, M. H. (1908). The California sage. *Glean. Bee Cult.* 36: 1202–3. [M, B

MENZER, G. & STOCKHAMMER, K. (1951). Zur Polarisationoptik der Fazettaugen von Insekten. *Naturwissenschaften* 38: 190–1. [B, T, 154/53

MERRILL, J. H. (1923). Value of winter protection for bees. *J. econ. Ent.* 16: 125–30.
—— (1924a). Observations on brood-rearing. *Amer. Bee J.* 64: 337–8 [M, B
—— (1924b). Sealed and unsealed brood. *Amer. Bee J.* 64: 424–5. [M, B
—— (1925a). Colony influence on brood-rearing. *Amer. Bee J.* 65: 172–4 [M, B
—— (1925b). The relation of stores to brood-rearing. *J. econ. Ent.* 18: 395–9.

MEYER, H. & GOTTLIEB, R. (1925).* *Die experimentalle Pharmacologie als Grundlage der Arzneibehandlung.* Berlin: Urban & Schwarzenberg.

MEYER, W. & ULRICH, W. (1952). Zur Analyse der Bauinstinkte unserer Honigbiene. Untersuchungen über die Kleinbauarbeiten. *Naturwissenschaften* 39: 264. [T

MILLEN, T. W. (1942). Bee breeding; laying workers and their progeny. *Indian Bee J.* 4: 94–5. [B

MILLER, C. C. (1901). Stray straws. *Glean. Bee. Cult.* 29: 502. [B
—— (1902). (No title). *Glean. Bee Cult.* 30: 136. [B

MILUM, V. G. (1930). Variations in time of development of the honeybee. *J. econ. Ent.* 23: 441–7. [M

MINDERHOUD, A. (1929a). Onderzoekingen over de wijze, waarop de honigbij haar voedsel versamelt. *Maandschr. Bijent.* 32: 109–12 (cited *Bee World* 12: 22–4). [B
—— (1929b). *Onderzoekingen over die wijze, waarop de honigbij haar voodsel verzamelt.* Wageningen: Veenmann & Zonen. [B
—— (1931). Untcrouchungen über das Betragen der Honigbiene als Blütenbestäuberin. *Gartenbauwiss.* 4: 342–62. [B, T

MINNICH, D. E. (1931). The sensitivity of the oral lobes of the proboscis of the blow fly, *Calliphora vomitoria* L., to various sugars. *J. exp. Zool.* 60: 121–39.
—— (1932). The contact chemoreceptors of the honeybee. *J. exp. Zool.* 61: 375–93.

MOMMERS, J. F. A. M. (1948). De plaatsvastheid der honigbijen. *Meded. dir. Tuinb.* 11: 529–39.

MONCRIEFF, R. W. (1944). *The chemical senses.* London: Leonard Hill.

MOORE EDE, W. E. (1947). Some notes on bee behaviour. *Brit. Bee J.* 75: 448–9. [M, B

MORGENTHALER, O. (1927). Beiträge zur Kenntnis der Bienenkrankheiten. I. Überwinterung und Nosema. *Arch. Bienenk.* 8: 145–70. [B

MORGENTHALER, O. (1931). Das 'Hobeln' der Bienen. *Schweiz. Bienenztg.* **54**: 210–14. [B, T
MORLAND, D. (1930). On the causes of swarming in the honeybee: an examination of the brood food theory. *Ann. appl. Biol.* **17**: 137–47. [M
—— (1935). De l'essaimage et de la division du travail dans la ruche. *Bull. Soc. Apic. Alpes-Marit.* **14**: 109–10. [B
—— (1938). Recent investigations into bee-keeping at Rothamsted. *J. roy. Soc. Arts* **86**: 394–402. [B
MOSKOVLJEVIC, VASILJA (1939). Reported *Bee World* **20**: 83, and **21**: 39–41. [M, B
—— (1940). Reported *Bee World* **21**: 39–41. [M, B
MÜLLER, E. (1931). Experimentelle Untersuchungen an Bienen und Ameisen über die Funktionsweise der Stirnocellen. *Z. vergl. Physiol.* **14**: 348–84.
MÜLLER, E. (1950). Ueber Drohnensammelplätze. *Bienenvater, Wien* **75**: 264–5. [B, 180/50
MÜLLER, H. (1881).* *Kosmos* **11**: 414–25.
—— (1882).* Versuche über die Farbenliebhaberei der Honigbiene. *Kosmos* **12**: 273–99.
MÜLLER, J. (1826).* *Zur vergleichenden Physiologie des Gesichtsinnes des Menschen und der Tiere.* Leipzig.
MÜSSBICHLER, A. (1952). Die Bedeutung äusserer Einflüsse und der Corpora Allata bei der Afterweiselentstehung von *Apis mellifica*. *Z. vergl. Physiol.* **34**: 207–21. [160/53
MYSER, W. C. (1952). Ingestion of eggs by honeybee workers. *Amer. Bee J.* **92**: 67. [M, B, 235/52
NEKRASOV, V. U. (1949). [The drifting of bees.] *Pchelovodstvo*: 177–84. [B, T, 152/52
NELSON, J. A. & STURTEVANT, A. P. (1924). The rate of growth of the honeybee larva. *Bull. U.S. Dep. Agric.* No. 1222: 1–24. [B
NELSON, F. C. (1927). Adaptability of young bees under adverse conditions. *Amer. Bee J.* **67**: 242–3. [M, B
NESTERVODSKY, V. A. (1939).* Cited Taranov (1947).
NEWELL, W. (1913). Investigations pertaining to Texas bee-keeping. *Bull. Tex. agric. Exp. Sta.* No. 158: 1–14. [M
NIXON, H. L. & RIBBANDS, C. R. (1952). Food transmission in the honeybee community. *Proc. roy. Soc. B* **140**: 43–50. [M, B, 147/53
NOLAN, W. J. (1925). The brood-rearing cycle of the honeybee. *Bull. U.S. Dep. Agric.* No. 1349: 1–56. [M, B
OERTEL, E. (1940). Mating flights of queen bees. *Glean. Bee Cult.* **68**: 292–3 [M, B
—— (1949). Relative humidity and temperature within the beehive. *J. econ. Ent.* **42**: 528–31. [B, 17/50
OETTINGEN-SPIELBERG, THERESE (1949). Ueber das Wesen der Suchbiene. *Z. vergl. Physiol.* **31**: 454–89. [B. 98/50
OPFINGER, ELIZABETH (1931). Ueber die Orientierung der Biene an der Futterquelle. *Z. vergl. Physiol.* **15**: 431–87.
—— (1949). Zur Psychologie der Duftdressuren bei Bienen. *Z. vergl. Physiol.* **31**: 441–53. [B, 56/50
ÖRÖSI-PÁL, Z. (1936). Die Rolle der Mandibeldrüsen der Honigbiene. *Dtsch. Imkerführer* **10**: 364–5. [B
PALMER-JONES, T. (1947). Use of pollen supplements in New Zealand. *N.Z. J. Agric.* **75**: 147–50. [M, B
PARK, O. W. (1922). Time and labor factors involved in gathering pollen and nectar. *Amer. Bee J.* **62**: 254–5. [M, B
—— (1923a). Flight studies of the honeybee. *Amer. J. Bee* **63**: 71. [M, B
—— (1923b). The temperature of the bee's body. *Amer. Bee J.* **63**: 232–4. [M, B
—— (1923c). Water stored by bees. *Amer. Bee J.* **63**: 348–9. [M, B
—— (1923d). Behaviour of water carriers. *Amer. Bee J.* **63**: 553. [M, B
—— (1925). The storing and ripening of honey by honeybees. *J. econ. Ent.* **18**: 405–10.
—— (1927). Studies on the evaporation of nectar. *J. econ. Ent.* **20**: 510–16.
—— (1928a). Further studies of the evaporation of nectar. *J. econ. Ent.* **21**: 882–7.
—— (1928b). Time factors in relation to the acquisition of food by the honeybee. *Res. Bull. Ia. agric. Exp. Sta.* No. 108: 183–225. [M, B
—— (1932). Studies on the changes in nectar concentration produced by the honeybee. I. Changes which occur between the flower and the hive. *Res. Bull. Ia. agric. Exp. Sta.* No. 151: 209–44. [B

PARK, O. W. (1933). Studies on the rate at which honeybees ripen honey. *J. econ. Ent.* 26: 188–93.
—— (1949). In *The hive and the honeybee*, Hamilton, Ill.: Dadant. [M, B
PARKER, R. L. (1926). The collection and utilization of pollen by the honeybee. *Mem. Cornell agric. Exp. Sta.* No. 98: 1–55. [M, B
PARKS, H. B. (1925). Critical temperatures in bee-keeping. *Beekeeper's Item* 9: 125–7. [M
PARRY, D. A. (1947). The function of the insect ocellus. *J. exp. Biol.* 24: 211–19.
PASEDACH-POEVERLEIN, KATHARINA (1940). Ueber das 'Spritzen' der Bienen und über die Konzentrationsänderung ihres Honigblaseninhalts. *Z. vergl. Physiol.* 28: 197–210. [T
PEARSON, P. B. & BURGIN, C. J. (1941). Pantothenic acid content of royal jelly. *Proc. Soc. exp. Biol. N.Y.* 48: 415–17.
PERCIVAL, MARY (1947). Pollen collection by *Apis mellifera*. *New Phytol.* 46: 142–73. [B
—— (1950). Pollen presentation and pollen collection. *New Phytol.* 49: 40–63. [B, 62/51
PEREPELOVA, L. I. (1926). [Biology of laying workers, I.] *Opuit. Pas.* (12): 8–10. See also *Bee World* 8: 90–1. [B, T
—— (1928a). [Biology of laying workers, II, III.] *Opuit. Pas.*: 6–10, 59–61. [B, T
—— (1928b). [Laying workers, the egg-laying activity of the queen, and the swarming.] *Opuit. Pas.*: 214–17. Transl. *Bee World* 10: 69–71. [M, B, T
—— (1928c). [Materials concerning the biology of the bee: the work of bees in the hive.] *Opuit. Pas.*: 492–502. [B, T
—— (1928d). [The nurse bees.] *Opuit. Pas.*: 551–7. [B, T
—— (1947). [Ways of increasing breeding in colonies.] *Pchelovodstvo* (4): 10–14. [B, T, 134/50
PERRET-MAISONNEUVE, A. (1927). Sécrétion et utilisation de la cire chez l'abeille. *C. R. Acad. Sci., Paris* 185: 1317–19.
PÉREZ, J. (1894). De l'attraction exercée par les odeurs et les couleurs sur les insectes. *Act. Soc. linn. Bordeaux* 47: Sér. 5, 7: 245–53.
PETERKA, V. (1929). A contribution to our knowledge of laying workers. *Bee World* 10: 5–6. [M, B
PHILIPP, P. W. (1928). Das Kittharz, seine Herkunft und Verwendung im Bienenhaushalt. *Biol. Zbl.* 48: 705–14. Cited *Amer. Bee J.* 70: 273.
PHILLIPS, E. F. (1915). *Beekeeping*, London: Macmillan.
—— (1923). Humidity in the beehive. *Glean. Bee Cult.* 51: 779. [M, B
PHILLIPS, E. F. & DEMUTH, G. S. (1914). The temperature of the honeybee cluster in winter. *Bull. U.S. Dep. Agric.* No. 93: 1–16. [M, B
PIRSCH, G. B. (1923). Studies on the temperature of individual insects, with special reference to the honeybee. *J. agric. Res.* 24: 275–87.
PIXELL-GOODRICH, HELEN (1920). Determination of age in honeybees. *Quart. J. micr. Sci.* 64: 191–206.
PLANTA, A. v. (1888–9). Ueber den Futtersaft der Bienen. *Hoppe-Seyl. Z.* 12: 324–54, 13: 552–61.
PLATEAU, F. (1876).* L'instinct des insectes peut-il être mis en défaut par les fleurs artificielles? *C. R. Ass. franç. Av. Sci.* 5: 535–40.
—— (1897a). Comment les fleurs attirent les insectes. Recherches expérimentales, III. *Bull. Acad. Belg.*; Sér. 3, 33: 17–41.
—— (1897b). Comment les fleurs attirent les insectes. Recherches expérimentales, IV. *Bull. Acad. Belg.*, Sér. 3, 34: 601–44.
—— (1901). Observations sur le phénomène de la constance chez quelques Hyménoptères. *Ann. Soc. ent. Belg.* 45: 56–83.
—— (1910). Recherches expérimentales sur les fleurs entomophiles peu visitées par les insectes, rendues attractives au moyen de liquides sucrés odorants. *Mém. Acad. roy. Belg., Cl. Sci.,* Sér. 2, 2: 3–55.
PRINGLE, J. W. S. (1938). Proprioception in insects. *J. exp. Biol.* 15: 101–31, 467–73.
PROTHEROE, J. (1923). Safe introduction. *Amer. Bee J.* 63: 303. [M, B
PUMPHREY, R. J. (1940). Hearing in insects. *Biol. Rev.* 15: 107–32.
—— (1950). Hearing. *Symp. Soc. exp. Biol.* 4: 3–18.
PUMPHREY, R. J. & RAWDON-SMITH, A. F. (1936). Sensitivity of insects to sound. *Nature, Lond.* 137: 990.
—— (1939). Frequency discrimination in insects; a new theory. *Nature, Lond.* 143: 806–7.
RAHMAN, KHAN A. (1945). The Indian honeybee (*Apis indica* F.) at Lyallpur. *Proc. roy. ent. Soc. Lond. A* 20: 33–42.

RAMDOHR, R. H. (1811).* Cited Cheshire (1886) and Bugnion (1928).
RATNAM, R. (1938).* Climatic conditions and bee behaviour. *Madras Agric. J.* **26**.
RAUSCHMAYER, F. (1928). Das Verfliegen der Bienen und die optische Orientierung an Bienenstand. *Arch. Bienenk.* **9**: 249–322. [B
REINHARDT, J. F. (1939). Ventilating the bee colony to facilitate the honey ripening process. *J. econ. Ent.* **32**: 654–60.
REITER, R. (1947). The coloration of anther and corbicular pollen. *Ohio J. Sci.* **47**: 137.
RHEIN, W. v. (1933). Ueber die Entstehung des weiblichen Dimorphismus im Bienenstaate. *Arch. Entw. Mech. Org.* **129**: 601–65. [B
—— (1951a). Ueber die Ernährung der Drohnenmaden. *Z. Bienenf.* **1**: 63–6.
[B, 96/52
—— (1951b). Ueber die Entstehung des weiblichen Dimorphismus im Bienenstaate und ihre Beziehung zum Metamorphoseproblem. *Verh. dtsch. zool. Ges.*: 99–101. [B, 141/53
RIBBANDS, C. R. (1949). The foraging method of individual honeybees. *J. Anim. Ecol.* **18**: 47–66. [M, B, 1/50
—— (1950a). Autumn feeding of honeybee colonies. *Bee World* **31**: 74–6.
[M, B, 109/52
—— (1950b). Changes in the behaviour of honeybees following their recovery from anaesthesia. *J. exp. Biol.* **27**: 302–10. [M, B, 138/53
—— (1951). The flight range of the honeybee. *J. Anim. Ecol.* **20**: 220–6. [M, B, 167/52
—— (1952a). The relation between the foraging range of honeybees and their honey production. *Bee World* **33**: 2–6. [M, B, 132/53
—— (1952b). Division of labour in the honeybee community. *Proc. roy. Soc. B* **140**: 32–43. [M, B, 162/53
—— (1953a). The inability of honeybees to communicate colours. *Brit. J. Anim. Behav.* **1**: 5–6. [M, B, 145/53
—— (1953b). The defence of the honeybee community. Prepared for *Proc. roy. Soc. B*.
—— (1953c). The scent perception of the honeybee. Prepared for *Proc. roy. Soc. B.*
—— (1953d). Not yet published.
RIBBANDS, C. R., KALMUS, H. & NIXON, H. L. (1952). New evidence of communication in the honeybee colony. *Nature, Lond.* **170**: 438–40 and *Bee World* **33**: 165–9. [M, B, 142/53
RIBBANDS, C. R. & SPEIRS, NANCY (1953a). The adaptability of the homecoming honeybee. *Brit. J. Anim. Behav.* **1**: 59–66. [M, B
—— (1953b). Not yet published.
RICHARDS, O. W. (1927). Sexual selection and allied problems in the insects. *Biol. Rev.* **2**: 298–364.
RICHTMYER, F. K. (1923). The reflection of ultraviolet by flowers. *J. Opt. Soc. Amer.* **7**: 151–68.
RISGA, P. (1931). When and how many times do queens leave the hive? *Bee World* **12**: 76–8. [M, B
—— (1936). The weight of bees, nectar and pollen loads they carry, and the causes for the changes in these weights. *Acta. Univ. latv. Agron.*, ord. Ser. **3**: 1–46.
RIVIÈRE, G. & BAILHACHE, G. (1921). Contributions à l'étude de la propolis. *Ann. Sci. agron., Paris* **38**: 82–6. [T
ROBERTS, W. C. (1944). Multiple mating of queen bees proved by progeny and flight tests. *Glean. Bee Cult.* **72**: 255–9, 303. [B
ROCKSTEIN, M. (1950a). Longevity in the adult worker honeybee. *Ann. ent. Soc. Amer.* **43**: 152–4. [B, 60/51
—— (1950b). The relation of cholinesterase activity to change in cell number with age in the brain of the adult worker honeybee. *J. Cell. comp. Physiol.* **35**: 11–24.
[B, 59/51
ROEPKE, W. (1930). Beobachtungen an Indischen Honigbienen, insbesondere an *Apis dorsata* F. *Meded. Wageningen* **34** (6): 1–26.
ROMANES, G. J. (1885). Homing faculty of Hymenoptera. *Nature, Lond.* **32**: 630.
ROOT, A. I. (1883–1950). *ABC and XYZ of Bee culture.* Medina, Ohio: Root. [M, B
ROOT, E. R. (1908). (No title.) *Glean. Bee Cult.* **36**: 830 and 868. [M, B
RÖSCH, G. A. (1925). Untersuchungen über die Arbeitsteilung im Bienenstaat, 1. Die Tätigkeiten im normalen Bienenstaate und ihre Beziehungen zum Alter der Arbeitsbienen. *Z. vergl. Physiol.* **2**: 571–631.
—— (1927). Ueber die Bautätigkeit im Bienenvolk und das Alter der Baubienen. Weiterer Beitrag zur Frage nach der Arbeitsteilung im Bienenstaat. *Z. vergl. Physiol.* **6**: 265–98.

REFERENCES

RÖSCH, G. A. (1930). Untersuchungen über die Arbeitsteilung im Bienenstaat. 2. Die Tätigkeiten der Arbeitsbienen unter experimentell veränderten Bedingungen. *Z. vergl. Physiol.* **12**: 1–71. [B, T

ROSOV, S. A. (1944). Food consumption by bees. *Bee World* **25**: 94–5. [M, B

ROSSER, J. H. (1934). On drone rearing. *Bee World* **15**: 58. [M, B

ROUBAUD, E. (1916). Recherches biologiques sur les guêpes solitaires et sociales d'Afrique. La génèse de la vie sociale et l'évolution de l'instinct maternel chez les vespides. *Ann. Soc. Nat. Zool.* **10**: 1–160.

ROUSSY, L. (1929). Biological experiments and a test of the thesis of M. Perret-Maissonneuve on the secretion of wax by bees. *Bee World* **10**: 4–5. [M, B

SALT, G. (1937). The egg parasite of *Sialis lutaria*: A study of the influence of the host upon a dimorphic parasite. *Parasitology* **29**: 539–53.

SANDER, W. (1933). Phototaktische Reaktionen der Bienen auf Lichten verschiedener Wellerlänge. *Z. vergl. Physiol.* **20**: 267–86.

SANTSCHLI, F. (1911). Observations et remarques critiques sur le méchanisme de l'orientation chez les Fourmis. *Rev. Suisse Zool.* **19**: 303–38.

SCHAEFER, C. W. & FARRAR, C. L. (1946). The use of pollen traps and pollen supplements in developing honeybee colonies. *Circ. U.S. Dep. Agric.* No. E-531 rev.: 1–7. [M, B

SCHENK, O. (1902). Die antennalen Hautsinnesorgane einiger Lepidopteren und Hymenopteren. *Zool. Jb.*, Abt. 2, **17**: 573–618.

SCHIEMENZ, P. (1883). Ueber des Herkommen des Futtersaftes und die Speicheldrüsen der Biene nebst einem Anhange über das Riechorgan. *Z. wiss. Zool.* **38**: 71–135.

SCHIRACH, A. G. (1770)*. [Natural history of the queen of the bees.] Leipzig ? [French transl. M

SCHMIDT, ANNALIESE (1938). Geschmacksphysiologische Untersuchungen an Ameisen. *Z. vergl. Physiol.* **25**: 351–78.

SCHÖN, A. (1911). Bau und Entwicklung der tibialen Chordotonalorgane bei der Honigbiene und der Ameisen. *Zool. Jb.*, Abt. 2, **31**: 439–72.

SCHOORL, P. (1936).* Vitamin E research. *Z. Vitaminforsch.* **5**: 246–53.

SCHUÀ, L. (1952a). Untersuchungen über den Einfluss meteorologische Elemente auf das Verhalten der Honigbienen. *Z. vergl. Physiol.* **34**: 258–77. [B

—— (1952b). Hat die Entfernung der Trachtquelle vom Stock einen Einfluss auf die Saugleistung der Bienen? *Z. vergl. Physiol.* **34**: 376–82. [B, 137/53

SCOTT, W. N. (1936). An experimental analysis of the factors governing the hour of emergence of adult insects from their pupae. *Trans. roy. ent. Soc. Lond.* **85**: 303–30.

SCRIVE, F. (1948). L'essaim suit-il la reine ou la reine suit-elle l'essaim ? *Apiculteur* **92**: 47–8. [B

SECHRIST, E. L. (1944). *Honey getting*. Hamilton, Ill.: American Bee Journal. [M, B

SENDLER, O. (1940). Vorgänge aus dem Bienenleben vom Standpunkte der Entwicklungsphysiologie. *Z. wiss. Zool.* **153**: 29–82.

SHAFER, G. D. (1917). A study of the factors which govern mating in the honeybee. *Tech. Bull. Mich. agric. Exp. Sta.* No. 34: 1–19. [M, B

SHARMA, P. L. & SHARMA, A. C. (1950). Influences of numbers in a colony on the honey-gathering capacity of bees. *Indian Bee J.* **12**: 106–7. [M, B, 57/51

SHAW, T. P. G. (1924). An investigation of Canadian propolis gum. *Canad. Chem. Metall.* **8**: 33–6.

SIMMINS, S. (1914). *A modern bee farm*. Heathfield, Sussex: Simmins. [M, B

SIMPSON, J. (1950). Humidity in the winter cluster of a colony of honeybees. *Bee World* **31**: 41–3. [M, B, 210/52

—— (1953). *Not yet published*.

SINGH, SARDAR, (1950). Behaviour studies of honeybees in gathering nectar and pollen. *Bull. Cornell agric. Exp. Sta.* No. 288: 1–59. [M, B, 91/51

SKAIFE, S. H. (1952). The yellow-banded carpenter bee, *Mesotrichia caffra* Linn., and its symbiotic mite, *Dinogamasus braunsi* Vitzthun. *J. ent. Soc. S. Afr.* **15**: 63–76.

SKRAMLIK, E. v. (1926). Physiologie des Geschmackssinnes. *Handb. normalen. Path. Phys.* **11**: 306–92.

SLADEN, F. W. L. (1901). A scent-producing organ in the abdomen of the bee. *Glean. Bee Cult.* 29: 639–40. [B
—— (1902). A scent-producing organ in the abdomen of the worker of *Apis mellifica*. *Ent. mon. Mag.* 38: 208–11.
—— (1911). How pollen is collected by the social bees and the part played in the process by the auricle. *Brit. Bee J.* 39: 491–514. [M, B
—— (1912a). Further notes on how the corbicula is loaded with pollen. *Brit. Bee J.* 40: 144–5. [M, B
—— (1912b). Pollen collecting. *Brit. Bee J.* 40: 164–6. [M, B
—— (1913). *Queen-rearing in England*. London: Madgwick, Houlston.
SMALLWOOD, W. M. & PHILLIPS, R. L. (1916). The nuclear size of the nerve cells of bees during the life cycle. *J. comp. Neurol.* 27: 69–75.
SMITH, J. (1923). *Queen rearing simplified*. Medina, Ohio: Root. [M, B
—— (1928). Young bees not nurse bees. *Amer. Bee J.* 68: 80. [M, B
—— (1949). *Better Queens*. Fort Myers, Florida: Smith. [M, B
SNELGROVE, L. E. (1940). *The introduction of queen bees*. Bleadon, Somerset: Snelgrove. [M, B
—— (1946). *Queen rearing*. Bleadon, Somerset: Snelgrove. [M, B
SNODGRASS, R. E. (1925). *Anatomy and physiology of the honeybee*. New York: McGraw-Hill. [M, B
SOUDEK, S. (1927). Hltanové zlázy vcely medonasné. *Bull. ecol. agron.*, Brno. 10: 1–63. [M, B
—— (1929.) Pollen substitutes. *Bee World* 10: 8–9. [M, B
SOUTHWICK, A. M. (1947). Surplus royal food. *Glean. Bee Cult.* 75: 584. [M, P
SPENCER, H. (1867). *Principles of biology*. London: Williams & Norgate.
SPITZNER, M. J. E. (1788). *Ausführliche Beschreibung der Korbbienenzucht im sächsischen Churkreise*. Leipzig: Junius. Cited *Bayer Bienenztg.* 42: 167–8.
[M, citation in B
STAPEL, C. (1934). Honningbier og rødkløverfrøavl. Kan man tvinge honningbierne til at bestøve rødkløveren? *Tidsskr. handbr. Planteavl.* 40: 301–13. [T
STEINHOFF, HILDTRAUT (1948). Untersuchungen über die Haftfähigkeit von Duftstoffen am Bienenkörper. *Z. vergl. Physiol.* 31: 38–57.
STURTEVANT, A. P. & FARRAR, C. L. (1935). Further observations on the flight range of the honeybee in relation to honey production. *J. econ. Ent.* 28: 585–9.
SVOBODA, J. (1940). Ueber den wert des Pollens als Nahrungsmittel für Bienen. [The role of pollen as a food for bees.] *Schweiz. Bienenztg.* 63: 206–9 and *Bee World* 21: 105–7. [M, B
SYNGE, ANN D. (1947). Pollen collection by honeybees. *J. Anim. Ecol.* 16: 122–38. [M, B
TARANOV, G. F. (1936).* *Pchelovodstvo* (1). Cited Natsin (1951), *Slov. Vcelar* 29: 132–4.
[Citation in 81/52
—— (1937).* *Pchelovodstvo* (1). Same citation.
—— (1946). [Distribution of beehives in the apiary.] *Pchelovodstvo* (8, 9): 37–41. [B
—— (1947). [The genesis and development of the swarming cycle.] *Pchelovodstvo* (2): 44–54. [B, T, 80/50
—— (1951). [Biological and economic characteristics of the high mountain grey Gruzinian bee.] *Pchelovodstvo* (1): 15–20 and·(2) : 28–36. [B, T, 102/52
TARANOV, G. F. & IVANOVA, L. V. (1946). [Observations upon queen behaviour in bee colonies.] *Pchelovodstvo* (2, 3): 35–9. [B, T, 65/50
THOMPSON, F. (1930). Observations on the position of the hexagons in natural comb building. *Bee World* 11: 107. [M, B
THOMPSON, W. R. (1923). Sur le déterminisme de l'aptérisme chez un Ichneumonide parasite. *Bull. Soc. ent. Fr.*: 40–2.
THORLEY, J. (1744). *Melisselogia; or, the female monarchy*. London: Thorley. [M
TODD, F. E. & BISHOP, R. K. (1940). Trapping honeybee-gathered pollen and factors affecting yields. *J. econ. Ent.* 33: 866–70. [B
—— (1941). The role of pollen in the economy of the hive. *Circ. U.S. Bur. Ent.* No. E-536: 1–9. [M, B
TODD, F. E. & BRETHERICK, O. (1942). The composition of pollens. *J. econ. Ent.* 35: 312–17. [B
TSCHUMI, P. (1950). Ueber den Werbetanz der Bienen bei nahen Trachtquellen. *Schweiz. Bienenztg.* 73: 129–34. [M, B, 143/53
TUENIN, F. (1928). [The wax glands of the worker bee.] *Opuit. Pas.* (11): 502–8 [B, T

TURNER, C. H. (1910). Experiments on colour-vision of the honeybee. *Biol. Bull., Wood's Hole* 19: 257–79.
—— (1911). Experiments on pattern-vision of the honeybee. *Biol. Bull., Wood's Hole* 21: 249–64.
VANSELL, G. H. (1934). Relation between the nectar concentration in fruit blossoms and the visits of honeybees. *J. econ. Ent.* 28: 943–5.
—— (1942). Factors affecting the usefulness of honeybees in pollination. *Circ. U.S. Dep. Agric.* No. 650: 1–31. [B
VANSELL, G. H. & BISSON, C. S. (1935). Origin of colour in western beeswax. *J. econ. Ent.* 28: 1001–2.
VEPRIKOV, P. N. (1936).* [*The pollination of cultivated agricultural plants.*] Moscow: Selchogis. German transl. Berlin: Reichsfachgruppe Imker.
VERLAINE, L. (1927). Le déterminisme de déroulement de la trompe et la physiologie du goût chez les Lépidoptères (*Pieris rapae* L.). *Ann. Soc. ent. Belg.* 67: 147–82.
—— (1929). L'instinct et l'intelligence chez les Hyménoptères. X. La reine des abeilles dispose-t-elle à volonté du sexe de ses œufs? *Bull. (Ann.) Soc. ent. Belg.* 69: 224–38.
VINOGRADOVA, T. V. (1950). [Acceleration of the development of the honeybee and increase in its fertility.] *Agrobiologiya*: 79–86. [T
—— (1951). [The effect of feeding colonies with yeast.] *Pchelovodstvo*: 17–18. [B, 11/53
VIVINO, A. E. & PALMER, L. S. (1944). The chemical composition and nutritional value of pollens collected by bees. *Arch. Biochem.* 4: 129–36. [B
VOGEL, BERTA (1931). Ueber die Beziehungen zwischen Süssgeschmack und Nährwert von Zuckern und Zuckeralkoholen bei der Honigbiene. *Z. vergl. Physiol.* 14: 273–347.
VOGEL, R. (1923). Zur Kenntnis des feineren Baues der Geruchsorgane der Vespen und Bienen. *Z. wiss. Zool.* 120: 281–324.
VOWLES, D. M. (1953*a*). The orientation of ants. I. The substitution of stimuli. *J. exp. Biol.* (in press).
—— (1953*b*). The orientation of ants. II. Orientation to gravity, light and polarized light. *J. exp. Biol.* (in press).
WADEY, H. J. (1948). Section de chauffe? *Bee World* 29: 11. [M, B
WAHL, O. (1932). Neue Untersuchungen über das Zeitgedächtnis der Bienen. *Z. vergl. Physiol.* 16: 529–89.
—— (1933). Beiträge zur Frage der biologischen Bedeutung des Zeitgedächtnisses der Bienen. *Z. vergl. Physiol.* 18: 709–17.
WALKER, C. R. (1945). At what temperature will bees fly? *Glean. Bee Cult.* 73: 452–3.
[M, B
WALSTROM, R. J. (1950). Pollen substitute tests in Nebraska. *Amer. Bee J.* 90: 118–19. [M, B, 13/51
WALSTROM, R. J., PADDOCK, F. B., PARK, O. W. & WILSIE, C. P. (1951). Red clover pollination. *Amer. Bee J.* 91: 244–5. [M, B, 181/52
WANKLER, W. (1924). *Die Königin.* Freiburg: Theodor Fisher. [M
WEAVER, N., ALEX. A. H. & THOMAS, F. L. (1953). Pollination of Hubam clover by honeybees. *Prog. Rep. Tex. Agr. Exp. Sta.* No. 1559: 1–3.
WEAVER, N. & FORD, R. N. (1953). Pollination of crimson clover by honeybees. *Prog. Rep. Tex. Agr. Exp. Sta.* No. 1557: 1–4.
WEDMORE, E. B. (1929). The building of honeycomb. *Bee World* 10: 52–5. [M, B
—— (1947). *The ventilation of bee hives.* Petts Wood, Kent: Bee Craft. [M, B
—— (1952*a*). Methods of introducing queens sent by post: preliminary inquiry. *Rep. Bee Res. Ass.* No. 70: 1–14. [B
—— (1952*b*). Experimental queen cages. *Rep. Bee Res. Ass.* No. 78: 1–8. [B
—— (1952*c*). Queen Introductions final report on 1951 introductions. *Rep. Bee Res. Ass.* No. 93: 1–8. [B
WEIPPL, T. (1934). Der Brutzelldeckel. *Arch. Bienenk.* 15: 46–50. [M, B
WERY, JOSÉPHINE (1904). Quelques expériences sur l'attraction des abeilles par les fleurs. *Bull. Acad. Belg. Cl. Sci.*: 1211–61.
WHEELER, W. M. (1918). A study of some ant larvae, with a consideration of the origin and meaning of the social habit among insects. *Proc. Amer. phil. Soc.* 57: 293–343.
—— (1928.) *The social insects.* London: Kegan Paul.
WHITCOMB, W. Jr. (1946). Feeding bees for comb production. *Glean. Bee Cult.* 74: 198–202, 247. [M, B

WIGGLESWORTH, V. B. (1950). *The principles of insect physiology*, 4th ed. London: Methuen. [B
WILL, F. (1885). Das Geschmacksorgan der Insecten. *Z. wiss. Zool.* **42**: 674–707.
WILSKA, A. (1935). Eine Methode zur Bestimmung der Hörschwellenamplituden des Trommelfells bei verschiedenen Frequenzen. *Skand. Arch. Physiol.* **72**: 161–5.
WILSON, H. F. (1922). Winter care of bees in Wisconsin. *Bull. Wis. agric. Exp. Sta.* No. 338: 1–26.
WILSON, H. F. & MILUM, V. G. (1927). Winter protection for the honeybee colony. *Res. Bull. Wis. agric. Exp. Sta.* No. 75: 1–47. [M, B
WILTZE, J. (1882). Work done by two quarts of bees, before 16 days old. *Glean. Bee Cult.* **10**: 596–7. [M, B
WOLF, E. (1926). Ueber das Heimkehrvermögen der Bienen, I. *Z. vergl. Physiol.* **3**: 615–91.
—— (1927). Ueber das Heimkehrvermögen der Bienen, II. *Z. vergl. Physiol.* **6**: 227–54.
—— (1931). Sehschärfeprüfung an Bienen im Freilandversuch. *Z. vergl. Physiol.* **14**: 746–62. [B
—— (1933a). The visual intensity discrimination of the honeybee. *J. gen. Physiol.* **16**: 407–22. [B
—— (1933b). Critical frequency of flicker as a function of intensity of illumination for the eye of the bee. *J. gen. Physiol.* **17**: 7–19. [B
—— (1933c). Das Verhalten der Bienen gegenüber flimmernden Feldern und bewegten Objekten. *Z. vergl. Physiol.* **20**: 151–61. [B
—— (1935). Der Einfluss von intermittierender Reizung auf die Reaktionen von Insekten. *Naturwissenschaften* **23**: 369–71. [B
—— (1937). Flicker and the reactions of bees to flowers. *J. gen. Physiol.* **20**: 511–18.
WOLF, E. & ZERRAHN-WOLF, GERTRUD (1935a). The effect of light intensity, area and flicker frequency on the visual reactions of the honeybee. *J. gen. Physiol.* **18**: 853–63. [B
—— (1935b). The dark adaption of the eye of the honeybee. *J. gen. Physiol.* **19**: 229–37.
WOLFE, E. A. (1950). The use of pollen supplements. *Rep. Ia. St. Apiar.* for 1950: 48–50. [M, B, 110/52
WOLFF, O. B. J. (1875).* Das Riechorgan der Bienen nebst Beschreibung des Respirationswerkes. *Nova Acta Lesp. Carol.* **38**: 1–251.
WOLSKY, A. (1933). Stimulationsorgane. *Biol. Rev.* **8**: 370–417.
WOODROW, A. W. (1932). The comparative value of different colonies of bees in pollination. *J. econ. Ent.* **25**: 331–6.
—— (1933). The comparative value of different colonies of bees for pollination. *Mem. Cornell agric. Exp. Sta.* **147**: 1–29.
—— (1934). The effect of colony size on the flight rates of honeybees during the period of fruit bloom. *J. econ. Ent.* **27**: 624–9.
—— (1935). Some effects of relative humidity on the length of life and food consumption of honeybees. *J. econ. Ent.* **28**: 565–8.
—— (1941). Some effects of temperature, relative humidity, confinement and type of food on queen bees in mailing cages. *Circ. U.S. Bur. Ent.* No. E-529: 1–13. [M, B
WOODS, E. F. (1950). Sounds in beekeeping. *Brit. Bee J.* **78**: 766–7, 804–6, 831–4, 873–5. [M, B, 110/51
WYKES, GWENYTH R. (1952a). An investigation of the sugars present in the nectar of flowers of various species. *New Phytol.* **51**: 210–15. [M, B, 149/53
—— (1952b). The preferences of honeybees for solutions of various sugars which occur in nectar. *J. exp. Biol.* **29**: 511–19. [M, B
ZANDER, E. (1925). Die Königinnenzucht im Lichte der Beckerschen Untersuchungen. *Erlanger Jb. Bienenk.* **3**: 224–46. [B
ZERRAHN, GERTRUD (1933). Formdressur und Formunterscheidung bei der Honigbiene. *Z. vergl. Physiol.* **20**: 117–50. [B
ZOUBAREV, A. (1883). A propos d'un organe de l'abeille non encore décrit. [Concerning an organ of the bee not yet described.] *Bull. Apic. suisse romande* **5**: 215–16 and *Brit. Bee J.* **11**: 296–7.

Index of Authors

ADAM, BROTHER, 289–90, 292–3
ALEX, A. H., 130
ALFONSUS, A., 233
ALFONSUS, E. C., 205, 209, 263
ALLEY, H., 289, 297
ALPATOV, V., 259
ALPATOV, W. W., 216–17
ALTENBERG, E., 13
ALTMANN, G., 253, 281
ANDERSON, E. J., 201, 213, 222–3
ANDERSON, J., 274
ANDREAE, E., 94
ARMBRUSTER, L., 68, 218, 224
AUTRUM, H., 16, 30, 53

BACHMETJEV, P., 218
BAILEY, L., 29–30
BAILHACHE, G., 209
BARLOW, H. B., 14
BARYKIN, D. J., 288
BATEMAN, A. J., 107, 113–14
BAUMGARTNER, F., 145
BAUMGARTNER, H., 13–15
BECKER, F., 245–6, 254
BEECKEN, W., 183
BELING, INGEBORG, 119, 133–5, 138
BERLEPSCH, A. v., 234
BERTHOLF, L. M., 15, 25, 28, 270
BETHE, A., 49, 77–8, 81, 88, 172, 179
BETTS, ANNIE D., 108–9, 143, 227
BEUTLER, RUTH, 34–5, 128, 132, 217
BISHOP, G. H., 143–4
BISHOP, R. K., 167, 238, 240, 265
BISSON, C. S., 207
BLUMENHAGEN, R., 265
BODENHEIMER, F. S., 235, 239, 265
BOETIUS, J., 132
BONNIER, G., 71, 100, 103–4, 109, 148, 161
BORDAS, L., 55, 59–62
BÖTTCHER, F. K., 229
BOZLER, E., 30
BRAUN, E., 228, 230, 237
BRETHERICK, O., 249
BRIAN, ANN D., 191–2
BRIAN, M. V., 191–2
BRIANT, T. J., 37
BRITTAIN, W. H., 70–1, 108
BRUMAN, F., 224
BRUN, R., 84
BRÜNNICH, K., 234–6

BRUNSON, M. H., 259
BUDDENBROCK, W. v., 31
BÜDEL, A., 214
BUGNION, E., 59
BULMAN, G. W., 107
BURGIN, C. J., 250, 252
BURTOV, V., 225
BUTLER, C., 62
BUTLER, C. G., 36, 70, 92, 105, 110, 115, 131–2, 180–3, 242, 263, 304
BUTTEL-REEPEN, H. v., 78, 92, 133, 143, 146, 172, 179, 182, 278, 285
BUXTON, P. A., 44
BUYSSON, R. du, 191
BUZZARD, C. N., 110

CALE, G. H., 180, 263, 293
CAMERON, A. T., 33, 35, 70
CASTEEL, D. B., 119–21, 198
CHADWICK, P. C., 226–7
CHANTRY, 291–2
CHELDELIN, V. H., 250
CHERIAN, M. C., 70
CHESHIRE, F., 13, 55, 58–9, 65, 142, 201–3, 290, 297
CHEWYREUV, I., 258
CIESIELSKI, T., 218
CLEMENTS, F. E., 93
COOPER, B. A., 105
CORKINS, C. L., 70, 89–90, 219, 223–4, 228, 271, 275
COWAN, T. W., 297
CRANE, E. EVA, 242
CRANE, M. B., 113
CRAWSHAW, L. S., 202
CURRIE, G. A., 240

DADANT, C., 278
DADANT, H. C., 202
DAHLBERG, A. C., 35
DARWIN, C., 104, 107, 198–201, 256
DAVIS, J. L., 287
DEMUTH, G. S., 218, 222, 260, 265–6, 268
DIETERICH, K., 209
DOBKIEWICZ, L. V., 133
DOLGOVA, L. P., 124
DÖNHOFF, E., 37, 279, 281–2, 296, 299
DOOLITTLE, G. M., 210, 287–90, 292
DREYLING, L., 62
DRÖSI, 279

341

INDEX OF AUTHORS

DUFOUR, L., 234
DUJARDIN, F., 147, 209
DUNHAM, W. E., 225–6, 237
DZIERZON, J., 257

ECKERT, A., 197
ECKERT, J. E., 72, 166–7, 234, 249, 289, 292
EGGERS, F., 49
ELLEMENT, N. E., 298
ELSER, E., 248–9
ELVEHJEM, C. A., 249–52
EMERSON, G. A., 249
EMERY, J., 147
EVANS, H. M., 249
EVENIUS, C., 208, 269, 274

FABER, A., 53
FABRE, J. H., 76
FAHN, A., 35
FARRAR, C. L., 72, 227, 229–30, 238, 240–1, 243, 274, 290
FIELDE, ADELE M., 172
FILMER, R. S., 124–5, 190
FINNEY, D. J., 70
FIRSSOV, J. G., 187
FISCHER, G., 55
FLANDERS, S. E., 259
FORD, R. N., 132
FOREL, A., 23, 32, 60, 78, 94, 106, 133, 135, 172
FRAENKEL, G., 24–5
FRANÇON, J., 150
FREE, J. B., 180–3, 304
FREUDENSTEIN, K., 208–9
FRINGS, H., 32, 39
FRISCH, K. v., 17, 23–4, 26, 28, 30, 33–6, 38–42, 49, 79–80, 88, 95–7, 104–5, 140, 147–61, 164–5, 173–5, 185–9, 239, 243
FYG, W., 143, 145, 272

GATAULIN, F., 190
GATES, B. N., 219–22, 225
GEIGER, J.-E., 228, 230
GERSTUNG, F., 56, 121, 203, 264, 268, 299
GILBERT, C. S., 228, 271, 275
GILLETTE, C. P., 128
GILTAY, E., 94, 109
GOETZE, G., 254
GONTARSKI, H., 92, 128, 248, 250, 252, 258, 307–8
GOTTLIEB, R., 136
GRABENSBERGER, W., 136–8
GRAHAM, C. H., 14
GROOT, A. P. de, 233
GROSDANIC, S., 183, 208
GUBIN, A. F., 104, 184–5, 187, 189
GWIN, C. M., 202

HAAN, J. A., BIERENS de, 26
HAMBLETON, J. I., 71

HAMILTON, W., 267
HANSSON, A., 54, 287
HARTRIDGE, H., 158
HAVILAND, G. D., 255
HAYDAK, M. H., 233–4, 240–1, 248–53, 279–82, 307
HAZELHOFF, E. H., 212, 227
HEALY, M. J. R., 237
HECHT, S., 14–15
HEIN, G., 153–4
HENKEL, C., 150
HERAN, H., 45–46, 160
HERROD-HEMPSALL, W., 13, 292
HERTZ, MATHILDE, 17–22, 24, 26–7, 44–5
HESELHAUS, F., 59–62
HESS, C. v., 15–16, 23, 28
HESS, GERTRUD, 277–82
HESS, W. R., 221, 224–6
HESSE, A., 186
HIMMER, A., 218–22, 224–6, 228, 269, 307–9
HIRSCHFELDER, H., 126, 234
HODGE, C. F., 274
HODGES, DOROTHY, 119–20
HOLLICK, F. S. J., 49
HOMANN, H., 30
HOWELL, D. E., 144
HUBER, F., 143, 197–200, 205–9, 212, 226, 245, 256–7, 260–1, 265–6, 278, 285, 287–8, 294, 296
HUISH, R., 291
HUMPHREYS, W., 297
HUNTER, J., 3, 62
HUNTER, W. S., 14
HUTCHINSON, W. Z., 288–90

ILLNER, W., 309
INGLESENT, H., 58–60
IVANOVA, L. V., 261, 268, 284

JACOB, N., 245
JACOBS, W., 63, 173
JANSCHA, A., 143, 261, 289, 293, 298, 317
JANET, C., 191
JAUBERT, G. F., 207, 209
JEFFREE, E. P., 110, 115, 131–2
JESSUP, J., 212–13
JONES, D. B., 249–51
JONES, G. D. GLYNNE, 36
JONESCU, C. N., 65–7
JONGBLOED, J., 216–17
JORDAN, R., 257–8, 309
JURINE, 245

KALMUS, H., 31, 110, 115, 129, 131–2, 135, 138–9, 174–8, 196
KALNITZKI, A., 144
KALTOFEN, R. S., 177
KAPUSTIN, 184
KATHARINER, L., 78–9

INDEX OF AUTHORS

KELSALL, A., 287
KERNER, A. v. MARILAUN, 94-5
KHALIFMAN, T. A., 161
KING, G. E., 89, 191-2
KITZES, G., 249-52
KLEBER, ELISABETH, 140
KLEIN, 245
KLEIST, F. V., 227
KNIGHT, T. A., 168, 171, 209
KNUTH, F., 23, 94
KOCH, P., 27, 90
KOCH, R. K., 265
KOCHER, V., 250
KÖHLER, ADRIENNE, 247-8, 269
KOMAROV, P. M., 58-61
KÖRNER, ILSE, 140
KOSMIN, NATALIE P., 58-61, 216-17
KOTOGYAN, A. M., 267
KOVTUN, F. N., 287-8
KRATKY, E., 55-61, 269
KRIJGSMAN, B. J., 44
KROGH, A., 208
KRÖNING, F., 51
KÜHN, A., 19, 23-26
KUNZE, G., 34, 38
KUPETZ, F., 145
KUWABARA, M., 233

LAIDLAW, H. H., 289, 292
LAMMERT, F., 224
LANDOIS, H., 54, 287
LANGER, J., 56, 249
LANGSTROTH, L. L., 256-7, 266, 288-9, 297
LATHAM, A., 297
LECOMTE, M. J., 180, 182, 195-6
LEFEBVRE, A., 38
LEHNART, A., 265
LEUCKART, R., 56
LEUENBERGER, F., 278
LEVIN, M. D., 233
LINDAUER, M., 71, 103, 125-6, 161-4, 168-71, 190-2, 200-1, 207, 231-2, 249, 252, 302-7, 310, 312
LINEBURG, B., 191, 198-9, 203, 231
LISSMANN, H. W., 49
LIVENETZ, T. P., 92
LOEW, E., 107
LONG, F. L., 93
LOTMAR, RUTH, 24, 270
LOVELL, J. H., 27, 95
LUBBOCK, J., 23, 27, 147
LUNDEN, W., 130-1
LUNDIE, A. E., 69-72
LUTZ, F. E., 23, 28

MAHADEVAN, V., 70
MANLEY, R. O. B., 292, 296-7
MANUILOVA, A. I., 251
MARCHAL, P., 253

MARSHALL, J., 38
MASON, K., 249
MATHER, K., 108, 113
MATHIS, M., 256
MAURIZIO, ANNA, 109, 124, 128, 251, 262, 270-2, 275
MAYER, A. M., 51
McCOOK, H. C., 172
McGREGOR, S. E., 132, 208-9, 247
McINDOO, N. E., 38, 49, 63
MECKEL, H., 55
MEHRING, J., 207, 255
MEINERT, F., 60
MELAMPY, R. M., 217, 247, 249-51
MELNIK, M. I., 288
MENDLESON, M. H., 166
MENZER, G., 30
MERRILL, J. H., 234-5, 237, 239
MEYER, H., 136
MEYER, W., 203
MILLEN, T. W., 277
MILLER, C. C., 129, 243
MILUM, V. G., 219-23, 225, 228, 232, 235
MINDERHOUD, A., 110, 131-2
MINNICH, D. E., 32, 35, 37-8
MOMMERS, J. F. A. M., 110-11
MONCRIEFF, R. W., 32
MOORE EDE, W. E., 166
MORGENTHALER, O., 207, 222
MORLAND, D., 260, 263-4, 268, 273
MOSKOVLJEVIC, VASILJA, 271, 309
MÜLLER, E., 16, 30-1, 143
MÜLLER, H., 27, 109
MÜLLER, J., 13
MÜSSBICHLER, A., 279-81, 286, 292
MYSER, W. C., 235

NEKRASOV, V. U., 90-2
NELSON, J. A., 232, 247-8
NELSON, F. C., 307-8
NERYA, A. BEN, 235, 239
NESTERVODSKY, V. A., 267
NEWELL, W., 266
NEWTON, DOROTHY E., 108
NIXON, H. L., 192-4, 259
NOLAN, W. J., 145, 235-9

OERTEL, E., 143-5, 213
OETTINGEN-SPIELBERG, THERESE, 27, 101-5
OPFINGER, ELISABETH, 86, 97-100, 103
ÖRÖSI-PAL, Z., 227

PADDOCK, F. B., 75
PALMER, L. S., 249-51
PALMER-JONES, T., 241
PARK, O. W., 35, 69, 75, 127-8, 130-2, 150, 162, 192, 210-12

PARKER, R. L., 117, 119–21, 124, 128, 130
PARKS, H. B., 70
PARRY, D. A., 31
PASEDACH-POEVERLEIN, KATHARINA, 60–1
PEARSON, P. B., 250, 252
PENCZEK, E. S., 35
PERCIVAL, MARY, 109, 124, 130
PEREPELOVA, L. I., 193, 231, 237, 259, 262, 267–8, 277–9, 282–4, 297, 300–3, 310
PERRET - MAISONNEUVE, A., 203
PÉREZ, J., 104
PETERKA, V., 279
PHILIPP, P. W., 208
PHILLIPS, E. F., 210, 218, 222, 229
PHILLIPS, R. L., 274
PIRSCH, G. B., 218
PIXELL-GOODRICH, HELEN, 274
PLANTA, A. v., 247–8
PLATEAU, F., 93
POHL, R., 23
PRINGLE, J. W. S., 48
PROTHEROE, J., 292
PUMPHREY, R. J., 51–3

RAHMAN, KHAN A., 317
RAMACHANDRAN, S., 70
RAMDOHR, R. H., 55
RATNAM, R., 70
RAUSCHENFELS, 129
RAUSCHMAYER, F., 89
RAWDON-SMITH, A. F., 51–2
REGEN, J., 53
REINHARDT, J. F., 211–12
REITER, R., 119
RESNITCHENSKO, M. G., 216–17
RHEIN, W. v., 246, 248, 252–3
RIBBANDS, C. R., 39–40, 42–4, 71, 73–5, 86, 92, 107, 110, 114–18, 121–3, 129, 131–2, 140, 150, 163, 174–8, 180–3, 192–4, 204, 213, 241–2, 259, 262, 271–4, 300–1, 310–12
RICHARDS, O. W., 142–3
RICHTMYER, F. K., 23
RISGA, P., 92, 129, 145
RIVIÈRE, G., 209
ROBERTS, W. C., 144–6
ROCKSTEIN, M., 273–4
ROEPKE, W., 247
ROMANES, G. J., 76–7
ROMASHOV, G. I., 104, 184
ROOT, A. I., 227, 257, 273, 288–91
ROOT, E. R., 148
RÖSCH, G. A., 56, 63, 121, 173–5, 208, 263–4, 272, 284, 299–304, 308–10
ROSOV, S. A., 233
ROSSER, J. H., 257
ROUBAUD, E., 191
ROUSSY, L., 203

SAF'YANOVA, V., 259
SALT, G., 245
SANDER, W., 28
SANTSCHLI, F., 84
SCHACHINGER, 243
SCHAEFER, C. W., 241
SCHENK, O., 65
SCHIEMENZ, P., 56, 59–60
SCHIRACH, A. G., 245
SCHMIDT, ANNALIESE, 35
SCHNEIDER, W., 53
SCHÖN, A., 52
SCHÖNTAG, ADELE, 132
SCHOORL, P., 249
SCHUÀ, L., 71, 129
SCHUETTE, H. A., 249–52
SCOTT, W. N., 138
SCRIVE, F., 261
SECHRIST, E. L., 289, 293
SENDLER, O., 195
SHAFER, G. D., 143
SHARMA, A. C., 243
SHARMA, P. L., 243
SHAW, T. P. G., 209
SIMMINS, S., 290
SIMPSON, J., 122, 207, 214, 225, 248, 257, 268
SINGH, SARDAR, 111–12, 115, 122, 131
SKAIFE, S. H., 86
SKRAMLIK, E. v., 33, 35
SLADEN, F. W. L., 63, 119, 172, 292
SMALLWOOD, W. M., 274
SMITH, J., 287–92
SNELGROVE, L. E., 289–90, 292
SNODGRASS, R. E., 51, 60, 62
SOUDEK, S., 56, 240
SOUTHWICK, A. M., 265
SPEIRS, NANCY, 86, 118
SPENCER, H., 247
SPITZNER, M. J. E., 147
STAPEL, C., 190
STEINHOFF, HILDTRAUT, 149
STOCKHAMMER, K., 30
STUMPF, HILDEGARD, 30
STURTEVANT, A. P., 72, 232, 247–8
SVOBODA, J., 251
SYNGE, ANN, D., 117, 124, 167

TANQUARY, M. C., 241
TARANOV, G. F., 92, 202, 238, 240, 260–3, 265, 268, 275, 284, 287, 303
THOMAS, F. L., 130
THOMPSON, F., 201
THOMPSON, W. F., 245
THORLEY, J., 289
TISCHER, J., 309
TODD, F. E., 132, 167, 238, 240, 249, 265
TSCHUMI, P., 153–4
TUENIN, F., 203–4

INDEX OF AUTHORS

TURNER, C. H., 16, 23

ULRICH, W., 203
USINGER, R. L., 144

VANSELL, G. H., 35, 117, 207
VEPRIKOV, P. N., 190
VERLAINE, L., 35, 277
VINOGRADOVA, T. V., 241
VIVINO, A. E., 249-52
VOGEL, BERTA, 34
VOGEL, R., 49
VOWLES, D. M., 48-9, 164

WADEY, H. J., 227
WAHL, O., 134-8
WALKER, C. R., 70, 228
WALSTROM, R. J., 75, 241
WANKE, L., 160
WANKLER, W., 292
WEAVER, N., 130, 132
WEDMORE, E. B., 201, 219, 290-1
WEIPPL, T., 203
WERY, JOSÉPHINE, 94-5
WHEELER, W. M., 149, 191, 197, 295
WHITCOMB, W. Jr., 202

WIERSMA, C. A. G., 216-17
WIGGLESWORTH, V. B., 46-7, 138, 142, 144, 217-18, 281
WILL, F., 33, 37
WILLIAMS, R. J., 250
WILLIS, E. R., 217, 247
WILSIE, C. P., 75
WILSKA, A., 53
WILSON, H. F., 70, 219-23, 228, 235
WILTZE, J., 307-8
WINDRED, G. L., 44
WOLF, E., 14-16, 20, 22, 49, 77, 79-85, 88
WOLFE, E. A., 241
WOLFF, O. B. J., 60
WOLSKY, A., 31
WOODROW, A. W., 72, 213, 292
WOODS, E. F., 54
WYKES, GWENYTH R., 34

ZANDER, E., 246
ZERRAHN, GERTRUD, 20
ZERRAHN-WOLF, GERTRUD, 16, 20
ZOUBAREV, A., 63

General Index

Absconding, 317
Adaptability, 46, 85–8, 96, 116, 118, 299
  age and, 86
  to crops, 116, 118
Akinesis, 183
Anaesthesia,
  on orientation, 92
  on length of life, 272
  on pollen gathering and hive duties, 122, 272
Analogy, inferences from, 94, 106
ANTENNAE
  amputation of, 38, 45–6, 81–4, 296
  in cell inspection, 205
  in communication, 155, 210, 285
  of drone, queen and worker, 65
  Johnston's organ, 47–50, 53
  in queen recognition, 265, 282, 284
  of gnats respond to sound, 51
  see dances; distance perception; gravity perception; orientation; smell; taste; temperature perception; touch; water perception
Anthropomorphism, 11, 143
Ants, 35, 48–9, 60, 136–8, 164, 172, 191, 318
Apiary siting, 75
Apis dorsata, 247, 316
  florea, 316
  indica, 243, 277, 316–17

Balling, 296–8
Bee balm, 208
BEE MILK
  composition of, 248–9
  food of drones and queens, 56–8, 259, 302
  and ovary development, 254, 281
  preparation of, 306, 310
  source of, 56, 315
  vitamins in, 249–53
  see brood food theory; pharyngeal glands
BEE SCENT
  attractiveness of, 172–6, 178
  differences not inherited, 175
  distinctiveness of, 174–8
  persistence of, 175–6
  source of, 177–8
  in clustering, 195–6
  in orientation, 86–7, 176
  in recognition of companions, 172–8

BEE SCENT (*contd.*)
  see defence; orientation; queen odour; scent gland; fanning and scent dispersal; mandibular gland
Bees, weight of, 128, 233
Biotin *see* vitamins
Brain, of drone, queen and worker, 65–7
Brood food, *see* bee milk
Brood food theory, 264–5, 268, 299
BROOD REARING, 231–44
  age of nurse bees, 301–2, 308–9
  food reserves for spring, 239, 269
  proportion of bees to brood, 265
  rate and yearly total, 234–5
  seasonal fluctuations in, 234–9
  time spent on, 231
  winter, 229–30
  colony size on, 235
  nectar or syrup on, 239–43
  pollen required for, 233–4
  pollen supply on, 238–40, 265
  pollen supplements on, 240–1
  queen on, 236
  temperature on, 232, 237–8
  temperature for, 225–6
  see bee milk; flying activity; larvae; length of life; metabolism; pollen collecting
Broodless colonies, 272, 279, 281, 289
Bumble bees, 55, 58–61, 68, 199, 253, 314–15

Campaniform sensillae, 47
Carpenter bees, orientation of, 86–7
CELL capping, 200–1, 231, 304, 305–7
  cleaning, 205, 231, 302–3, 305–7, 310
  varnishing, 206–8
Choice, exercise of, 20, 111
Chordotonal sensillae, 47
CLUSTERING, 195–6, 217
  the winter cluster, 210, 214–15, 218–24
  glands in winter cluster, 270
Colony odour, *see* bee scent
Colony shape, 227
Colony size, *see* brood rearing; flying activity; honey production
COLOUR, not communicated, 150
  memory of, 96
  in foraging, 94–7, 105
  in orientation, 78–80, 94–8, 106

GENERAL INDEX

COLOUR, preferences, 27–8
  stimulation by, 28
COLOUR VISION, 23–7
  after-image phenomena, 26
  colour triangle, 26–7
  complementary colours, 24, 26
  range of, 25
  shades of colour, 24–5
  simultaneous contrast phenomena, 26
  ultraviolet, 23–8
COMB building, 198–200, 264, 300, 305–7, 310
  colour, 205–7
  dimensions, 201–2
  drone, 255–6
  reinforcement, 205–6
  spacing, 267
  see wax; cell
COMMUNICATION, see antennae; bee scent; crops; dances; direction; distance; food transmission; home selection; queen recognition; recruitment; scents
Congestion of broodnest, see swarming, causes of
Constancy to a crop, 107–9
Co-ordination, 48, 67
Corpora allata, 281
CROPS, change of, 107, 115, 117–18, 121
  COMMUNICATION ABOUT, 104, 147–67
  constancy to, 107–9
  FINDING AND RETURNING TO, 93–106
  learning recognition marks, 97–100
  memory of scent and colour, 96
  working two at once, 107–9
  see directing bees to crops
Crossing see pollination

DANCES, use of antennae in, 53, 148–9, 155
  discovery of, 148
  description of, 148–9
  round, 148, 150, 153–4
  sickle, 153
  waggle or figure-of-eight, 53, 149–50, 153–5
  individual variation, 158, 163
  on horizontal combs, 140, 157
  orient to gravity, 156, 158
  orient to plane of polarization of light, 140, 156–8
  origin of, 163–4
  stimuli for, 50, 161–3
  nectar flow on, 104, 161–2
  and scenting, 164
  scents and tastes on, 161–3
  wind on, 160
  of swarming bees, 168–70
  attached bees respond, 164–5
  facilitate change of crop, 102

DANCES (*contd.*)
  see communication; distance; direction; gravity; plane of polarization of light
Defence against beekeepers, 183
DEFENCE AGAINST ROBBERS, 179–83, 316
  age of guard bees, 304, 307
  nectar flows on, 181–3, 296
  parallel to queen introduction, 295–6
  recognition of enemies, 179–83
  bee scent in, 180–2
Defence against usurpers, 295, 316
Defence by immobility, 183
DIRECTION, communication of, 151–3, 155–61
  perceived under clouds, 140, 158
  in orientation, 76, 88
  of foraging, adjusted to passing of time, 140–1
DIRECTING BEES TO CROPS, 184–90, 239
  methods, 184–6, 190
  honey yields, 187–90
  seed yields, 184, 187–90
DISTANCE, communication of, 53, 150–1, 153–5, 159–60
  perception, 49–50, 160
  in orientation, 88
Distinctive colony odour see bee scent; recognition
Diurnal rhythm, see time perception
DIVISION OF LABOUR, 64, 299–312
  age and occupation, 299–301, 309
  arrangement, 310–12
  abnormal colonies, 307–9
  extent of, 309–10
  food supplies on, 311–12
  among nurse bees, 301–2, 309–10
  among foragers, 121–3
Dominant and submissive bees, 182
DRIFTING, 89–92
  anaesthesia on, 92
  estimates of, 89–92
  of drones, 92
  of queens, 89
  honey production and, 27, 90–2
DRONES
  antennae and eyes of, 65–7
  congregation of, 143
  flight range, 146
  flight statistics, 144
  feeding of, 193, 259, 302, 315
  glands of, 55, 59–63
  PRODUCTION OF, 255–9
  rearing and swarming, 256, 264, 266
  specializations of, 64, 142
  worker behaviour toward, 256
  see brain; comb; drifting; length of life; mating; vision

Egg laying, 225, 235–7, 264–5, 267, 282
 in drone cells, 256–9
Eggs, hatching of, 225
 eating of, 235
Enzymes, 58–60
EVOLUTION, 255, 295, 313–18
Exteroceptors, 47
Eyes *see* vision

FANNING to cool hive, 212, 226–7
 noxious gases on, 212
 to ripen nectar, 212
FANNING AND SCENT DISPERSAL
 of caged bees, 196
 on flowers, 173
 at the hive, 86–7, 170, 172–3
 on syrup dishes, 173–6
Fat bodies, 269–72, 274–6, 309
Feeding colonies, pollen supplements, 240–1
 sugar syrup, 213, 241–2
 for scent-directing, 184–90
Feeding *see* food transmission; larvae; queen
'Fixed' bees, 115
FLIGHT, age at first flight, 300
 cleansing, 222
 food consumption in, 216–17
 orientation, 86, 98–9, 305
 speed, 127
Flowers, per acre, 132
 amount of nectar in, 132
 ultraviolet, 23
FLYING ACTIVITY, brood rearing on, 243–4
 colony size on, 72
 foraging range on, 72–5
 light intensity on, 70
 nectar supply and, 69, 104
 in rain and storms, 70–1
 temperature on, 69–70, 72
 ultraviolet on, 70
 weather on, 69–75
 wind on, 70–1
FOOD, consumption in flight, 216–17
 consumption in winter, 227–30
 exchange, 191–3
 storage, 316
 taken out by foragers, 128
 see metabolism
FOOD SHARING AND TRANSMISSION, 191–4, 312, 315–18
 communication by, 194, 312, 315–18
 extent of, 192–3
FORAGING age, 300–1, 308
 direction, 166
 direction adjusted to passage of time, 140–1
 dissimilar by neighbouring colonies, 166–7, 315
 reduced before swarming, 262, 275

FORAGING (*contd.*)
 see flying activity; nectar; pollen; propolis
FORAGING AREAS, 109–14
 competition on, 114
 movement of, 111–12
 origin of, 111, 114
 pollination and, 110–14
 size of, 110–14
FORAGING METHOD, 111, 115–18
 basis of, 115–16, 118
 when working 2 crops, 116–17
FORAGING RANGE, 72–5
 for pollen gathering, 74
 weather on, 73–4
 see flying activity; honey production; pollination
FORAGING STATISTICS, 127–32
Form, perception of, *see* vision

Glands are indexed individually
GRAVITY, dances orienting to, 156, 158
 perception of, 48–9
 relation to light, 163–4
Guard bees, *see* defence

Hair sensillae, 47
HEARING, 51–4, 196, 287
 arbitrary distinction from vibration perception, 51
 displacements perceived, 52
 through solids, 52–4
 sounds bees might perceive, 51–2, 285
 of crickets, locusts and gnats, 51–3
Hives, coloured, 27, 78
HOME FINDING, 76–88
HOME SELECTION, 168–71
Honeydew, sugars in, 34
Honeysac, 61, 128, 161, 192, 204, 211
Honey, water content of, 210
HONEY PRODUCTION, of colonies with queen cells, 262
 drifting on, 27, 90–2
 flying time on, 73–4
 foraging range on, 72–5
 pollen supplements on, 241
 age of queen on, 236–7
 size of colony on, 230, 243–4, 288
 winter food consumption on, 230
Hormones, 253, 280–2
HUMIDITY, 210–15
 on adult bees, 213
 in the hive, 213–15

Individual variation, 33, 103, 118, 163, 173, 182
Insulated colonies, *see* humidity in the hive; wintering
Irritability, 182–3

Johnston's organ, 47–50, 53

LARVAE
   duration of immaturity, 232 (queens), 254
   evaporation from, 215
   feeding of, 56, 191–2, 231, 233–4, 300–2, 305–7, 313–15
   food on rate of development, 232–3
   queen and worker diets, 245–53
   temperature on development, 232
   vitamins accumulated in, 252
   see bee milk; brood rearing; metabolism

LAYING WORKERS, 254, 272, 277–83
   behaviour of and to, 282–3, 289
   pharyngeal glands of, 278–9
   production of, 254, 277–8

Learning *see* crops; orientation; scent; temperature perception; time perception; vision

LENGTH OF LIFE, brood rearing on, 271–2
   drones, 144
   workers, 271–4
   seasonal differences in, 269–76

Light-compass reaction, 84–5, 140–1
Light intensity, *see* flying activity
Light, relation to gravity, 163–4
Lunch intervals? 71

Mandibular glands, 60–1
Mass provisioning, 313
MATING, 12, 142–6, 297
   age, 143–4
   flight duration, 144
   flight range, 146
   of laying queens, 145
   multiple, 145
   position in, 143

Melipona, combs of, 199
Memory, *see* colour; communication; crops; directing bees; drifting; foraging method; orientation; recognition; scent; taste; time perception; vision

METABOLISM, bees and brood, 216–17
   larvae and pupae, 217
   queen and worker larvae, 247
   summer and winter bees, 271, 275
   and time perception, 136–8
   in winter, 228

Movement, perception of, *see* vision
Multiple-queen colonies, 287–8
Mushroom bodies, 67
Mutual attraction, 195–6

Nasanov gland *see* scent gland
Natural selection, 107, 199, 255
NECTAR, gatherers, trips per day, 131
   flower visits for 1 load, 129

NECTAR (*contd.*)
   load, gathering-time, 129–31
   weight and uptake of, 128–9
   quantity secreted, 132
   sugar(s) in, 34–5
   supply on foraging, 104
   supply *see* brood rearing; defence; queen introduction; swarming
   transfer, 210–11

NECTAR RIPENING, 210–12, 226, 300, 310
   cost of, 212–13
   fanning on, 212
   ventilation on, 211

NURSE BEES, *see* brood rearing; brood food theory; division of labour; pharyngeal glands

Nutritional castration, 253

Ocelli, 30–1
Odour *see* orientation; scents; smell
Oecotrophobiosis, 191
ORIENTATION, 76–88
   through angles, 77
   antennae in, 81–4
   circling on departure, 86, 98–9
   colour in, 78–80, 94–8, 106
   to crops, 93–106
   to displaced hives, 77–82, 86–7
   to height, 77
   to landmarks, 76–8, 85–6, 98–100
   without landmarks, 82–5
   to marks at hive entrance, 89
   odour in, 79–80, 86–7, 93–6, 100, 106, 176
   effect of rotation on, 50, 81–4
   in relation to sunlight, 84–5, 140–1
   see direction; distance; plane of polarization

Ovary enlargement *see* bee milk; laying workers; swarm bees
Oxygen consumption *see* metabolism

Packed colonies, *see* wintering
Pantothenic acid, *see* vitamins
Patterns, recognition of, *see* vision
PHARYNGEAL GLANDS, 55–8
   pollen and substitutes on, 56, 240, 251
   development of, 56, 275
   and foraging, 56, 58, 275, 308
   functions of, 56, 61
   of laying workers, 278–9
   of broodless bees, 271–2
   of queenless bees, 272, 278–9
   of swarming bees, 262, 272
   regeneration of, 308–9
   relation to longevity, 269
   retrogression of, 56, 58, 61, 308, 310
   source of bee milk, 56
   source of enzymes, 58
   structure of, 55

PHARYNGEAL GLANDS (contd.)
  of wax-producers, 304
  of winter bees, 269–70, 274–5
  see bee milk; brood food theory
Piping of queens and workers, 54, 287
'Planing' of bees, 207
POLARIZED LIGHT, PLANE OF, 28–30
  effect on dances, 140, 157–8
  orientation to, 30, 140
POLLEN, autumn consumption of, 270
  quantity required, 234
  reserves, 238
  supplements, 240–1, 251, 253
  variation in quality, 251
  vitamins in, 249–51
POLLEN COLLECTING, 119–23
  age of bees, 122–3, 308
  anaesthesia on, 122
  and brood rearing, 124–5, 238–9, 265
  daily rhythm of, 117, 124
  deliberate and incidental, 122
  flower visits per load, 129
  foraging range for, 74
  proportion of foragers, 124
  trips per day, 132
POLLEN LOADS, gathering time, 129–31
  mixed, 108–9, 116
  packing of, 119–21
  ratio to nectar loads, 124
  weight of, 128–9
POLLINATION, crosses between varieties, 107–8, 113–14
  and foraging areas, 110–14
  and foraging range, 75
  nectar supplies and flower density on, 132
  orchard arrangement, 110
  lucerne (alfalfa), 185, 190
  red clover, 75, 184, 186–90
  see direction to crops
Polymorphism, 245
Postcerebral glands, 57, 59, 61
Postgenal glands, 57, 60
Progressive provisioning, 314
PROPOLIS, analysis and sources of, 208–9
  uses of, 205–9
  collection and manipulation, 205–9
Proprioceptors, 47, 49

QUEEN, antennae and eyes of, 65–6
  balling of, 296–8
  behaviour of workers to, 284
  on drone comb, 257
  eggs laid by, 58, 234–7, 254, 264–5
  feeding of, 56–8, 193, 284, 302, 315
  flights and mating flights, 144
  glands of, 55, 59–60, 63
  grafting, 245–7, 254

QUEEN (contd.)
  laying of drone eggs, 256–9
  odour of, 79, 143, 267, 285, 291, 293–8
  use of distinctive odour, 294–6
  rearing, 254
  retinue of, 261, 284
  sex regulation by, 257–9
  specializations of, 64
  treatment of rivals, 263, 287
  see antennae; brain; brood rearing; drifting; mating; scent gland; swarming; vision
QUEEN INTRODUCTION, 288–96
  methods of, 290–2
  nectar supply on, 288, 296
  theories of, 293
QUEEN PRODUCTION, 245–54
QUEEN RECOGNITION, 284–6
  Queen cages, 290–2
  Queen cells, 245, 247, 261–3, 265, 280, 285–6
  Queen cell introduction, 292–3
  Queenless bees, behaviour of, 284–5
  see laying workers; queen introduction

RECOGNITION OF COMPANIONS, 172–8
  see bee scent; defence; queen recognition; smell
Recognition marks, see crop-finding; orientation
RECRUITMENT TO CROPS, 103, 147–67
  discovery of, 147–8
  see bee scent; communication; dances
Red clover pollination, 75, 184, 186–90
Re-orientation see orientation
Robber bees, see defence
Royal jelly, 56, 248–52
  and see bee milk
Rotation, orientation disturbed by, 50, 76, 81–4

Saliva, 58, 60–1, 191
Salivary glands, 55–62
  see enzymes; pharyngeal glands
Saturation deficiency, see humidity
SCENTS, communication of, 149
  effect on dances, 162
  in foraging, 93–7, 105
  learning of, 100
  memory of, 96
  see directing to crops; orientation; smell
SCENT GLAND
  origin and structure, 63
  dancing and exposure of, 164
  exposed on flowers, 173
  amount of syrup and exposure, 173

SCENT GLAND (*contd.*)
  in orientation, 81, 86–7
  queens, 63, 143.
  sealing of, 152, 173
  see bee scent; fanning
SCOUT BEES, age of, 101–2
  discovery of crops by, 100–6
  proportion of, 101–3
Senility, 273
Sequence of duties, *see* division of labour
Sex differences, 64–8
Sex regulation, 257–9
SMELL, 38–43
  antennae as receptors, 32, 38–9, 65
  contact odours, 32
  discrimination between scents, 40–2
  experimental technique, 39–40
  of mandibular gland secretion, 60
  recognition of companions by, 172–8, 316
  relation to taste, 32
  repellent odours, 41
  scentless flowers, 41
  spatial distribution of odours, 32
  threshold values, 42–3
  see orientation; scents
SOCIAL LIFE, adaptations for, 63–4
  evolution of, 12, 312–18
Solitary bees, 63, 86, 295, 310, 313–14
Sound perception, *see* hearing
Sound production, 54
Stress perception, 48
Stimuli, substitution of, 163–4
Stings and sting glands, 57, 62, 68, 183
Subgenual organ, 47, 53
Sublingual gland, 62
Sugars *see* taste
Supersedure, 263, 287
Sun, orientating to direction of, 28–30, 84–5, 140–1
  dancing angle and direction of, 156
SWARMING, 260–8, 288
  causes of, 264–8
  conditions which aggravate, 264, 266–7
  control of, 265–8
  and drone rearing, 256, 264, 266
  relation to nectar supply, 264, 266
  evolution of, 316–17
  preparations for, 260–2
  prime swarms and casts, 261
  and supersedure, 263
  swarm bees, 262
  hive temperature on, 267
  hive ventilation on, 264, 266–7
  value of, 317
  weather on, 263–4, 266
SWARMS, ages of bees in, 263
  choice of home by, 168–71
  joining of, 171, 293

TASTE, 32–8
  acid, bitter and sour substances, 33, 36, 38
  experimental technique, 33–4, 36–7
  individual variations, 33
  receptors, antennae, 32, 37
    legs, 37
    mouth, 33
  sugars, chemical composition and taste, 35
  concentration to produce dances, 161–2
  discrimination between concentrations, 35
  food value and taste, 34
  relative attractiveness, 34–5
  threshold values, 33, 36, 38
TEMPERATURE of individual bees, 218
  perception and preferences, 45, 220
  see brood rearing; flying activity; larvae; time perception
TEMPERATURE REGULATION, 216–30
  in broodnest, 225–6
  effect of stores, 222
  at high temperatures, 226–7
  in winter, 218–24
'Thanatosis', 183
Thoracic glands, 59–61
TIME PERCEPTION, 133–41
  drugs on, 136–8
  and food metabolism, 136–8
  proof of, 133–4
  temperature on, 136, 138
  value of, 124, 139
  adaptation to weather, 140
Touch perception, 49, 65
Trips per day, 131–2
Trophallaxis, 191

Ultraviolet *see* flying activity; flowers; vision

Ventilation, *see* fanning; nectar ripening; swarming
Vibration perception, *see* hearing
Vibratory stimuli in clustering, 195–6
VISION, 13–31
  acuteness, 13–15, 22
  astigmatic, a vertical receptor, 15, 22
  attractiveness of contrasts, 18
  brightness discrimination, 15–16, 31
  choice of patterns, 18
  darkness adaptation, 16
  eye facets (ommatidia), 13–15, 65
  eyes of drone, queen and worker, 65
  eye structure, 13
  no focusing mechanism, 15
  form and pattern, 16–22
  movement and flicker, 16, 20, 22

GENERAL INDEX

VISION (*contd.*)
   shades of grey, 23, 26
   see plane of polarization; orientation; colour vision
Vitamins, 241, 249-53, 280

Wasps, 59-61, 68, 191, 226, 253
WATER, choice of, 36
   gathering, to cool hive, 226-7
      time taken, 131
      trips per day, 132
   perception, 44-5
   see humidity; nectar ripening
WAX, age of workers, 300, 303-4, 308-9
   adulteration of, 203
   cost of, 202-3
   glands, 62-3, 304
   redevelopment of glands, 309
   PRODUCTION AND HANDLING OF, 59, 62, 197-204, 304, 310

WAX (*contd.*)
   properties of scales, 197
   repeated use of, 203
   secretion, 197
   see cell capping; comb building
WEATHER and flight, 69
   and foraging range, 73-5
   and swarming, 263-4, 266
   and time perception, 140
Weight of a bee, 128, 233
Wind on flight, 127
Wing beats, 54
Winter bees, 269-71, 273-6
Winter cluster, 210, 214-15, 218-24, 270
WINTERING, 215, 222, 227-30, 237
   in cellars, 219, 222, 228
   insulated colonies, 213, 228, 237
   food consumption, 227-30
   pollen reserves, 238
Worker bees, production of, 231-54
   origin of, 314